Finite-Element Modelling of Structural Concrete

Short-Term Static and Dynamic Loading Conditions

Finite-Element Modelling of Structural Concrete

Short-Term Static and Dynamic Loading Conditions

Michael D. Kotsovos

National Technical University of Athens

CRC Press
Taylor & Francis Group
Boca Raton London New York

CRC Press is an imprint of the
Taylor & Francis Group, an **informa** business
A SPON PRESS BOOK

CRC Press
Taylor & Francis Group
6000 Broken Sound Parkway NW, Suite 300
Boca Raton, FL 33487-2742

First issued in paperback 2017

© 2015 by Taylor & Francis Group, LLC
CRC Press is an imprint of Taylor & Francis Group, an Informa business

No claim to original U.S. Government works

ISBN-13: 978-1-4987-1230-9 (hbk)
ISBN-13: 978-1-138-74926-9 (pbk)

Visit the Taylor & Francis Web site at
http://www.taylorandfrancis.com

and the CRC Press Web site at
http://www.crcpress.com

In memory of Jan Bobrowski

Contents

Preface

Over the years, in collaboration with Milija N. Pavlovic, the author has steadily conducted a programme of research aimed at rationalising the analysis and design of structural concrete. This programme of work was preceded by a decade of experimental research at the material level which furnished the key data input for structural considerations, namely the fundamental behaviour of concrete materials under multiaxial stress conditions. Thus, the latter provided the starting point for a comprehensive investigation into the response of concrete to loading up to collapse, a study that encompassed the following three basic structural aspects: numerical modelling by the finite-element method; laboratory testing of structural members and a consistent design methodology.

This book is concerned with the first of these three fundamental approaches to structural behaviour; it extends the range of application of the part of the work carried out in the mid-1990s (Kotsovos M. D. and Pavlovic M. N., *Structural Concrete: Finite-Element Analysis for Limit-State Design*, Thomas Telford, 1995) so as to cover structural concrete behaviour within the whole spectrum of short-term loading ranging from static (monotonic and cyclic) to dynamic (seismic and impact). The characteristic feature of the work described is that it contrasts widely accepted tenets in that concrete is considered to be a brittle material and it is demonstrated that the ductility of concrete structures or members of structures is dictated by triaxial stress conditions, the latter invariably developing in concrete when its strength is approached, rather than strain-softening material properties. Some elements of the remaining two structural aspects, namely laboratory testing of structural members and a consistent design methodology, will also be mentioned whenever the necessary practical supporting evidence to the theoretical findings may be deemed appropriate; a full outline of the proposed unified design methodology (also based on the concepts underlying the numerical modelling of structural concrete) has formed the subject of a recent publication (Kotsovos M. D., *Compressive Force-Path Method: Unified Ultimate Limit-State Design for Concrete Structures*, Springer, 2014).

This book, therefore, not only provides the theoretical background of, and justification for, the latter design approach for ultimate strength but also affords a powerful tool for both analysis and design of the more complex structural elements for which hand calculations and/or simplified design rules are not sufficient and which therefore demand a more formal, rigorous and, indeed, sophisticated computational model such as the one based on non-linear finite-element analysis described here.

The book is divided into eight chapters. The need for a reappraisal of the concepts that underlie the methods adopted in practical structural analysis and design is demonstrated in Chapter 1 through the use of typical study cases highlighting the significance of valid experimental information on the behaviour of concrete under triaxial stress conditions for interpreting structural behaviour. Such information is presented in Chapter 2, where the techniques developed for obtaining valid test data are also discussed, and used for modelling concrete behaviour as described in Chapter 3. The model has been developed by regression analysis based on the *internal stress* concept which allows for the effect of micro-cracking on deformation. The modelling of the steel properties as well as the interaction between concrete and steel are also discussed in Chapter 3, whereas the numerical techniques developed for incorporating the material models into non-linear finite-element analysis for the case of short-term static (monotonic and cyclic) loading are presented in Chapter 4.

The objectivity and generality of the numerical scheme presented in Chapter 4 are investigated in Chapter 5 where, by merely specifying the uniaxial cylinder compressive strength for concrete and the yield stress of the steel, it is shown to be capable of producing realistic predictions of the behaviour (load-carrying capacity, deformation and mode of failure) for a wide range of structural concrete configurations. The numerical techniques adopted for extending the use of the numerical analysis scheme for the solution of dynamic problems are presented in Chapter 6 and the validity of the resulting package is investigated in Chapters 7 and 8 where it is used to predict the response of a wide range of structural concrete configurations respectively to seismic and impact excitations. In all cases, the proposed numerical model is found to provide a realistic description of the experimentally established response.

The author is indebted to many colleagues and past students who, throughout the years have contributed to the development of the present unifying approach to structural modelling. Among the former, mention must be made of Jan Bobrowski who encouraged the author with his salutary – albeit somewhat irreverent comments – on codes of practice and Dr. J. B. Newman who initiated the research on triaxial testing that led eventually to the material model adopted herein. Mr. D. Hitchings kindly provided the linear package FINEL, which formed the basis of the program for the non-linear modelling of concrete structures, and he was particularly generous with his advice on the use of the former. At the early stages of the work, a time of rapid development in hardware, many useful suggestions for the choice and implementation of the computing equipment were made by Dr. K. Anastasiou. The actual numerical implementation of the finite-element model and its application to specific problems was carried out with the help of the following collaborators: Dr. C. Bedard (two-dimensional modelling); Dr. F. Gonzalez Vidosa (three-dimensional modelling); Dr. D. M. Cotsovos (three-dimensional modelling under dynamic modelling and studies of structural concrete under seismic and impact loading); Dr. G. M. Kotsovos (study of punching failure of slabs).

Finally, the author wishes to express his sincere appreciation to Tony Moore of CRC Press and Spon Press for encouraging him to pursue this project and the Taylor & Francis team, especially Jennifer Ahringer, Andrea Dale and Syed Mohamad Shajahan who patiently and painstakingly guided the editing and production process throughout.

Author

After conducting extensive research into the triaxial behaviour of concrete materials, **Dr. Michael D. Kotsovos** has pioneered its application to the analysis and design of structural concrete members. He has authored over 100 journal papers and 3 books in these fields, and has been engaged in various consultancies covering the design of concrete structures, their assessment and upgrading under earthquake conditions. He is a former professor and director of the laboratory of concrete structures at the National Technical University of Athens where he is currently engaged in research and post-graduate teaching.

Chapter 1

Need for a reappraisal

Most methods widely used for the ultimate limit-state analysis and design of concrete structures place emphasis on modelling post-peak material characteristics such as, for example, strain softening, tension stiffening, shear-retention ability and so on, coupled with stress- and/or strain-rate sensitivity when blast or impulsive types of loading are considered. Such modelling is based on a variety of theories, such as, for example, endochronic (Bazant and Bhat 1976), plasticity (Chen and Chen 1975; Fardis et al. 1983; Yang et al. 1985), plastic fracturing (Bazant and Kim 1979), viscoplasticity (Cela 1998), damage mechanics (Mazars 1986; Mazars and Pijaudier-Cabot 1989; Suaris et al. 1990) and so on, all invariably developed within the context of continuum mechanics (Truesdell 1991).

The implementation of such modelling in finite-element (FE) analysis has led to the development of numerical packages each found to produce realistic solutions to particular, rather to a wide range of, structural problems such as, for example, FE modelling of reinforcement with bond (Jendele and Cervenka 2006), reinforced-concrete (RC) columns under monotonic and cyclic loading (Kwon and Spacone 2002), nuclear reactor containment RC vessels (Jendele and Cervenka 2009), RC walls (Agrawal et al. 1981; Ile and Reynouard 2000) and frames (Mochida et al. 1987; Lee and Woo 2002) under earthquake loading, plain-concrete prisms or cylinders (Tedesco et al. 1997; Thabet and Haldane 2001), RC beams (Dube et al. 1996), slabs (Cela 1998) and plates (Sziveri et al. 1999) under impact loading and so on. This apparent lack of generality has been attributed to poor material modelling due to misinterpretation of the observed behaviour of concrete at both material and structural levels (Kotsovos 2014). In fact, it has been suggested that such misinterpretations have also resulted in complex design methods which cannot safeguard the code requirements for structural performance.

In view of the above it is considered that there is a need for reappraising the currently prevailing views regarding the modelling of concrete behaviour (Kotsovos 2012) and the present chapter is intended to demonstrate this need. The main concepts underlying the modelling of structural concrete are outlined and typical examples of the shortcomings resulting from their implementation into methods developed for predicting structural behaviour are presented. Emphasis is placed on the physical modelling underlying practical structural design rather than the constitutive modelling underlying analysis since, unlike the latter, there is ample published work on the shortcomings of current design methods.

1.1 PHYSICAL MODELLING OF STRUCTURAL CONCRETE

1.1.1 Typical models

Any structural form comprising linear elements may be seen as an assemblage of simply supported elements extending between successive points of zero bending moment (simple

Figure 1.1 Typical crack pattern of RC beams just before suffering flexural failure under a transverse point load applied at mid-span.

supports, points of contraflexure or inflection), the simplest form of such elements being simply supported beams. Therefore, an indication of the physical modelling of RC structures may be provided by reference to the modelling of simply supported RC beams without any loss of generality.

Figure 1.1 shows the typical crack pattern of an RC beam at its ultimate limit state, just before the occurrence of flexural failure, under a transverse point load acting at mid-span. From the figure, it can be seen that the crack pattern is characterised by inclined cracking of the web throughout the beam span; these cracks become visible when the applied load exceeds a value of around 50% of the load-carrying capacity of the beam and, with increasing load, they progressively extend towards the load point. The load reaches its peak value just before horizontal splitting of the compressive zone which transforms the beam into a mechanism and leads to loss of load-carrying capacity.

When the beam reaches its ultimate limit state, it is widely considered that the manner in which the applied load is transferred to the supports is realistically described by truss (Ritter 1899; Collins and Mitchell 1980; Morsch 1902) or strut-and-tie mechanisms (Schlaich et al. 1987), which underlie the development of current methods for structural concrete design (Eurocode 2 [EC2] 2004). The simplest form of a truss that can be used to represent the physical state of a simply supported beam-like RC element such as that in Figure 1.1 is shown in Figure 1.2. In fact, such a structural element is considered to start behaving as a truss once inclined cracking occurs; the compressive zone and the flexural reinforcement form the longitudinal struts and ties, respectively, the stirrups form the transverse ties, whereas the cracked concrete of the element web is assumed to allow the formation of inclined struts.

Adopting a truss model for describing the function of an RC beam at its ultimate limit state implies that the inclined struts and the transverse ties form the path along which the applied load is transferred from its point of application (at the truss-member joints at the upper face of the beam-like RC element) to the supports. The load transfer occurs as

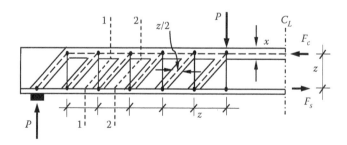

Figure 1.2 Truss model of a beam-like RC element.

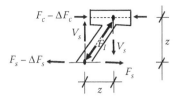

Figure 1.3 Portion of truss in Figure 1.2 between cuts 1-1 and 2-2.

indicated in Figure 1.3. The figure shows, in isolation, the portion of the truss between two successive cuts (1-1 and 2-2 in Figure 1.2) on either side of an inclined strut. The resultant ΔF_c of F_c and $F_c - \Delta F_c$ (i.e., the forces developing within the horizontal strut due to the bending of the beam-like RC element) combines with the force V_s (= P) transferred to the upper side of the truss by the right-hand side vertical tie, and, through the inclined strut, the resulting force F_l is transferred to the lower end of the strut; there, the vertical component (V_s) of F_l is transferred to the vertical tie at this end, whereas its horizontal component balances the action of the resultant ΔF_s of F_s and $F_s - \Delta F_s$ (i.e., the forces developing within the horizontal ties due to the bending of the beam-like RC element). Through this vertical tie, V_s is transferred to the upper end of the inclined strut adjacent to the one considered in the figure, and, in this manner, the load transfer continues until the applied load reaches the support.

1.1.2 Underlying concepts

The fundamental prerequisite for adopting a truss model for the description of the physical state of a beam-like RC element at its ultimate limit state is that, after visible cracking, concrete is capable of making a significant contribution to the load-carrying capacity of the beam. Such a prerequisite implies strain softening material characteristics (i.e., a gradual loss of load-carrying capacity once the peak-load level is reached), since, as it will be discussed in the next chapter, it has been established by experiment that the formation of visible cracking begins at the peak-load level under any state of stress.

1.1.2.1 Load transfer

As discussed in Section 1.1.1, in spite of the formation of visible cracking in the beam's web, concrete allows the formation of inclined struts, the latter enabling load transfer. The formation of inclined struts implies load transfer across the crack faces, since it is inevitable for the directions of inclined struts and cracks to intersect as a result of the close spacing of the cracks forming within the web at the ultimate limit state of the element. Such a load transfer, however, also implies a shearing movement of the crack faces which is resisted by frictional forces developing on the crack faces.

Therefore, it appears from the above that, since strain softening (which characterises 'cracked' concrete behaviour) is indicative of material instability, it is predominantly the development of frictional forces resisting the shearing movement of the crack faces which enables load transfer by allowing the formation of inclined struts through cracked concrete.

1.1.2.2 Load-carrying capacity

It should also be noted that the post-peak (strain softening) characteristics of concrete in uniaxial compression are used to formulate the failure criterion of the horizontal strut of

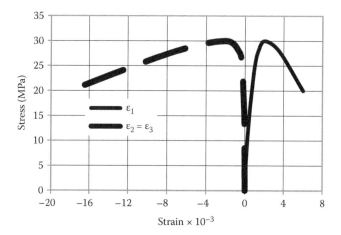

Figure 1.4 Stress–strain curves for a typical concrete in uniaxial compression.

the truss in Figure 1.2 (compressive zone of the beam in Figure 1.1). The experimentally established stress–strain curves for a typical concrete in uniaxial compression are shown in Figure 1.4 (Barnard 1964). A usually overlooked feature of these stress–strain curves is that not only the rate of increase but also the magnitude of the transverse strains (strains orthogonal to the axial compression) increase at a rate significantly higher than that of the axial strain when the stress approaches its peak value. In fact, at the peak stress level, the values of axial and transverse strains are numerically nearly equal, that is, $\varepsilon_1 = |\varepsilon_2| = 0.002$.

Failure of the longitudinal strut is considered to occur when the compressive strain of concrete reaches the limiting value $\varepsilon_c = 0.0035$ (EC2 2004) corresponding to a post-peak stress of the order of $0.85 f_c$, where f_c is the uniaxial cylinder compressive strength of concrete. Since visible cracking of concrete in uniaxial compression appears just before ε_c reaches the value 0.002 (i.e., the value corresponding to f_c), the above failure criterion implies that the 'cracked' concrete of the longitudinal strut makes a significant contribution to load-carrying capacity of the beam before flexural capacity is exhausted.

1.1.3 Shortcomings

As discussed in Section 1.1.1, the significance placed by current design methods on the post-peak characteristics of concrete is reflected in the use of various types of truss or strut-and-tie models for describing the function of RC linear elements at their ultimate limit state. The use of such models often leads to large deviations of the calculated from the measured values of shear capacity, whereas the assumptions underlying the assessment of flexural capacity are in sharp contrast with true structural behaviour. In what follows, emphasis is placed on presenting cases where shear capacity is overestimated and/or flexural capacity underestimated, since these are cases with grave implications for the safety of structures.

1.1.3.1 Shear capacity

Figure 1.5 shows the design details of an RC beam with an overhang investigated by Kotsovos and Michelis (1996); the beam was subjected to two transverse point loads, one at mid-span and the other near the end face of the overhang. The former of the two loads was applied first; it was increased to a predefined value, close to that corresponding to flexural

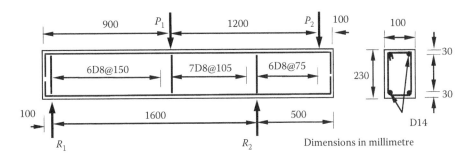

Figure 1.5 Design details of an RC beam with an overhang (beam denoted as BIGRC) designed in compliance with the truss model and subjected to sequential transverse loading. (From Kotsovos M. D., and Michelis P., 1996, *ACI Structural Concrete Journal*, 93(4), 428–437.)

capacity, where it was maintained constant. Then, the latter load was increased monotonically to failure.

The beam was designed in accordance with the provisions of the European Code for the design of concrete structures (EC2 2004) so as to eventually fail when the flexural capacity of its overhang is exhausted. In accordance with these provisions, shear failure can be prevented through the provision of a sufficient amount of transverse reinforcement assessed on the basis of the assumption that the beam behaves as a truss when the ultimate limit state is reached.

And yet, as indicated in Figure 1.6, beam failure occurred well before the flexural capacity had been exhausted. In fact, the beam failed in a brittle manner due to severe inclined cracking within its shear span adjacent to the right support under a shear force which was around 35% smaller than the calculated value of shear capacity. Such failure is indicative of failure of the inclined struts which the provisions of current codes for shear design, and, in particular, those considered safeguarding against failure of inclined struts, were apparently found unable to prevent. The results obtained were subsequently confirmed by Jelic et al. (2004).

The truss model also underlies the design of the column shown in Figure 1.7a; the column was designed in accordance with the earthquake-resistant design clauses of the European Codes (EC2, 2004; Eurocode [EC8] 8 2004) so as to exhibit a flexural mode of failure that safeguards ductile behaviour. And yet, as indicated in Figure 1.7b, the column, as in the case of the beam in Figure 1.5, failed prematurely in a brittle manner caused by severe inclined cracking within its critical end regions, in spite of the dense stirrup arrangement which was expected to allow the development of a shear force more than three times larger than the shear force corresponding to flexural capacity.

Another unexpected type of failure that may be suffered by columns is that indicated in Figure 1.8, which shows two slender and one short (in the middle) column of buildings damaged by the earthquake that hit Athens in 1999 (Kotsovos and Pavlovic 2001). In all cases, the truss model underlies design; however, in contrast with the left-hand side slender column which was designed in accordance with the earthquake-resistant design clauses of the old generation of codes based on the permissible stress philosophy (Deutsche Industrie Normen 1045 [DIN1045] 1959), the other two were designed in accordance with the earthquake-resistant design clauses of more recent codes of practice based on the limit-state philosophy (EC2; EC8 2004). And yet, in spite of the assumed design improvements (in the form of a significantly denser stirrup arrangement) introduced by the latter design codes, in all cases, failure occurred unexpectedly at mid-height, rather than at the column end regions which are widely considered as critical.

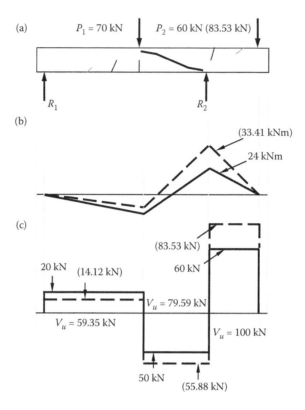

Figure 1.6 Experimental results for the RC beam in Figure 1.5 under sequential loading: (a) crack pattern at failure; (b) bending moment diagram and (c) shear force diagram. (The diagrams correspond to the experimentally established [continuous line] and the calculated [dashed line] load-carrying capacities, with the shear force diagram also including the shear capacity [V_u] of the various beam regions.)

As for the cases of the RC beam and column elements discussed above, the code adopted methods for the design of beam–column joints also rely on modelling the joint as a truss; such a truss, schematically represented in Figure 1.9b, combined with the development of a diagonal strut mechanism, schematically represented in Figure 1.9a, is assumed to resist the action of the forces transferred to the joint by the adjacent beam and column elements. However, in spite of the considerable research work on the behaviour of RC beam–column joints (such as that shown in Figure 1.10) carried out to date, the adopted methods have been found unable to satisfy the code performance requirements. This is because, as indicated by the crack pattern shown in Figure 1.11, not only did the typical joint shown in Figure 1.10 suffered considerable cracking before the formation of a plastic hinge in the adjacent beam, but also, such cracking occurred at early load stages and thus violated the assumption of 'rigid joint', which underlies the methods adopted in practice for structural analysis (Kotsovou and Mouzakis 2011). Results similar to those in Figure 1.11 have been obtained from a number of investigations (see e.g., Ehsani and Wight 1985a,b; Hwang et al. 2005; Tsonos 2007).

The investigation of the causes of deviations between the intended and the observed structural behaviour, such as those discussed above, forms the subject of extensive research work which appears in numerous publications, many of which are summarised elsewhere

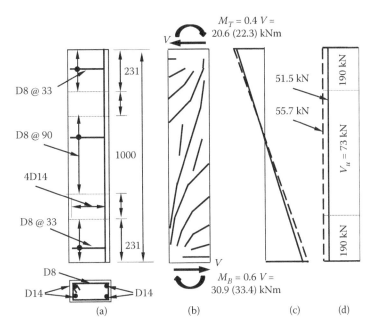

$M_T = 0.4\ V = 20.6\ (22.3)$ kNm

D8 @ 33

D8 @ 90

4D14

D8 @ 33

231

1000

231

51.5 kN

55.7 kN

190 kN

$V_u = 73$ kN

190 kN

D8

D14

D14

$M_B = 0.6\ V = 30.9\ (33.4)$ kNm

(a)   (b)   (c)   (d)

Figure 1.7 Experimental results from tests on RC columns under combined flexure and shear: (a) design details; (b) crack pattern at failure; (c) bending moment diagram and (d) shear force diagram. (The diagrams correspond to the experimentally established [continuous line] and the calculated [dashed line] load-carrying capacities, with the shear force diagram also including the shear capacity [V_u] of the various beam regions.)

Figure 1.8 Typical mode of failure suffered by columns of buildings damaged by the earthquake that hit Athens in September 1999. The column at the left-hand side was designed in compliance with the permissible stress philosophy, rather than the limit-state philosophy that formed the basis of the design of the two other columns.

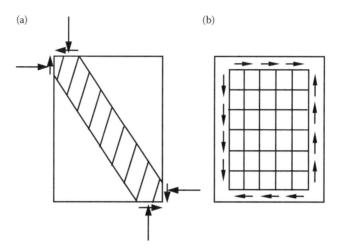

Figure 1.9 (a) Diagonal strut and (b) truss mechanisms of beam–column joint resistance.

(Kotsovos and Pavlovic 1995, 1999; Kotsovos 2014). From this work, it becomes clear that the apparent inability of current methods to produce design solutions safeguarding the code specified requirements for structural performance reflects the incompatibility between the concepts underlying the development of the methods and fundamental concrete properties, the latter forming the subject of the next chapter.

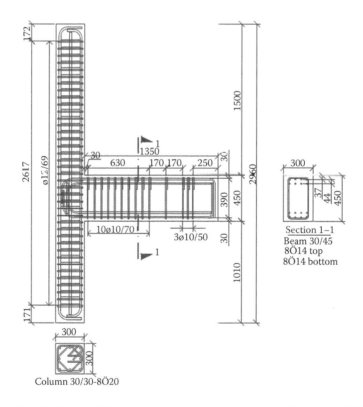

Figure 1.10 Design details of a typical beam–column joint designed in accordance with EC2 and EC8.

Figure 1.11 Failure mode of beam–column joint shown in Figure 1.10.

1.1.3.2 Flexural capacity

Amongst the assumptions underlying the assessment of flexural capacity is that the behaviour of concrete in the compressive zone is adequately described by $\sigma - \varepsilon$ curves, comprising both an ascending and a gradually descending branch, established from tests on cylinders or prisms in uniaxial compression. This assumption, on the one hand attributes the strains, of the order of 0.35%, measured at the extreme compressive fibre of an RC beam at its ultimate-limit state in flexure, to strain softening, which, as discussed in Section 1.1.2, characterises the behaviour of 'cracked' concrete, and, on the other hand, implies that the transverse stresses, which invariably develop in any RC structural element, have an insignificant effect on concrete behaviour.

And yet, there is easily reproducible experimental information which shows that the above assumptions are not correct. Figure 1.12 shows the details of an RC beam designed to fail in flexure under the action of a two-point transverse load (Kotsovos 1982). The material characteristics of both concrete and steel together with the values of the design and measured

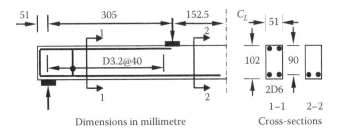

Figure 1.12 RC beams under two-point loading: design details. (From Kotsovos M. D., 1982, *Materials and Structures, RILEM*, 15(90), 529–537.)

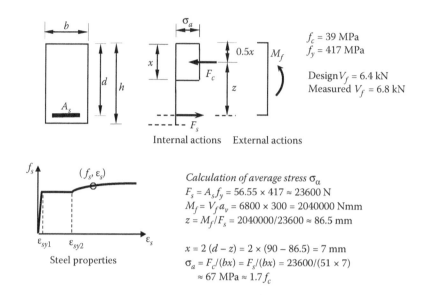

Figure 1.13 Assessment of average stress in compressive zone based on the measured values obtained from testing the beam in Figure 1.12. (From Kotsovos M. D., 1987, *ACI Structural Journal, Proceedings*, 84(3), 266–273.)

load-carrying capacity are provided in Figure 1.13 where they are used to assess the average stress in the compressive zone corresponding to the measured load-carrying capacity. The calculated average axial stress in the compressive zone is found to be $\sigma_a = 67$ MPa, that is, 70% higher than the uniaxial cylinder compressive strength of the concrete (f_c) and about 150% higher than the design stress, assuming a safety factor equal to 1.5. Such a large stress can only be sustained if the stress conditions in the compressive zone are triaxial compressive (Kotsovos 1987). Moreover, horizontal cracks appeared in the compressive after the peak load had been reached which is an indication that concrete in the compressive zone is described by strain-hardening, rather than strain-softening, material characteristics. A full description of the behaviour of concrete in the compressive zone of RC beams exhibiting a flexural mode of failure is provided in Section 2.4.2.

1.2 CONSTITUTIVE MODELLING

1.2.1 Underlying concepts

The concepts discussed in the preceding sections also underlie the constitutive models of concrete behaviour incorporated in most packages currently used for the numerical analysis of concrete structures. Although emphasis is placed on the description of the post-peak material characteristics which essentially describe the behaviour of a discontinuous material, since they reflect the effect of cracking on deformation, the material modelling relies on continuum mechanics theories. Within the context continuum mechanics, a theory (e.g., endochronic, damage, plasticity, etc.) or combination of theories (plastic fracturing, viscoplasticity, etc.) considered as the most appropriate for the description of the phenomenological features of the behaviour of concrete under load is selected for formulating analytical expressions describing the stress–strain and strength characteristics of concrete under

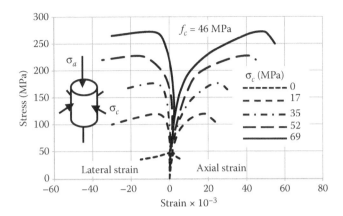

Figure 1.14 Stress–strain curves for a typical concrete in triaxial axisymmetric compression ($\sigma_a > \sigma_c$).

generalised (triaxial) stress states. The formulation of analytical expressions is followed by calibration through the use of experimental data. Such typical data obtained from tests on concrete under axisymmetric (Kotsovos and Newman 1980) and biaxial (Kupfer et al. 1969) states of stress are shown in Figures 1.14 through 1.16 and Figures 1.17 and 1.18, respectively.

Figure 1.14 presents stress–strain curves obtained from tests on concrete cylinders under triaxial axisymmetric compression, that is under the combined action of an axial compressive stress, σ_a, and a confining pressure, σ_c, such that $\sigma_a > \sigma_c$ assuming compression as positive. The cylinders were first subjected to a hydrostatic pressure ($\sigma_a = \sigma_c$) which was increased to a predefined value; then, σ_c was maintained constant during the subsequent application of the displacement controlled σ_a which was increased monotonically until the cylinder suffered significant loss of load-carrying capacity.

The figure shows that all curves exhibit similar trends of behaviour which are independent of the applied σ_c. Both axial stress–axial strain and axial stress–lateral strain curves comprise ascending and gradually descending branches. It is important to note, however, that, when the σ_a approaches its peak value, the rate of increase of the lateral strain (i.e., the strain in the direction of σ_c ($<\sigma_a$)) becomes significantly larger than the rate of the axial strain (i.e., the strain in the direction of σ_a).

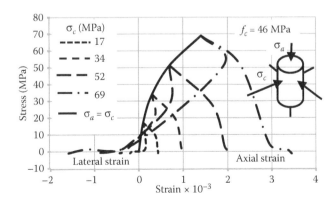

Figure 1.15 Stress–strain curves for a typical concrete in triaxial axisymmetric extension ($\sigma_a < \sigma_c$).

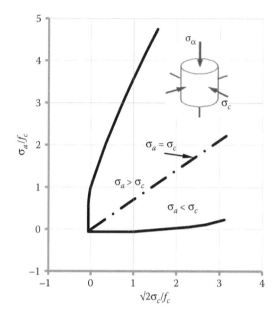

Figure 1.16 Strength envelope of concrete under axisymmetric states of stress.

Figure 1.15 presents stress–strain curves obtained from tests on concrete cylinders under triaxial axisymmetric extension (i.e., under the combined action of σ_a and σ_c such that $\sigma_a < \sigma_c$). As for the case of the triaxial compression tests, the cylinders were first subjected to a hydrostatic pressure ($\sigma_a = \sigma_c$) increasing to a predefined value; then, σ_c was maintained constant during the subsequent application of a displacement-controlled axial stress counteracting the vertical component of the hydrostatic pressure until the specimen suffered a complete loss of load-carrying capacity.

The figure shows that all curves exhibit similar trends of behaviour which are independent of the applied σ_c, but, unlike the stress–strain curves in Figure 1.14, these curves have only ascending branches; when the axial stress reaches a critical value, the cylinder suffers a complete and immediate loss of load-carrying capacity. However, as for the curves obtained

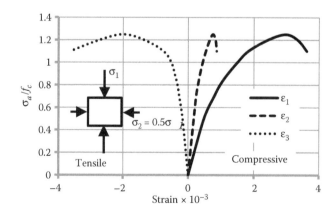

Figure 1.17 Stress–strain curves for a typical concrete in biaxial compression.

from the triaxial compression tests, the rate of increase of the strain in the direction of the smaller stress, σ_a is significantly larger than that of the strain in the direction of the larger stress, σ_c when the axial stress approaches its peak value.

Figure 1.16 shows the variations with confining stress, σ_c, of the maximum and minimum values of the axial stress, σ_a, sustained by concrete for the cases of triaxial axisymmetric compression ($\sigma_a > \sigma_c$) and extension ($\sigma_a < \sigma_c$), respectively. The figure shows that, for the case of triaxial compression, σ_a increases sharply with σ_c, even for a small increase of the latter; similarly, for the case of triaxial extension, σ_c increases sharply with σ_a, even for a small increase of the latter. On the other hand, the presence of small tensile stress is sufficient even to reduce to zero the load-carrying capacity of the material in the orthogonal direction.

The stress–strain curves in Figure 1.17 have been obtained from tests on square concrete plates under a plane state of stress σ_1, σ_2 such that $\sigma_2 = 0.5\sigma_1 > 0$. As for the case of the stress–strain curves of concrete under axisymmetric compression, the stress–strain curves under biaxial compression comprise an ascending and a gradually descending branch with the descending branch, which describes the out-of-plane deformational response (ε_3), exhibiting a significantly smaller slope than the slopes of the in-plane descending branches (ε_1, ε_2).

As regards the strength of concrete under biaxial (plane) stress conditions, Figure 1.18 shows that, while the presence of a compressive stress up to f_c in any of the two principal directions (σ_1, σ_2) leads to an up to 25% increase of the compressive strength of concrete in the orthogonal direction, the presence of a small tensile stress (smaller than the uniaxial tensile strength of concrete) rapidly diminishes the compressive strength of concrete in the orthogonal direction to zero.

The calibration of the analytical formulations of the constitutive models proposed to date appears to place emphasis only on the use of stress–strain data (such as those presented above) describing the deformational response of concrete in the direction of the maximum principal compressive stress. (It should be noted that the directions of the axes of symmetry of the specimens tested for establishing the stress–strain behaviour of concrete are the directions of the principal stresses.) On the other hand, the shapes of the stress–strain curves in the directions orthogonal to the direction of the maximum principal compressive stress (i.e., the directions of the intermediate and minimum principal stresses) are dictated by the continuum mechanics theory adopted for the formulation of the analytical expressions.

Although the post-peak material characteristics as described by the constitutive models proposed to date also describe the effect of cracking on deformation, only the formulation of

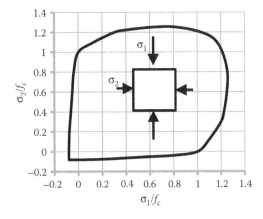

Figure 1.18 Strength envelope of concrete under biaxial states of stress.

the deformational response of concrete in compression is considered to adequately describe this effect. Under a state of stress with at least one of the principal stress components being tensile, the numerical description of the effect of cracking on deformation is complemented with the use of numerical techniques (such as e.g., the smeared-crack approach [Ngo and Scordelis 1967] with fixed [Cervera and Chiumenti 2006] or free-rotating [Jirásek and Zimmermann 1998] crack axes and the discrete-crack approach [Saouma and Ingraffea 1981]) which allow for the effect of cracking on deformation through the implementation of modifications in the geometry of the structure analysed. Such techniques place emphasis on the shear resistance considered to be provided by friction developing across the crack interfaces through the introduction of shear-resistance parameters such as the shear-retention factor (Scotta et al. 2001).

1.2.2 Inconsistencies of concepts underlying constitutive modelling

RC structures subjected to in-plane loading, such as, for example, frames, structural walls and so on are often analysed assuming plane-stress conditions. Even if in-plane loading were possible (since there always exist unintended eccentricities of the applied load), the assumption of plane-stress conditions is unrealistic; this is because the development of out-of-plane actions is inevitable due to variations in the transverse expansion resulting from the non-uniform distribution of the internal stresses. As discussed in Section 2.4.2, such variations in transverse expansion inevitably result in the development of small transverse stresses, for purposes of transverse deformation compatibility, the effect of which is considerable in strength, as indicated in Figure 1.16. Ignoring the development of transverse stresses on account of their small magnitude leads to misinterpretations of the available experimental information and assumptions which divert attention from the true causes of observed and measured structural response.

A typical misinterpretation links the causes of the 'size effect' phenomenon (i.e., the dependence of the behavioural characteristics of certain concrete members, such as, beams without transverse reinforcement, on the actual member dimensions) with intrinsic material properties (Bazant and Oh 1963; Hillerborg 1985; Gustafsson and Hillerborg 1988). On the other hand, it has been suggested by Kotsovos and Pavlovic (1994, 1997) that size effects are due to out-of-plane actions resulting from non-symmetrical cracking caused by unintended eccentricities of the applied load and/or the heterogeneous nature of concrete. In fact, it has been shown that realistic predictions of the size effect on the load-carrying capacity of RC beams without transverse reinforcement can be obtained by numerical analysis which allows for the formation of non-symmetrical cracking (Kotsovos and Pavlovic 1997).

As discussed in Section 1.1.2.2, the post-peak stress–strain behaviour of concrete under any state of stress is significantly affected by visible cracking which first occurs when the peak-stress level is reached. As it will be fully discussed in the following chapter, visible cracking predominantly affects the strains measured in the direction orthogonal to that of the crack plane, that is, the strains in the direction of σ_c in Figures 1.14 and 1.15 and strains ε_3 in Figure 1.17. This effect is reflected on the rate of increase of these strains which exceeds the rate of increase of the strains in the other directions by an amount significantly larger than the amount that could be described by a continuum mechanics theory. As a result, constitutive models developed on the basis of a continuum mechanics theory cannot provide a realistic description of the post-peak deformational behaviour of concrete as a material.

The inability of constitutive models based on continuum mechanics concepts to provide a realistic description of the post-peak concrete characteristics is more pronounced for the case of states of stress with at least one tensile principal stress component. In such cases, the description of cracking on deformation is complemented through the use of numerical techniques (e.g., smeared- or discrete-crack approach) which modify the geometry of the structure so as to account for the effect of cracking in excess of that which is accounted for by the adopted constitutive model.

Moreover, although the large rate of increase of the strains orthogonal to the crack plane implies void formation due to the lack of contact of the crack faces, it is often further assumed that some resistance to the shearing movement of the crack faces is possible to develop due to friction. This resistance is usually described by assuming that 'cracked' concrete retains a portion of the shear rigidity of 'uncracked' concrete.

As a result of the above inconsistencies characterizing the constitutive models, the applicability of the analysis packages which incorporate them appears to be limited only to particular structural elements. In fact, there has been no evidence presented to date on the ability of the analysis packages to provide realistic predictions of the behaviour of a wide range of structural configurations without a suitable modification of the adopted constitutive model of concrete behaviour. The above implies that the ability of the analysis packages for realistic predictions is linked with the use of a constitutive model dependent on the type of structural element analysed; however, there has not as yet been any criterion suggested for selecting a 'suitable' constitutive model. This apparent lack of generality of the analysis packages, as a result of the lack of objective criteria for adopting a particular constitutive model, represents a major drawback of the use of numerical methods for the analysis of RC structures. Moreover, even if such criteria existed, linking the ability of realistic predictions of structural behaviour to the use of a particular constitutive model does not appear to be rational, since such a link implies that concrete possesses some sort of intelligence allowing it to adapt its behaviour to the needs of particular structures.

1.3 CONCLUDING REMARKS

The methods widely used for the ultimate limit-state analysis and design of concrete structures under load make use of models whose development places emphasis on the contribution of concrete to load-carrying capacity after the formation of visible cracking.

The use of such models has proven to date ineffective. As regards design, current methods have been found unable to always produce solutions which satisfy the code requirements for structural performance. This shortcoming appears to reflect the conflict between the concepts underlying the modelling of structural elements and the mechanisms dictating the observed and/or measured structural response.

As regards analysis, the applicability of most numerical packages developed to date appears to be limited only to particular structural elements, as there has not as yet been published evidence on the ability of the packages to provide realistic predictions of the behaviour of a wide range of structural configurations. The causes of this apparent lack of generally appears to be due to the use of constitutive models which, although developed on the basis of continuum mechanics theories, are used to describe the behaviour of an essentially discontinuous material such as concrete.

REFERENCES

Agrawal A. B., Jaeger L. G. and Mufti A. A., 1981, Response of reinforced concrete shear walls under ground motions, *Journal of the Structural Division, Proceedings of the ASCE*, 107, 395–411.

Barnard P. R., 1964, Researches into the complete stress–strain curve for concrete, *Magazine of Concrete Research*, 16(49), 203–210.

Bazant Z. P. and Bhat P. D., 1976, Endochronic theory of inelasticity and failure of concrete, *Journal of the Engineering Mechanics Division, Proceedings of ASCE*, 102, 701–722.

Bazant Z. P. and Kim S. S., 1979, Plastic-fracture theory for concrete, *Journal of the Engineering Mechanics Division, Proceedings of ASCE*, 105, 407–428.

Bazant Z. P. and Oh B. H., 1963, Crack band theory for fracture of concrete, *Materials and Structures, RILEM*, 16, 155–177.

Cela J. J. L., 1998, Analysis of reinforced concrete structures subjected to dynamic loads with a visco-plastic Drucker–Prager model, *Applied Mathematical Modelling*, 22(7), 495–515.

Cervera M. and Chiumenti M., 2006, Smeared crack approach: Back to the original track, *International Journal for Numerical and Analytical Methods in Geomechanics*, 30(12), 1173–1199.

Chen A. C. T. and Chen W.-F., 1975, Constitutive relations for concrete, *Journal of the Engineering Mechanics Division, Proceedings of ASCE*, 101, 465–481.

Collins M. P. and Mitchell D., 1980, Shear and torsion design of prestressed and non-prestressed concrete beams, *Prestressed Concrete Institute*, 25(5), 32–100.

Deutsche Industrie Normen 1045 (DIN 1045), 1959, *Bauwerke aus Stahlbeton*, Ausgabe.

Dube, J. F., Pijaudier-Cabot, G. and La Borderie, C., 1996, Rate dependent damage model for concrete in dynamics, *Journal of the Engineering Mechanics Division, Proceedings of ASCE*, 122, 359–380.

Ehsani M. R. and Wight J. K., 1985a, Effect of transverse beams and slab on behaviour of reinforced concrete beam–column connections, *ACI Journal*, 82, 188–195.

Ehsani M. R. and Wight J. K., 1985b, Exterior reinforced concrete beam–column connections subjected to earthquake-type loading, *ACI Journal*, 82, 492–499.

Eurocode 2 (EC2), 2004, Design of concrete structures. Part 1-1: General rules and rules of building, British Standards.

Eurocode 8 (EC8), 2004, Design of structures for earthquake resistance. Part 1: General rules, seismic actions and rules for buildings, British Standards.

Fardis M. N., Alibe B. and Tassoulas J. L., 1983, Monotonic and cyclic constitutive law for concrete, *Journal of the Engineering Mechanics, Proceedings of ASCE*, 109, 516–536.

Gustafsson P. J. and Hillerborg A., 1988, Sensitivity in shear strength of longitudinally reinforced concrete beams to fracture energy of concrete, *ACI Structural Journal*, 85, 286–294.

Hillerborg A., 1985, The theoretical basis of a method to determine the fracture energy G_f of concrete, *Material and Structures, RILEM*, 18, 291–296.

Hwang S.-J., Lee H.-J., Liao T.-F., Wang K.-C. and Tsai H.-H., 2005, Role of hoops on shear strength of reinforced concrete beam–column joints, *ACI Structural Journal*, 102, 445–453.

Ile N. and Reynouard J. M., 2000, Nonlinear analysis of reinforced concrete shear wall under earthquake loading, *Journal of Earthquake Engineering*, 4, 183–213.

Jelic I., Pavlovic M. N. and Kotsovos M. D., 2004, Performance of structural-concrete members under sequential loading and exhibiting points of inflection, *Computers and Concrete*, 1(1), 99–113.

Jendele L. and Cervenka J., 2006, Finite element modelling of reinforcement with bond, *Computers and Structures*, 84(28), 1780–1791.

Jendele L. and Cervenka J., 2009, On the solution of the multi-point constraints – Application to FE analysis of reinforced concrete structures, *Computers and Structures*, 87(15–16), 970–980.

Jirásek, M. and Zimmermann, T., 1998, Analysis of rotating crack model, *Journal of Engineering Mechanics, ASCE*, 124(8), 842–851.

Kotsovos M. D., 1982, A fundamental explanation of the behaviour of reinforced concrete beams in flexure based on the properties of concrete under multiaxial stress, *Materials and Structures, RILEM*, 15(90), 529–537.

Kotsovos M. D., 1987, Consideration of triaxial stress conditions in design: A necessity, *ACI Structural Journal, Proceedings*, 84(3), 266–273.

Kotsovos M. D., 2012, Structural concrete analysis and design: Need for a sound underlying theory, *Archives of Applied Mechanics* 82(10), 1439–1459.

Kotsovos M. D., 2014, *Compressive Force–Path Method: Unified Ultimate Limit-State Design of Concrete Structures*, Springer, London, UK, 221pp.

Kotsovos M. D. and Michelis P., 1996, Behavior of structural concrete elements designed to the concept of the compressive force path, *ACI Structural Concrete Journal*, 93(4), 428–437.

Kotsovos M. D. and Newman J. B., 1980, Mathematical description of the deformational behaviour of concrete under generalised stress beyond ultimate strength, *ACI Journal, Proceedings*, 77(5), 340–346.

Kotsovos M. D. and Pavlovic M. N., 1994, A possible explanation for size effects in structural concrete, *Archives of Civil Engineering (Polish Academy of Sciences)*, 40(2), 243–261.

Kotsovos M. D. and Pavlovic M. N., 1995, *Structural Concrete: Finite-Element Analysis for Limit-State Design*, Thomas Telford, London, UK, 550pp.

Kotsovos M. D. and Pavlovic M. N., 1997, Size effects in structural concrete: A numerical experiment, *Computers and Structures*, 64(1–4), 285–295.

Kotsovos M. D. and Pavlovic M. N., 1999, *Ultimate Limit-State Design of Concrete Structures: A New Approach*, Thomas Telford, London, UK, 164pp.

Kotsovos M. D. and Pavlovic M. N., 2001, The 7/9/99 Athens earthquake: Causes of damage not predicted by structural-concrete design methods, *The Structural Engineer*, 79(15), 23–29.

Kotsovou G. M. and Mouzakis H., 2011, Seismic behaviour of RC external joints, *Magazine of Concrete Research*, 33(4), 247–264.

Kupfer H., Hilsdorf H. K. and Rusch H., 1969, Behavior of concrete under biaxial stress, *ACI Journal Proceedings*, 66(8), 656–666.

Kwon M. and Spacone E., 2002, Three-dimensional finite element analysis of reinforced concrete columns, *Computers and Structures*, 80, 199–212.

Lee H.-S. and Woo S-W., 2002, Seismic performance of a 3-story RC frame in a low-seismicity region, *Engineering Structures*, 24, 719–734.

Mazars J., 1986, A description of micro and macroscale damage of concrete structures, *Engineering Fracture Mechanics*, 25, 729–737.

Mazars, J. and Pijaudier-Cabot, G., 1989, Continuum damage theory – Application to concrete, *Journal of the Engineering Mechanics, Proceedings of ASCE*, 115(2), 345–365.

Mochida A., Mutsuyoshi H. and Tsuruta K., 1987, Inelastic response of reinforced concrete frame structures subjected to earthquake motion, *Concrete Library of JSCE*, 10, 125–138.

Morsch E., 1902, *Versuche uber Schubspannungen in Betoneisentragen*, Beton und Eisen, Berlin, 2(4), 269–274.

Ngo D. and Scordelis A. C., 1967, Finite element analysis of reinforced concrete beams, *ACI Journal Proceedings*, 64, 152–163.

Ritter W., 1899, Die bauweise hennebique, *Schweisserische Bauzeitung*, 33, 59–61.

Saouma V. E. and Ingraffea A. R., 1981, Fracture mechanics analysis of discrete cracking, *Proceedings of IABSE Colloquium on Advanced Mechanics of Reinforced Concrete*, Delft, June, pp. 413–436.

Schlaich J., Schafer K. and Jennewein M., 1987, Toward a consistent design of structural concrete, *Prestressed Concrete Institute*, 32(3), 74–150.

Scotta R, Vitaliani R., Saetta A., Oñate E. and Hanganu A., 2001, A scalar damage model with a shear retention factor for the analysis of reinforced concrete structures: Theory and validation, *Computers and Structures*, 79(7), 737–755.

Suaris W., Ouyang C. and Fernado V., 1990, Damage model for cyclic loading of concrete, *Journal of the Engineering Mechanics, Proceedings of ASCE*, 116, 1020–1035.

Sziveri J., Topping P. H. V. and Ivanyi P., 1999, Parallel transient dynamic non-linear analysis of reinforced concrete plates, *Advances in Engineering Software*, 30, 867–882.

Tedesco J. W., Powell J. C., Ross, A. C. and Hughes, M. L., 1997, A strain-rate-dependent concrete material model for ADINA, *Computers and Structures*, 64, 1053–1067.

Thabet A. and Haldane D., 2001, Three-dimensional numerical simulation of the behaviour of standard concrete test specimens when subjected to impact loading, *Computers and Structures*, 79, 21–31.

Truesdell C., 1991, *A First Course in Rational Continuum Mechanics*, Vol. 1, 2nd edition, Academic Press Ltd, New York, USA, 381pp.

Tsonos A. G., 2007, Cyclic load behavior of reinforced concrete beam–column subassemblages of modern structures, *ACI Structural Journal*, 104, 468–478.

Yang B.-L., Dafalias Y. F. and Herrmann L. R., 1985, A bounding surface plasticity model for concrete, *Journal of the Engineering Mechanics, Proceedings of ASCE*, 111, 359–380.

Chapter 2

Main behavioural characteristics of concrete

It is clear from the preceding chapter that concrete behaviour is an important constituent of the overall input required for the structural analysis of concrete structures. However, the modelling of concrete behaviour is fraught with difficulties. The first of these is the necessity of using triaxial-test data, namely that the material description should refer to the response of concrete under generalised (i.e., three dimensional) states of stress. As discussed in the preceding chapter, the reason for this is the large effect that small stresses, usually ignored in plane-stress analysis, have on concrete behaviour; this will become apparent throughout the following chapters of the book. The second source of difficulties associated with the establishing of actual properties of concrete materials relates to the scatter – and, thus, the reliability – of available experimental results. Not surprisingly, this apparent discrepancy of the failure envelopes and constitutive relations obtained by various laboratories and research groups working in this important field raises the question of whether or not a reliable model of concrete is at all possible, such that consistent and repeatable results might be obtained among a set of nominally identical specimens.

In an attempt to resolve the problem of data scatter, an international cooperative project was set up with subsequent publication of its findings on strength (Gerstle et al. 1978) and constitutive response (Gerstle et al. 1980). In the recognition that the large scatter in past studies of the strength and response of plain concrete (PC) under multiaxial stress states could be attributed to two principal factors – namely variation of materials tested and variation of the test methods themselves – the cooperative study concentrated on eliminating at least the first of these factors. One of the principal findings of this study, which are summarised in Kotsovos and Pavlovic (1995), was the identification of test methods capable of producing consistent and reliable data by inducing definable states of stress in specimens. The data discussed in the present chapter have been obtained from such a method which employs the triaxial cell developed by Newman (1974) for testing cylinders as described in the following section. Although this device is capable of reproducing only axisymmetric loading cases, such a limitation does not pose a practical problem for the reasons which will be outlined later. Moreover, unlike the data resulting from other test methods, those obtained by employing the triaxial cell have been found to provide a comprehensive description of the key features of concrete behaviour under generalised stress.

2.1 CYLINDER TEST

2.1.1 Underlying considerations

The triaxial cell device developed by Newman provides a simple and dependable technique for gathering triaxial data. Through the cell's steel ram, the cylinder is subjected to axial concentric load P_a at its ends combined with hydraulically imposed pressure p on its curved edges. However, due to the highly heterogeneous nature of concrete, even under such simple loading, it is only the mean values of the stresses developing within the cylinder that it is possible to determine accurately. Thus, the average axial stress developing due to P_a is $\sigma_a = 4P_a/\pi D^2$, where D is the diameter of the cylinder's cross section, whereas the mean value of the radial compressive stress developing due to p is $\sigma_c = p$. Since these normal stresses are the only stresses developing under the loading conditions considered, σ_a and σ_c are principal stresses.

It is important to note, however, that the state of stress within a specimen is affected by the means used to subject the cylinder to the intended loading. The axial load is applied through the rigid steel ram; it is inevitable, therefore, that frictional forces develop at the concrete–steel interface due to the incompatible mechanical properties of concrete and steel. As will be discussed in detail in a following section, these forces lead to the development of an indefinable stress field which restrains significantly the transverse expansion of concrete within the end regions of the cylinder. However, in accordance with St. Venant's principle, this indefinable stress field fades away with the distance from steel–platen interfaces and its effect on deformation becomes negligible at a distance of about D from the cylinder's ends. It is considered, therefore, that, for a cylinder with $H/D > 2$ (where H is the cylinder height), the middle portion of the specimen is free of the above end restraint effects. On the other hand, the transverse confining pressure is imposed hydraulically and, therefore, the fluid–concrete interaction does not lead to the development of any internal stress state other than that caused by the imposed pressure p. Therefore, only the middle portion of the cylinder under the combined action of an axial force and a transverse confining pressure is subjected to the definable state of stress σ_a, σ_c, and, hence, it is within this middle portion that the deformational response of concrete should be measured.

Moreover, it should be reminded that concrete comprises aggregate particles of various sizes bound together with cement paste. It has been found by experiment that, for the measured response to be representative of concrete, rather than isolated aggregate particles or cement paste, deformation should be measured over a length of around three times the size of the maximum aggregate of the concrete mix investigated (Newman and Lachance 1964). It appears from the above, therefore, that in order to both induce definable states of stress in concrete and obtain realistic measurements of material response, the test specimens should have a height-to-diameter ratio of 2.5. In fact, assuming a maximum size of aggregate of around 20 mm, the specimens selected for testing in the triaxial cell are 100 mm diameter × 250 mm height cylinders.

2.1.2 Test results and required clarifications

Typical results expressing both the deformational and strength characteristics of concrete already have been shown in Figures 1.14, 1.15, 1.17 and Figures 1.16, 1.17, respectively. As discussed in Section 1.2.1, a characteristic feature of the stress–strain relationships of Figures 1.14 and 1.17 is that, for stress levels beyond the peak value, the tensile strain increases at a rate which is very much higher than that of the compressive strain (Barnard 1964). Poisson's ratio values, which describe such behaviour, may vary from a value of approximately 1 at the peak-load level to values as large as 10 for stress levels beyond this level (Kotsovos 1983).

Now, it should be recalled that an isotropic continuum (which concrete is assumed to be throughout the whole range of testing) cannot exhibit Poisson's ratio values in excess of 0.5. In fact, this is implicit in the design of the testing devices used to obtain stress–strain relations such as those shown in Figures 1.14 and 1.17. Therefore, existing testing techniques assume that the deformational behaviour of concrete is compatible with that of a continuous medium up to and beyond the maximum sustained load. There is a historical justification for such an assumption, in the sense that the various test methods were devised originally with the aim of obtaining reasonably accurate estimates of the strength of concrete. In this respect, loading devices such as brush and flexible platens are considered to induce negligible frictional restraint at the specimen–platen interfaces when designed to allow displacements in the direction orthogonal to loading compatible with tensile strains calculated on the basis of Poisson's ratio values up to 0.5 (Gerstle et al. 1978).

Now, as shown in Figure 2.1, such values characterise concrete behaviour up to a level close to, but not beyond, the peak-load level. Therefore, it is evident that, as the maximum load is approached, and the Poisson's ratio becomes increasingly large, the concrete specimen is no longer a continuum but is beginning to be affected markedly by internal fracture processes. Despite this, an *average* load–deformation path can still be recorded for the specimen as a whole, and this implies that considerable frictional restraint between platen and specimen must, at these late stages of the deformational response, come into effect.

The importance of frictional forces between specimen and machine has been recognised as early as 1882, when Mohr criticised Bauschinger's results of compression tests made with cubic specimens, on the grounds that the friction of the cube surfaces in contact with the machine test-plates: 'must have a great effect upon the stress distribution so that the results are not those of a simple compression test' (Timoshenko 1953, p. 287). Similar observations regarding the importance of interface frictional forces were made by Föppl when working with cubic specimens of cement; he explored various ways of reducing such effects, showing that the usual tests of cubic specimens give exaggerated values of the compressive strength of the material (Timoshenko 1953).

As mentioned earlier, most of the attempts made subsequently to eliminate frictional effects between specimen and machine were aimed at obtaining realistic *strength* values. In addition to reducing these friction forces to varying effects, experimentalists also increased the aspect ratio of specimens (i.e., the length to cross-sectional dimensions ratio) in order to minimise end effects. As pointed out above, all these measures seem to have been successful in terms of strength-data accuracy (a typical example being the cylinder test). On the other

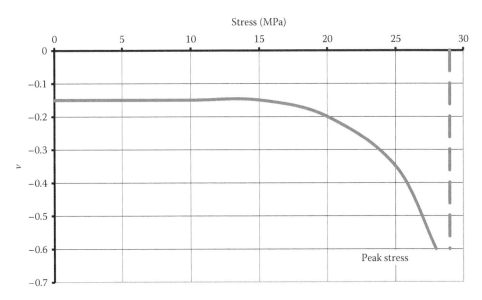

Figure 2.1 Typical variation of Poisson's ratio with increasing stress obtained from tests on cylinders under uniaxial compression.

hand, the larger deviation of the measured values of Poisson's ratio (i.e., $\nu \gg 1$) from their assumed counterparts (i.e., $\nu < 0.5$ used to design the experimental set-up that would minimise end restraint) at load levels beyond strength indicates that the post-peak tensile deformation of the specimens in the direction orthogonal to that of the maximum applied load (see Figures 1.14 and 1.17) is, in fact, significantly affected by the end restraint provided by the testing device, and that the experimental set-up cannot fulfil its objective of eliminating end friction. Clearly, the actual behaviour of a concrete specimen in uniaxial compression as the peak load is approached (and beyond) can be obtained only if the frictional effects are truly eliminated. One such investigation was successfully undertaken in order to assess the effect that reduction – and eventual removal – of the end friction would have on existing uniaxial-compression data on (a) strength and (b) post-peak characteristics (Kotsovos 1983). The investigation's findings, which were subsequently confirmed by other researchers (van Mier 1986; van Mier et al. 1997), are outlined in the next section.

2.2 POST-PEAK BEHAVIOUR

2.2.1 Uniaxial-compression tests

2.2.1.1 Behaviour of a test specimen under compressive load

Complete stress–strain relationships for concrete under uniaxial compression have been obtained to date by loading cylinders at a constant rate of displacement through a 'stiff' testing machine either by using a loading system capable of releasing almost instantaneously any load in excess of that which can be sustained by the specimen at any time (Ahmad and Shah 1979) or by loading a steel specimen in parallel with the concrete specimen in a manner such that, as the load-carrying capacity of concrete beyond ultimate strength (US) is reduced, the concrete–steel system transfers the excess load from concrete to steel so as to maintain the internal equilibrium of the overall system (Wang et al. 1978). Such relationships describe the response of the central zone of the cylinders, which is generally accepted to be subjected to a near-uniform *uniaxial* compressive stress in contrast to the complex and *indefinable* compressive stress state imposed on the end zones by frictional restraints resulting from the interaction between specimen and loading device (see Figure 2.2).

At a load level close to the maximum load-carrying capacity of the specimens, cracks aligned in the direction of loading appear in the central zone, and the maximum load-carrying capacity is reached when the US of this zone is attained. At this stage, the end zones remain stressed below their US capacity on account of the confining (triaxial) stress state (see Figure 2.3), without suffering any visible cracking.

The voids caused by the cracks of the central zone result in a dramatic increase of the lateral expansion of the zone which is incompatible with the much smaller lateral expansion of the 'uncracked' end zones. This incompatibility gives rise to *internal* forces acting so as to (a) *restrain* the lateral expansion of the central zone and (b) *increase* the lateral expansion of the end zones (see Figure 2.4). With increasing deformation, the internal forces acting on the end zones progressively reduce the effects of boundary frictional restraints and eventually create a state of stress in the end zone with at least one of the principal stress components tensile (see Figure 2.3). When the US of the end zones is attained, the cracks of the central portion propagate into the ends of the specimen, and the latter collapse under an applied load which induces, in the end regions, a state of stress similar to that indicated in Figure 2.3.

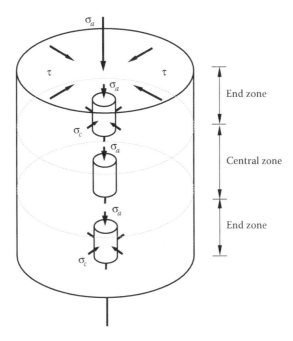

Figure 2.2 Schematic representation of the effect of boundary frictional restraint (τ) on the state of stress within cylinders under uniaxial compression.

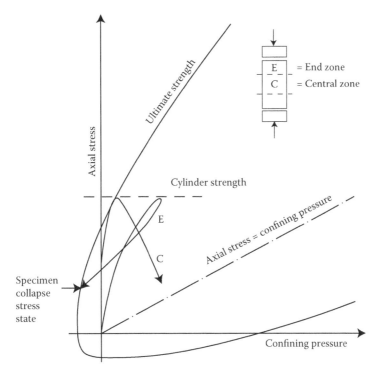

Figure 2.3 Schematic representation of stress paths induced in the central and end zones of cylinders under increasing uniaxial compression.

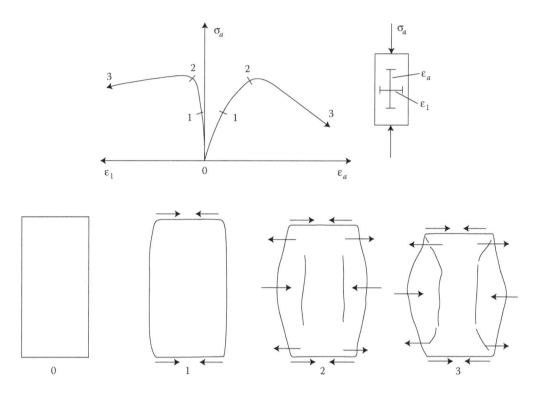

Figure 2.4 Stages of behavior of concrete cylinders under uniaxial compression.

It appears from the above, therefore, that the post-ultimate stress–strain relationships obtained from the tests describe behaviour under a *complex and indefinable* state of stress which is clearly induced by the frictional restraint (developing under increasing load) at the specimen–loading device interfaces. It therefore seems that the only reason for the central zone of the specimen to respond as a unit during the post-ultimate stage is the restraining action of the end zones. This suggests that, if the state of stress was the same throughout the specimen, no such restraining action would exist and the specimen would completely collapse as a result of the formation of continuous cracks at the US level. It is considered realistic to propose, therefore, that concrete suffers a complete and immediate loss of its load-carrying capacity when US is exceeded. Clearly, such a proposal implies that, while the end restraints affect the post-ultimate deformational response of concrete specimens, they have little influence on the maximum recorded load level since overall collapse immediately follows the onset of cracking in the central zone.

2.2.1.2 Experimental evidence for the brittle nature of concrete

An unequivocal experimental proof of the validity of the proposed behaviour of concrete can be obtained only by using testing techniques which completely eliminate any frictional restraint on the specimen–testing device interfaces. Although this is unlikely to be achieved by using existing testing techniques, an indication of the effect of frictional restraint on post-ultimate deformation can be obtained from uniaxial-compression tests by employing testing devices which induce varying degrees of boundary frictional restraint on the specimens.

For this purpose, an experimental programme was carried out (Kotsovos 1983) which involved the testing of two types of specimen: 250 mm height × 100 mm diameter cylinders and 100 mm cubes. Two mixes were used, the strength of which at the time of testing were: for the cylinders, 50.0 MPa (Mix 1) and 29.0 MPa (Mix 2); for the cubes, 60.0 MPa (Mix 1) and 37.7 MPA (Mix 2). The specimens were subjected to varying degrees of frictional restraint across their loaded surfaces. This was achieved by placing various types of 'anti-friction' media between the specimen and the hardened steel subsidiary load platens, the latter having a thickness of 25.4 mm and the same cross-sectional dimensions as the nominal cross section of the specimens tested. Three cylinders and three cubes, from each of the two concrete mixes used, were tested for every one of the following anti-friction media:

a. A layer of *synthetic rubber* (neoprene) 0.45 mm thick.
b. An *MGA pad* consisting of 0.008 mm thick hardened aluminium steel placed adjacent to the specimen; Molyslip grease (containing 3%MoS$_2$); and a Melinex polyester film, gauge 100, placed against the steel platen. (The higher-strength concrete specimens were tested by using previously unused MGA pads; the same pads were then used to test the lower-strength concrete specimens.)
c. A *brush platen* developed by splitting a steel platen longitudinally and transversely to form a large number of individual 'bristles'.
d. No 'anti-friction' medium, that is plain steel platens.
e. In addition to the above techniques, three cylinders made of the higher-strength concrete were subjected to an *active* restraint induced by 'Hi-Torque' hose clamps placed at a distance of 3 mm from the loaded surfaces and tightened by a small torque load of 1000 Nmm.

For the cubes, only the strength was determined, while for the cylinders both strength and deformational response were established. The complete deformational behaviour was measured within the central zone of the cylinders by attaching 60 mm electrical resistance strain gauges in the axial and circumferential directions. Two gauges placed diametrically opposite each other were used for each direction. The overall axial deformation of the specimens was also measured by using linear voltage displacement transducers (LVDTs).

The main results of the tests appear in Figures 2.5 through 2.10. Figure 2.5 shows the variation of the cube and cylinder strengths (normalised with respect to the cylinder strength (f_c) obtained from the tests using plain steel platens) with the various techniques adopted

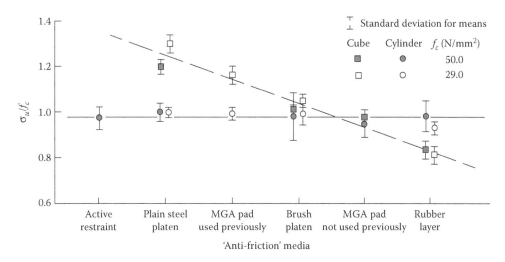

Figure 2.5 Variation of cube and cylinder normalised strength with the various 'anti-friction' devices adopted.

to reduce frictional restraint. The variation of cylinder strength with the technique used is also shown in Figure 2.6 which, in addition, includes the corresponding axial and lateral strains and the tangent values of Poisson's ratio. Figures 2.7 and 2.8 show the axial stress–axial strain and load–displacement relationships, respectively, obtained from the cylinder tests; for comparison purposes, these relationships are expressed in a normalised form with respect to both the maximum sustained stress (load) and corresponding strain (displacement). The typical deformational response of the cylinders under increasing load established

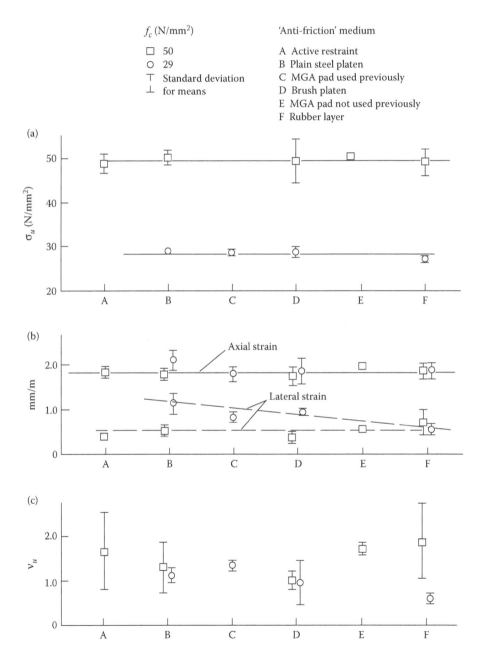

Figure 2.6 Variation of cylinder-tests' parameters at ultimate load with the various 'anti-friction' devices adopted: (a) strength; (b) axial and lateral strains; (c) Poisson's ratio.

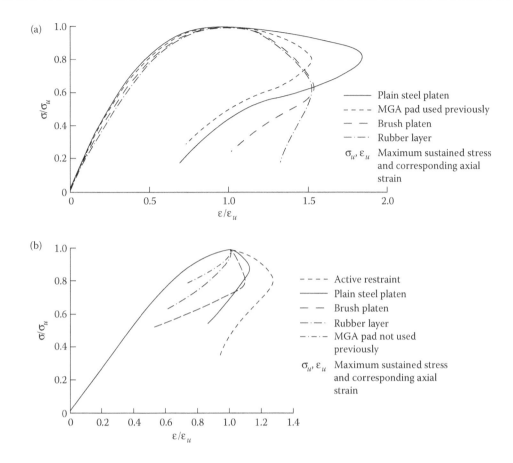

Figure 2.7 Axial–strain relationships established from cylinder tests: (a) $f_c = 29$ MPa; (b) $f_c = 50$ MPa.

from the strain-gauge measurements is compared with that established from the LVDT measurements in Figure 2.9. Finally, Figure 2.10 presents typical fracture modes of the cylinders subjected to the various degrees of boundary frictional restraint imposed by the testing techniques used.

The frictional restraint induced on the specimens by the testing techniques adopted may be quantified in terms of the strength of the cubes tested in compression. Since it is generally accepted that the difference between cube and cylinder strength reflects the influence of significant boundary restraints on the cube, it is considered realistic to accept also that, for a given concrete, the larger this difference, the higher the boundary restraints. On the basis of this argument, Figure 2.5 shows that the restraint is significantly reduced if increasingly efficient 'anti-friction' media are employed.

The normalised cylinder-strength data presented in Figure 2.5 are in agreement with those obtained in a previous investigation (Newman and Lachance 1964), which has shown that the uniaxial-compression strength of the cylinders with a height-to-diameter ratio of 2.5 is essentially independent of the frictional-restraint conditions across the loaded surface (a point already stressed in earlier parts of this chapter). It is of interest that the results obtained by using brush platens confirm the findings of similar investigations which have indicated that the use of such platens leads to the cube and cylinder strengths being nearly equal (see Kotsovos 1983). It may also be interesting to note in Figure 2.5 that, when a rubber layer is used, the cube strength is smaller than the cylinder strength. This is considered

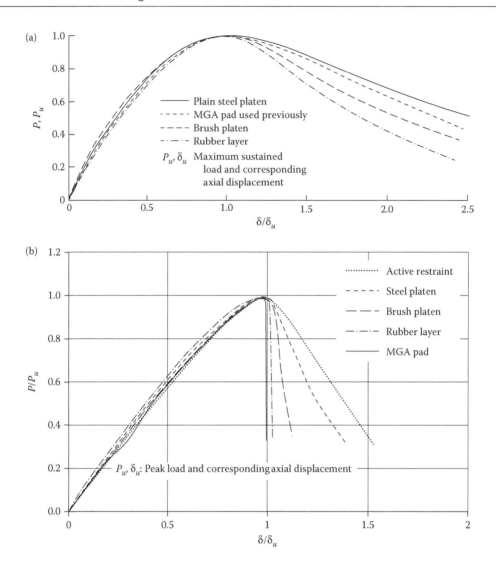

Figure 2.8 Load–displacement relationships established from cylinder tests: (a) $f_c = 29$ MPa; (b) $f_c = 50$ MPa.

to indicate that, as soon as the load increases to US, the rubber layer causes 'tensile' rather than 'compressive' stresses to develop across the loaded surface and that such stresses produce lateral expansion rather than induce restraint. (However, under stress levels beyond US, the capacity of the rubber for lateral expansion is much smaller than that of concrete, and, as a result, the 'tensile' stresses become 'compressive' and thus restrain further lateral expansion of the specimens, which then tend to be held together by friction.)

The cylinder-strength data, from which the normalised values of Figure 2.5 are derived, are shown in Figure 2.6 which also includes the corresponding data for axial and lateral strains. The figure indicates that, as for the strength data, the axial-strain values at US appear to be essentially independent of the boundary frictional restraint conditions. Furthermore, they are shown to be independent of the concrete strength. On the other hand, the lateral strains appear to decrease with increasing concrete strength, and while those of the higher concrete strength are also essentially independent of frictional restraint conditions, those of the lower concrete strength specimens seem to decrease with decreasing frictional restraint.

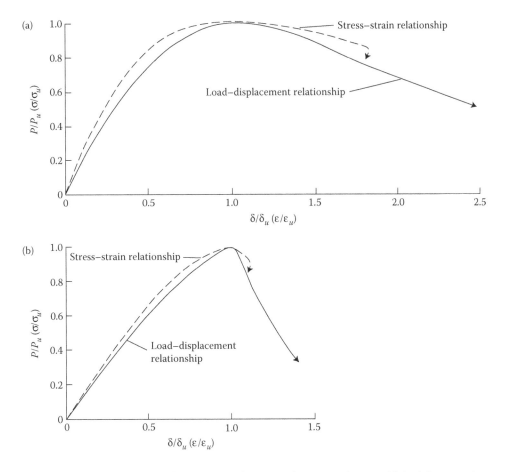

Figure 2.9 Deformation response of cylinders under uniaxial compression established from strain-gauge and LVDT measurements: (a) $f_c = 29$ MPa ; (b) $f_c = 50$ MPa.

(The reason for this difference in behaviour will become apparent later, when the inclination of the post-ultimate branch of the load–displacement relationships for the two types of concrete tested is discussed.)

The tangent values of Poisson's ratio at US are shown in Figure 2.6c which indicates that, for all cylinders tested, the values are much greater than 0.5. Such behaviour is compatible with earlier reported experimental information (Barnard 1964).

Figures 2.7 and 2.8 show that, for stress levels up to US, the deformational behaviour of the cylinders as established by both strain-gauge and LVDT measurements is practically the same for any of the boundary frictional conditions induced by the testing techniques adopted. On the other hand, for stress levels beyond US, both figures indicate a very pronounced dependency of the specimen response on the frictional-restraint conditions at their boundaries.

However, it is interesting to note that, while the load–displacement relationships of Figure 2.8 exhibit continuous increase of displacement throughout the whole of the loading path, the stress–strain relationships of Figure 2.7 show that the post-ultimate strain increases numerically but that the corresponding stresses decrease to a level between approximately 60% and 80% of US. At this level the trend reverses and the strain decreases continuously to complete disruption of the specimens. This late change of trend in behaviour in

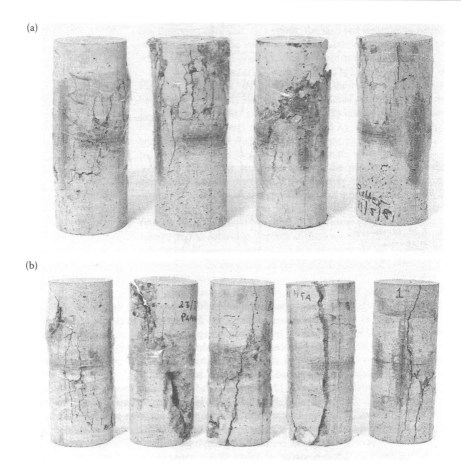

Figure 2.10 Typical fracture modes of cylinders (standard size 10 × 4 in/254 × 102 mm). 'Anti-friction' media used from left to right: (a) for lower-strength cylinders: plain-steel platen, brush platen, MGA pad, rubber layer; (b) for higher-strength cylinders: active restraint, plain-steel platen, brush platen, MGA pad, rubber layer.

the characteristics of Figure 2.7 is considered to reflect the effect on deformation of the occurrence of continuous axial cracks which subdivide the specimen in such a way that the applied load is predominantly supported by the core rather than the outer layers of the specimen, as indicated in Figure 2.11 (see also stage 3 in Figure 2.4). As a result, the stresses within the outer layers decrease and this causes an 'elastic recovery' (in fact, the outer layers are no longer load carrying) which is not typical of the overall specimen behaviour since it does not reflect the deformational response of the (ungauged) *core* of the specimen.

It can also be seen that, before the above 'elastic recovery' (i.e., unloading) occurs, the relationships of Figure 2.7 exhibit trends of post-ultimate behaviour similar to those of the relationships of Figure 2.8. There are, however, quantitative differences, demonstrated in Figure 2.9, which reflect the fact that, while the strain-gauge measurements describe essentially the deformational response of the central zone of the specimen (see Figure 2.2), the displacement measurements describe the overall response of the specimen–loading system. It can be shown, however, that, as the loading frame becomes stiffer, the displacement measurements become more representative of the specimen behaviour (Ahmad and Shah 1979). The relatively good correlation of the relationships shown in Figure 2.9 may be taken as an indication that, through the use of the relatively stiff loading frame, the loading-system

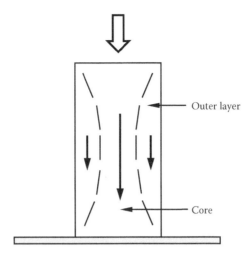

Figure 2.11 Distribution of post-ultimate load-carrying capacity of cylinders.

effects on the displacement measurements have been minimised to the extent that the resulting load–displacement relationships of Figure 2.8 provide a realistic description of the specimen behaviour.

Figure 2.8 indicates that, for all the boundary conditions investigated, the inclination of the post-ultimate branch of the relationships obtained for the lower concrete-strength cylinders appears to be less steep than that of the relationships obtained for the higher concrete-strength cylinders. This difference in behaviour is considered to reflect the difference in size of the central zone of the specimens within which continuous longitudinal cracks appear when the maximum load-carrying capacity of the specimens is approached (see Figure 2.4). For the higher concrete-strength specimens, the size of the above zone was observed to be substantially larger than that of the lower concrete-strength specimens, indicating a more dynamic type of crack propagation for the former as against the relatively gradual extension of the cracking exhibited by the latter; as a result, the size and restraining effect on lateral expansion of the end zones was smaller. Furthermore, it was also observed that cracking of the higher concrete-strength specimens always extended in both directions with increasing deformations (see Figure 2.10), thus reducing further the size and the restraining action of *both* end zones. In contrast, for the lower concrete-strength cylinders, cracking extended in one direction only and, as a result, the size and restraining action of at least one of the end zones was maintained throughout the whole length of the post-ultimate branch (see Figure 2.10).

However, a characteristic feature of the relationships shown in Figure 2.8 is that, for both concretes used, their descending portion becomes significantly steeper as the frictional restraint decreases. In fact, for the higher concrete-strength cylinders, the reduction of boundary frictional restraint achieved by some of the testing techniques adopted was sufficient to cause complete breakdown of the specimens as soon as US was exceeded. Such trends of behaviour appear to support the proposal that, if the boundary frictional restraints are completely eliminated, the post-ultimate behaviour of the specimens will be characterised by a complete and immediate loss of load-carrying capacity. At the same time, and as mentioned earlier, it is evident that the secondary effects due to interface friction at the ends of a cylinder having a height-to-diameter ratio of 2.5 are unimportant up to about the maximum sustained load, so that the uniaxial concrete strength is practically independent of the boundary restraint provided by the testing device.

It is important to stress that, even though the above conclusion regarding the brittle nature of concrete materials when tested uniaxially in compression has been reached on the basis of a particular experimental study (Kotsovos 1983), such a view is supported by the findings of a number of other investigators. In these, however, one rarely – if ever – finds an unambiguous statement that the post-ultimate deformation branch is attributable wholly to the interaction between specimen and loading platens, and is not a characteristic of the material (e.g., see Shah and Sankar 1987). Instead, such a belief (whether explicit or implicit) in the existence of the strain-softening regime of concrete is still widespread, despite evidence to the contrary, and it is often found even among those workers whose results lend support to the view that concrete suffers an abrupt and total loss of load-carrying capacity as soon as its peak strength is attained.

2.2.2 Triaxial compression tests and the effect of tensile stresses

In the preceding section, it was argued that a specimen of concrete under truly uniaxial compression does not exhibit a post-ultimate load–deformation branch. Since concrete in a structure is generally under the action of a multiaxial stress state, it is important to ascertain whether or not the mode of failure of the material remains brittle when subjected to these more complex loading conditions.

Considerable insight into the type of failure exhibited by concrete under multiaxial compression has been obtained by means of tests on concrete cylinders with a height-to-diameter ratio of 2.5 which, have been subjected to an axial compression (σ_a) and a lateral confining pressure (σ_c) (Kotsovos 1974; Kotsovos and Newman 1978). These actions were combined in such a way that the state of stress within the specimen was either $\sigma_a > \sigma_c$ (triaxial 'compression') or $0 < \sigma_a < \sigma_c$ (triaxial 'extension'). The axial compression was applied by using a loading method similar to that employed for uniaxial testing (i.e., no attempt was made to reduce the friction at the ends of the cylinder), while the confining pressure was hydrostatic (with the result that the curved surface of the specimen was essentially friction-free). In all tests, the entire specimens were subjected initially to a given hydrostatic pressure and then the axial compression was either increased (triaxial 'compression') or decreased (triaxial 'extension') to failure (see Figure 2.12a).

Typical stress–strain relationships obtained from the above tests are shown in Figure 2.12b which indicates that, while under triaxial 'compression' concrete exhibited a gradual reduction of load-carrying capacity for stress levels beyond US, under triaxial 'extension' concrete suffered an immediate and complete loss of load-carrying capacity. This difference in material behaviour is considered to reflect the effect of frictional restraint on the fracture processes of the specimens. According to the fracture mechanism of uniaxially tested concrete (discussed in Section 2.2.1.1), under triaxial 'compression' crack propagation occurs in the axial direction and, therefore, when cracking spreads in the end zones of the specimen, the frictional restraint will affect the specimen behaviour as for the uniaxial case. On the other hand, under triaxial 'extension' crack propagation takes place in the lateral direction (i.e., perpendicular to the cylinder's axis) and hence the fracture processes that occur in the central zone of the specimen are not affected by the frictional restraint that exists at the specimen–platen interfaces. The above findings strongly suggest that a truly 'unrestrained' specimen of concrete under arbitrary (but compressive) stress conditions would suffer a complete loss of load-carrying capacity on attaining its maximum strength. Therefore, it seems reasonable to postulate that brittle failure is a characteristic of concrete behaviour at a material level under any state of three-dimensional compression.

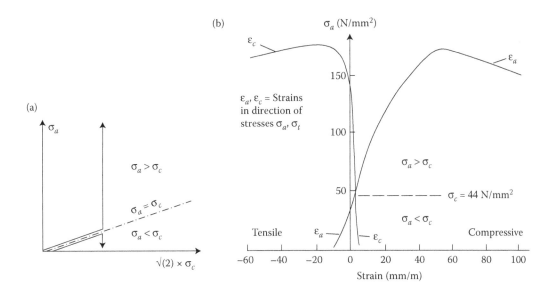

Figure 2.12 Triaxial compression tests on cylinders: (a) stress paths used; (b) typical stress–strain relationships obtained from 'compression' and 'extension' tests on cylinders with $f_c = 31.7$ MPa.

The effect of tensile stresses on the post-ultimate behaviour of concrete remains to be considered. The case when a cylinder is acted upon by confining circumferential pressure σ_c and an axial tension σ_a may be considered to be an extrapolation of the triaxial 'extension' case discussed above. In fact, such a specimen would be expected to exhibit an even more explosive type of failure than the case when σ_a is compressive. By reference to the particular instance of uniaxial compression, a similar lack of strain softening may be anticipated when σ_a is compressive while σ_c is tensile, provided that frictional end effects are eliminated. Finally, it seems unlikely that the presence of a fully tensile set of principal stresses will lead to an increase in ductility in the material. On the contrary, such a stress combination in concrete represents an inherently brittle system, despite gradually descending post-ultimate branches proposed by some investigators, as the latter strain softening, far from being a characteristic of the material, is a direct result of the control of crack propagation imposed by the machine in the course of tensile testing. In this respect, it is significant that strain-softening branches in direct tensile tests can be observed only if the stiffness of the testing machine is steeper than the steepest portion of the falling branch; otherwise, a sudden failure is deemed to occur (Evans and Marathe 1968). Evidently, the interaction between such testing machines and the specimen, which dictates the fracture processes in the material, has no relevance for concrete within a structure, as the latter is surrounded by material having the same (if PC) or similar (if reinforced concrete [RC]) stiffness.

Typical failed cylinders for the various triaxial-test types just described, showing the relevant fracture mechanisms, may be seen in Figures 2.13 and 2.14. For triaxial 'compression' (see Figure 2.13) (Newman 1973, 1979), where the principal stress is maintained compressive or is increased in compression, failure – under low values of confining pressure – can be observed to occur along planes parallel to the maximum compressive stress in a manner similar to uniaxial compression. For larger values of confining pressure, on the other hand, the failure planes become less distinct as the behaviour – although still brittle, since failure coincides with the peak of the stress–strain curve – becomes more

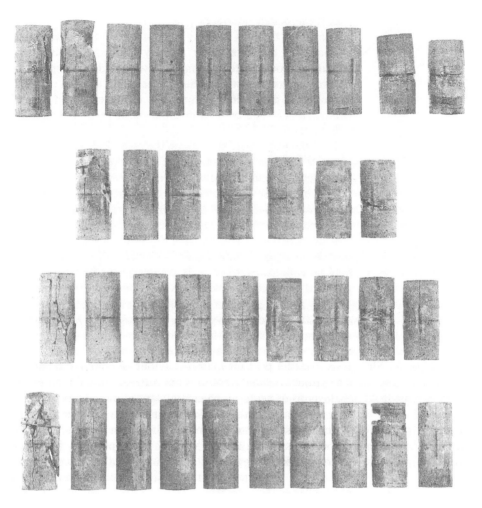

Figure 2.13 Typical concrete specimens (for four mix types) after failure under triaxial 'compression', with confining pressure increasing from left to right (all standard-size cylinders [standard size 10×4 in/254 $\times 102$ mm]).

'ductile' in the sense that larger strains now occur. The use of a conventional hydraulic coaxial cell to apply the loads gives rise to 'bulged' profiles of specimens under high confining pressures, allied to the considerable shortening in the axial direction (Newman 1973). In contrast, for the 'extension' or compression–compression–tension (C–C–T) test types (Figure 2.14) (Kotsovos 1974; Newman 1979), failure occurs abruptly on planes orthogonal to the minimum compressive stress. (In decreasing the axial stress to failure, the latter occurs in a 'tensile mode' under a resultant axial stress which may be either compressive [triaxial 'extension'] or tensile [C–C–T, or triaxial 'tension'], depending on the value of the maximum confining pressure in the test.)

2.2.3 Concrete: A brittle fracturing material

The test results discussed in previous sections shed considerable doubt over the validity of the current widely accepted view that a gradually decreasing post-ultimate load-carrying

Figure 2.14 Typical concrete specimens (for a given mix) after failure under triaxial 'extension' and C–C–T (all standard-size cylinders [standard size 10 × 4 in/254 × 102 mm]).

capacity for concrete is an essential part of the mechanism of stress redistribution within RC structures under increasing load. The validity of this view has, for a long time, been considered self-evident and in recent years a significant amount of research work has been concentrated on the development of testing apparatus capable of yielding stress–strain relationships which exhibit a gradually descending post-ultimate portion. Although such research has reached the stage when the shape of stress–strain relationships exhibiting 'softening' is readily obtainable, the results of the work presently described indicate that the time has come to reappraise existing testing methods in an attempt to establish to what extent the obtained relationships really represent the behaviour of an element of concrete *in a structure.*

The experimental evidence presented in the preceding sections points unequivocally to the conclusion that PC exhibits a sudden mode of failure under both uniaxial and the more general triaxial stress states. Such a brittle nature at the material level, however, is not incompatible with the fact that, when properly designed, a concrete structure or member can exhibit a ductile type of failure.

Ductility at the structural level is often due to the fact that material failure occurs locally and in a gradual manner, thus allowing stress redistribution to take place. Another obvious source of ductility is the presence of steel reinforcement. However, there is a more fundamental explanation as to why structural concrete is capable of failing in a ductile fashion, and this concerns the interaction of various elements of concrete within the structure (Kotsovos 1982; Kotsovos and Cheong 1984). This interaction gives rise to complex (triaxial) stress states which result in at least some of the principal stresses being very large: individually, these may be well in excess of the uniaxial strength of concrete but, when taken together, they represent a stress system below the *triaxial* strength of the material. Such large stresses are associated with large local strains which allow the formation of what one might term local 'hinges' that account for overall member ductility. This point will become evident later in the chapter, once the triaxial strength of concrete has been presented and the characteristic failure mechanism in a structure discussed.

In view of the above, it should be clear that, even though not consistent with the notion of concrete as a brittle material, plasticity concepts may often be used at a structural level, although they do not always guarantee the correct failure-mechanism and/or ultimate-load prediction. In the latter instances, the reason does not lie so much in the brittle nature of concrete. After all, Heyman (1982) has shown how, for brittle masonry structures, plasticity concepts can be applied. However, the stresses in the latter type of structure are very low and failure occurs through the formation of a mechanism. Concrete structures, in contrast, tend to be subject – albeit locally – to high triaxial stresses as their ultimate load is approached, and it is these large stresses which cause the strains necessary to produce ductile behaviour. Therefore, the latter is compatible with the brittle failure that occurs at the material level, and is not due to the existence of a strain-softening branch that is often argued to be necessary in order to explain the observed ductility of structural concrete.

2.3 FRACTURE PROCESSES IN CONCRETE

2.3.1 Non-linear behaviour of concrete materials

The response of concrete at a material level is characterised by a distinctly non-linear behaviour. Such a characteristic is already evident in the early portion of the stress–strain relations, becoming more pronounced as the US is approached. If non-linear computer-based techniques are to be employed to analyse rigorously the response of concrete structures to the imposed loading conditions, information must be expressed in a mathematical form suitable for use with such techniques. Before this is done, however, the origin and nature of material non-linearities should be properly understood.

The non-linear properties of concrete are generally accepted to reflect the fracture processes which occur under stress, and hence knowledge of these processes provides a suitable basis for the mathematical description of concrete non-linearity (Kotsovos 1979a). The term fracture process is used to denote the structural changes which a brittle material undergoes under increasing stress; in concrete, these structural changes are considered to be primarily those caused by cracking. A fracture mechanism (i.e., the law or group of laws which controls the above fracture processes) for concrete, valid under any state of stress and which can be used to predict observed deformational behaviour, may be postulated on the basis of theoretical considerations regarding the extension of a crack within a brittle medium (which, concrete is, as shown in Section 2.2). This fracture mechanism forms the basis for an analysis of experimental triaxial stress–strain data which will lead to the mathematical description of the deformational behaviour of concrete under static short-term loading conditions. Before such an analytical description is presented, it is helpful to give, in a unified form, the complete *qualitative* description of the structural changes of concrete under any type of increasing stress.

2.3.2 Causes of fracture

It is generally accepted that the cause of fracture and failure of concrete is the proliferation of flaws or micro-cracks which exist within the body of the material even before the application of load. These flaws are attributable to a number of causes; the main ones are

a. Discontinuities in the cement paste matrix resulting from its complex morphology (such flaws range in size from a number of angstrom units at the gel-lattice level, to several microns (or above) for isolated or continuous capillary pores).
b. Voids caused by shrinkage or thermal movements due to incompatibility between the properties of the various phases present in concrete.
c. Discontinuities at the boundary between the aggregate particles and the paste or mortar matrix caused by segregation.
d. Voids present in concrete as a result of incomplete compaction.

The above pre-existing flaws can be considered as randomly distributed and oriented within the material and to exhibit a range of shapes and sizes.

The stress and strain applied to the boundary of an element of a composite material, such as concrete, generates a strain field within the material which is dependent on the distribution of the component phases (i.e., aggregate particles and cement-paste matrix), and the size, shape and distribution of flaws. Local strain concentrations, therefore, develop throughout the material as a result of the incompatible deformation of the constituent phases. Such strain concentrations are further intensified to far higher orders of magnitude owing to the presence of flaws, particularly those with high aspect ratios (Griffith 1921). It is primarily these flaws that are considered to be the potential sources of any load-induced cracking.

2.3.3 Fracture mechanism of concrete

The existence of a flaw or micro-crack within a *brittle elastic material* under stress disturbs the stress and strain field around the micro-crack and causes *high,* predominantly *tensile,* stress and strain concentrations (and thus strain-energy concentrations) within small regions near the micro-crack tips. The energy capacity of these small regions must have a limit dependent on the material properties which, if exceeded, results in spontaneous crack extension (Griffith 1921) due to the initiation of branches. This process is followed by stable propagation of these branches, eventually becoming unstable and leading to ultimate collapse (Brace and Bombolakis 1960; Hoek and Bieniawski 1965). The above crack extension and propagation processes have been found to occur in the direction of the *maximum principal compressive stress* applied to the boundaries of a specimen (or orthogonal to the direction of the maximum principal tensile stress) (Brace and Bombolakis 1960; Hoek and Bieniawski 1965) and takes place in order to release any excessive strain energy fed within the material by external loads. (The cracking can be visualised by reference to a series of parallel planes of fracture containing the largest and intermediate principal compressive stresses, the planes extending along the former [i.e., maximum] stress direction as cracking propagates.)

For a tensile stress field, crack extension is insufficient to reduce the strain-energy concentrations to a level that can be contained within small regions near the crack tips. As a result, the fracture process is unstable in that crack extension continues to complete disruption even if the load is maintained constant.

On the other hand, for a predominantly compressive stress field, the strain energy which is released during the fracture process reduces the strain-energy concentrations to a level below the energy capacity of the material in the vicinity of the crack tips and, as a result, the crack ceases to extend. Crack extension is resumed when energy fed into the material by external loads tends to increase again the strain-energy concentrations to values above the capacity of the local energy-storing regions. The fracture process, therefore, occurs in discrete steps, and is stable since crack extension ceases when the load is maintained constant. Eventually, however, further increases in load lead to the stage when no more energy can be stored within the material; any additional energy induced by the applied load is released immediately by crack extension which, in turn, reduces further the energy capacity of the system and creates a state of instability (Kotsovos and Newman 1981b).

The energy release which occurs during the above fracture processes results in a reduction of the stress and strain concentrations which exist near the crack tips orthogonal to the path of crack extension. Furthermore, these fracture processes create voids within the material (see Figure 2.15). The reduction of these high tensile stress and strain concentrations tends to cause a contraction of the material in the direction normal to the crack-extension path, while void formation tends to cause an extension. The overall effect on deformation, therefore, will be either a contraction or an extension, depending on whether the effect of reduction of the tensile stress and strain concentrations or the effect of void formation predominates.

Based on the research of previous investigators, which has been concerned with the study of the fracture processes of concrete at both the microscopic and the phenomenological levels, it has been postulated that the fracture mechanism of this material is qualitatively similar to the fracture mechanism described above. The salient features stemming from the work of these investigators may be summarised as shown in Table 2.1 (Newman 1973; Kotsovos 1979a). (It should be noted that the deformational behaviour denoted in the table by the term 'quasi-elastic' and 'plastic' could be described more accurately as non-linearly elastic and inelastic [i.e., exhibiting permanent deformation upon unloading] respectively.) A feature of the fracture process being discussed is that its mechanism is specific, in the sense that, as mentioned previously, crack extension occurs in the direction of the 'applied'

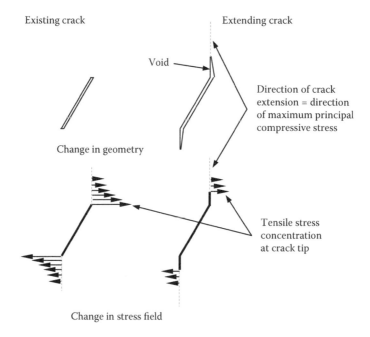

Figure 2.15 Schematic representation of changes in crack geometry and stress field associated with crack extension.

Table 2.1 Fracture processes of concrete when concrete is subjected to load

Observed material behaviour under increasing stress	Deduced changes in internal structure	Deduced changes in deformational behaviour	Approximate stress level	Influencing characteristics
Presence of cracks detected microscopically and from x-rays	Pre-existing cracks		Prior to load	Shrinkage, thermal effects, and so on
From structural investigations:				
Bond cracking and matrix cracking increases depending on spacing of particles of aggregate, with cracking aligned in direction of maximum compressive stress as load increases		Quasi-elastic behaviour	30%–40% ultimate	Spacing of particles for a given mix and so on
From phenomenological investigations:				
Decrease in ultrasonic pulse velocity, start of significant acoustic emissions, measurable void formation	Start of localised stable cracking	Plastic behaviour	45%–90% ultimate	Method of detection, type of aggregate, type of specimen. state of stress, rate of loading, curing conditions, and level decreases with: increase with w/c ratio; increase in volume fraction of aggregate: decrease in age; decreasing roughness and angularity of aggregate
From structural investigations:				
Formation of continuous crack patterns through matrix and around aggregate particles	Start of continuous cracking		70%–90% ultimate	Spacing of aggregate particles for a given mix
From phenomenological investigations:				
Increase in volume, marked increase in acoustic emission, and change in stress–strain relationship	Unstable behaviour caused by reduction in number of load paths due to cracking (leading to complete disruption)		70%–90% ultimate	As for localised stable cracking (detected phenomenologically)

Source: Newman J. B., 1973, Deformational behaviour, failure mechanisms and design criteria for concretes under combinations of stress, PhD thesis, University of London.

maximum principal compressive stress, and affects predominantly the deformation in the direction orthogonal to the crack-extension paths, while the effect on deformation along these paths is negligible. Clearly, it is implicitly assumed here that the many micro-cracks that exist within concrete under stress extend in a mode *qualitatively similar* to that of a single micro-crack within a stressed brittle medium. Such a postulate is reasonable, but it must be borne in mind that, in a heterogeneous material such as concrete, the crack extension paths will be orientated in the direction of the *local* rather than the 'applied' maximum principal compressive stress. Owing to this heterogeneous and complex nature of concrete, the direction of *local* principal stresses is indefinable, and hence the orientation of fracture paths is difficult, if not impossible, to establish. Nevertheless, it seems realistic to assume that, under a deviator stress (i.e., any stress state which is not purely hydrostatic), the directions of most of these *local* maximum principal compressive stresses practically coincide with the direction of the 'applied' maximum principal compressive stress. This is supported by a comprehensive investigation of the fracture processes of concrete specimens subjected to near-uniform states of multiaxial stress (Newman 1973; Kotsovos 1974; Kotsovos and Newman 1977), which has indicated that the fracture processes which lead to ultimate collapse of the material are, indeed, those that occur in the direction of the maximum applied compressive (principal) stress. Therefore, it would appear that, for concrete under deviator stress, the effect of the fracture processes on deformation may be reflected in the relationship between the deformation of the material in the direction of the 'applied' maximum principal compressive stress and the deformation in the orthogonal direction under increasing stress; such a relationship could thus form the basis for the investigation of the fracture processes of concrete under increasing load. On the other hand, it is found that hydrostatic stress states are associated with an essentially random orientation of fracture processes.

In view of the above, it is convenient to classify the fracture processes of concrete into the following two categories.

Category A: Processes with a unique orientation. These are easily definable since the propagation path coincides with the direction of the 'applied' maximum principal compressive stress. Such fracture processes may occur under any type of non-hydrostatic state of stress but they are unlikely to take place under pure hydrostatic stress.

Category B: Processes with a 'random' orientation (in the sense that the propagation direction is indefinable). Such fracture processes may occur under any stress state but they are considered to be reduced to a minimum under a pure deviatoric stress.

It follows from the above considerations, therefore, that a deviatoric stress state causes predominantly the fracture processes of category A, whereas a hydrostatic stress causes only the fracture processes of category B. A detailed description of these two categories of fracture processes is given below.

2.3.3.1 Fracture processes under deviatoric stress

It is assumed here that, under a deviatoric state of stress, the principal directions of both *local* and *applied* states of stress coincide. For this reason, no distinction is made in the following between local and applied stress states.

The various stages of the fracture processes – Comprehensive investigations into the behaviour of concrete under multiaxial stress (Newman 1973; Kotsovos 1974) have indicated that there are at least four stages in the process of crack proliferation under increasing stress. These are as follows.

Stage 1. When the load is first applied, micro-cracks additional to those pre-existing in the material may be formed at isolated points where the tensile stress concentrations due to the incompatible deformations of the aggregate and cement-paste phases are highest. During this stage, the micro-cracks do not propagate but remain stable.

Stage 2. As the load is increased, high tensile strain concentrations gradually develop near the tips of the micro-cracks as a result of the micro-crack geometry and/or orientation. The stage is reached, therefore, when the initially stable micro-cracks begin to initiate branches in the direction of the maximum principal compressive stress. This branching process tends to relieve the strain concentrations and, once strain redistribution has occurred, the individual crack configurations remain stable during further increases of applied stress. Although this process may produce voids, the reduction of strain concentrations along the crack branches is such that it results in contraction of the material in localised zones near the crack tips which, in turn, causes the rate of increase of the tensile strain in the direction at right angles to that of branching to be reduced with respect to the rate of increase of the strain in the direction of branching. The start of such deformational behaviour has been termed *local fracture initiation* (LFI) and is considered to mark the start of the branching-initiation process.

Stage 3. When the load is increased to a higher level, a stage is reached at which the branched cracks start to propagate. During this crack-propagation stage, each crack of the system extends in a relatively stable manner, in that, if the applied load is held constant, the process ceases. Although the relief of strain concentrations continues during this process, void formation is such that it causes the *rate of increase* of the strain at right angles to the direction of branching to increase with respect to the rate of increase of the strain in the direction of branching. The start of such deformational behaviour is considered to mark the start of the stable crack-propagation process and has been termed *onset of stable fracture propagation* (OSFP).

Stage 4. The degree of cracking eventually reaches a more severe level, after which the crack system becomes unstable and failure occurs even if the load remains constant. The start of this stage has been termed *onset of unstable fracture propagation* (OUFP). Under compressive stress states, this level is easily defined since it coincides with the level at which the overall volume of the material becomes a minimum. Under predominantly tensile stress states, it is marked by a rapid increase in the overall volume of the material and can be detected as described elsewhere (Kotsovos 1974).

The above four stages of crack extension and propagation are illustrated schematically in Figure 2.16 by reference to concrete under compressive stress. Such a representation may be thought of as possibly corresponding to a uniaxial cylinder test in which the frictional

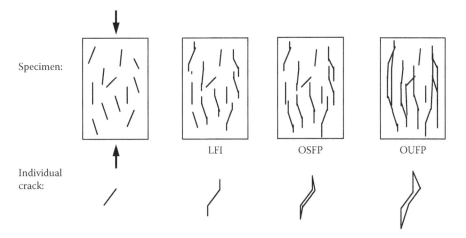

Figure 2.16 Stages in the process of crack extension and propagation for concrete under compressive stress.

effects between specimen and platens have been removed, or to the central portion of such a specimen if the end effects are present.

A simplified qualitative description of micro-cracking and macro-cracking processes: critical levels of concrete behaviour – It is important to emphasise that the above four stages of the fracture process are not all as clearly defined as might have been implied by the subdivision adopted. In particular, there is no easily detectable limit separating the local and essentially random process of stage 1 from the aligned cracking of stage 2. However, an analysis of triaxial stress–strain data can lead to a detection of reasonably distinct levels of change in the behaviour under increasing stress. As noted earlier, these levels are considered to represent the start of the stages within which crack branching (LFI), stable crack propagation (OSFP), and unstable crack propagation (OUFP) occur. (It is interesting to note that it has not been possible to detect the LFI level under triaxial 'compression' [C–C–C] of cylinder specimens when the maximum principal compressive stress is applied axially so that $\sigma_a > \sigma_c$. However, an indication of this level may be obtained by using the LFI levels detected under triaxial 'extension' [C–C–C but with the maximum principal stress applied laterally, i.e., $\sigma_a < \sigma_c$] and triaxial 'tension' [C–C–T i.e., σ_a tensile] – see Kotsovos and Newman 1981b.)

The above considerations suggest that, in attempting to establish failure criteria for concrete, three critical levels might be considered. The significance of these levels will depend on the damaging effect which the ensuing fracture process has on the structure of the material. Since branching-crack initiation appears to induce stabilisation of the material owing to the resulting relief of the stress concentrations (and, as intimated above, it is not always readily detectable), the OSFP and OUFP levels could be considered suitable for use as bases for lower-bound and upper-bound failure criteria in conformity with the limit-state requirements of serviceability and US. The validity of this consideration is supported by experimental evidence which has indicated a very close correlation between: (a) the OSFP level and the 'fatigue strength' of concrete, that is that level below which concrete does not suffer distress under repeated applications of load; and (b) the OUFP level and the 'long-term strength' of concrete, that is, that level below which concrete does not collapse under sustained load.

For structural purposes, however, the number of repeated loadings at the OSFP level which are required to cause fatigue failure of concrete is considered to be impractically high. For this reason, it seems that it is mainly the OUFP level which is relevant in structural applications where the governing criterion is that of static strength. In this respect, it is found convenient to simplify the cracking processes conceptually by considering these to consist of two major stages, which occur at the microscopic and macroscopic levels of observation, respectively. Accordingly, the first of these can be denoted generically by the term micro-cracking, and encompasses stages 1–3 (inclusive) described above. Micro-cracking can be said to be the underlying cause of the non-linear behaviour at the material level and thus determines the constitutive relations of concrete. One way of describing it would be as a static process in the sense that crack extension stops when the load is maintained constant. The second major stage in the fracture process is that of macro-cracking, which coincides with stage 4 discussed earlier. In contrast to micro-cracking, macro-cracking indicates material failure in localised regions within a structure and is a dynamic phenomenon in that crack extension continues even if the load is maintained constant. It stops when equilibrium, which is disturbed by material failure in the region of the macro-crack, is re-established through stress redistribution (elsewhere) in the structure. Due to material breakdown, macro-cracking causes local discontinuities in the original geometry of a structure. Clearly, unlike micro-cracking, macro-cracking affects concrete at the *structure level* and hence defines the *failure criterion* (or the failure 'envelope') of concrete.

Failure envelopes – Figure 2.17 shows typical variations in stress and strain space of LFI, OSFP, OUFP and US. (The actual values correspond to a concrete with $f_c = 46.9$ MPa subject

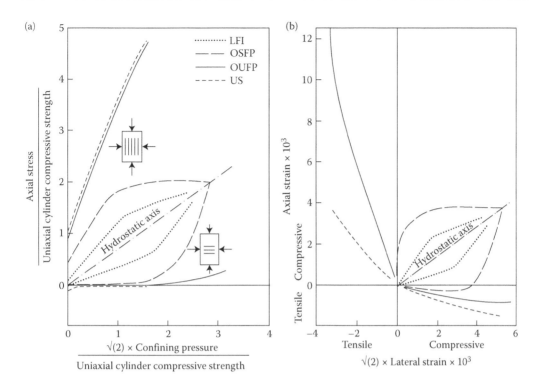

Figure 2.17 Typical LFI, OSFP, OUFP and US envelopes for concrete (with $f_c = 46.9$ MPa) subjected to axisymmetric triaxial stress states using the stress path 3 defined in Figure 2.18: (a) stress space; (b) strain space.

to axisymmetric triaxial conditions [Kotsovos and Newman 1977, 1981a,b; Kotsovos 1979a] in which the stress path 3 [Kotsovos 1979a,b] depicted in Figure 2.18 was followed.) The figure indicates that the OSFP forms a closed envelope in both stress and strain space. It may also be noted that, in both strain space and the wholly compressive portion of the stress space, the OSFP envelope is nearly symmetrical about the hydrostatic axis.

Such behaviour implies that, for stress and strain states enclosed by the envelope, the material may be considered isotropic. Since the orientation of crack branching under a triaxial 'compression' state of stress is perpendicular to that under triaxial 'extension' or C-C-T (see the schematic sketches in Figure 2.17a), such isotropic behaviour indicates that the fracture process up to the OSFP level causes insignificant disruption in the structure of the material. In view of this isotropy, it is possible that the LFI envelope within the triaxial 'compression' zone may be represented by the reflection of the LFI envelope in the triaxial 'extension' and C-C-T zones with respect to the hydrostatic axis, as shown in the figure.

In contrast to the LFI and OSFP envelopes, the OUFP and US envelopes are open-ended for the range of stresses used in the tests, and are non-symmetrical with respect to the hydrostatic axis. This latter observation suggests that, for stress and strain states outside the OSFP envelope, the material becomes anisotropic, because crack extension occurs along a particular direction dictated by the maximum principal compressive stress. However, allowing for the scatter of results that is encountered in processing the deformational stress–strain data within the OUFP envelopes for a wide variety of concrete mixes (this scatter will be discussed in the next chapter), it seems reasonable to postulate that the description of the stress–strain relations may be approximated by means of an isotropic model up to the

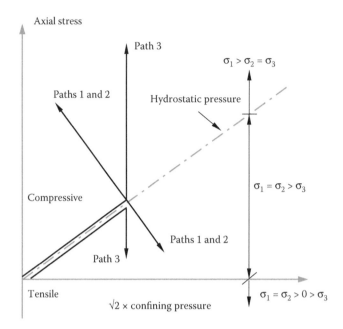

Figure 2.18 Schematic representations of various stress paths used in the triaxial testing of concrete cylinders.

OUFP level corresponding to macro-cracking. Moreover, it is found that the OUFP stress level forms a surface in stress space similar in shape and size with that of the ultimate-strength surface. As the strength of the concrete increases, the two surfaces become practically identical, but even for lower-strength concrete these surfaces are quite close to each other (Kotsovos 1974). Therefore, it seems reasonable to consider the OUFP and US levels as being essentially the same, at least for practical purposes. This may be represented schematically, as in Figure 2.19, by reference to both restrained and unrestrained concrete specimens under uniaxial compression.

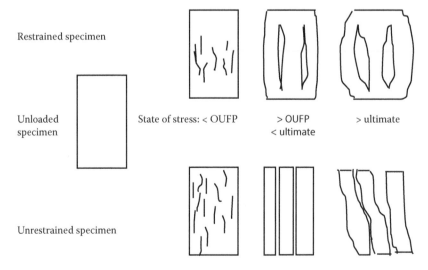

Figure 2.19 Schematic representations of the fracture processes for restrained and unrestrained concrete specimens under increasing compressive stress.

2.3.3.2 Fracture processes under hydrostatic stress

This section can be summarised as follows.

2.3.3.2.1 Description of the fracture processes

Except for the 'random' orientation (which, as a result of its local character, cannot be measured), the fracture processes under hydrostatic stress may be considered qualitatively similar to those described by the stages 1–3 of the fracture processes under deviatoric stress. Owing to this 'random' orientation, a characteristic feature of these processes is that they increase the likelihood that a crack will be situated in the path along which a potential crack may propagate under some future applied state of stress. The former crack will act as a crack-propagation 'inhibitor' which will tend to increase the energy required to start the crack-propagation process (Cook and Gordon 1964). The number of such crack 'inhibitors' will increase with the level of applied stress, and it is considered realistic to assume that the crack extension and propagation processes will eventually diminish under hydrostatic stress. Furthermore, under such a state of stress the fracture processes described above are considered capable of reducing the internal stress and strain concentrations to such small values that a very high load, well outside the capacity of currently used testing machines, will be required to restart a fracture process which will lead to ultimate collapse of the material.

The application of hydrostatic stresses, which cause a reduction in stress concentrations, increases the isotropy of concrete at the macroscopic level. Under such conditions, concrete can withstand very high hydrostatic loading (as mentioned above, considerably outside the capacity of current testing machines), a well-known feature of isotropic media.

2.3.3.2.2 Effect of fracture processes on failure envelopes

Figures 2.20 through 2.22 show the failure envelopes obtained from the tests on a typical concrete with $f_c = 31.7$ MPa subjected to various stress paths (see Figure 2.18) comprising both *hydrostatic* and subsequent *deviatoric* portions. Path 2 is similar to path 1 except that, after applying the required hydrostatic pressure, the latter is maintained constant until the rate of increase of deformation through 'creep' has reduced to a negligible value (as compared with the rate of deformation under increasing deviatoric load as assessed from the results of previous tests using path 1). This path has been investigated for values of hydrostatic pressure higher than $0.7f_c$ since, for lower values, 'creep' under hydrostatic pressure is so small that the path coincides with path 1.

The figures indicate that the stress path has a similar effect on all envelopes but this effect appears to be smaller for OUFP and US, particularly within the $\sigma_1 = \sigma_2 > \sigma_3$ region where it seems to be insignificant. For values of the hydrostatic component of the stress states at OSFP, OUFP and US up to approximately $0.8f_c$, the envelopes are unaffected by the stress path and therefore can be expressed by a single envelope. For higher values of the hydrostatic component the envelopes appear to be stress-path dependent, such that, for a given hydrostatic level, OSFP, OUFP and US occur under deviatoric stress levels which increase with the type of stress path, varying in the order 3, 1, 2 for $\sigma_1 > \sigma_2 = \sigma_3$ and 1, 2, 3 for $\sigma_1 = \sigma_2 > \sigma_3$. It is clear that the deviatoric stress levels increase by both increasing the maximum value of the pure hydrostatic stress of the test and sustaining this value until the rate of deformation becomes negligible in comparison with that occurring under the subsequent deviatoric stress. Such behaviour has been attributed to the fracture processes of 'random' orientation which occur under hydrostatic stress and, in particular, to the delaying effect which the crack propagation 'inhibitors' have on the fracture processes that occur under the deviatoric portions of the stress path. The number of such 'inhibitors' increases as a result of

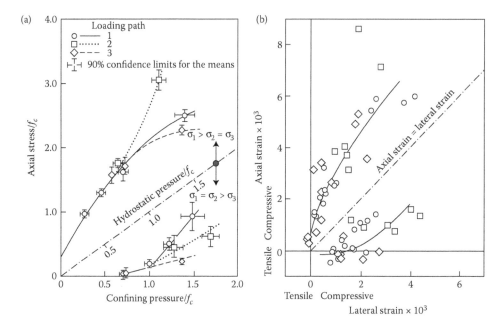

Figure 2.20 Typical OSFP envelopes ($f_c = 31.7$ MPa): (a) stress space; (b) strain space.

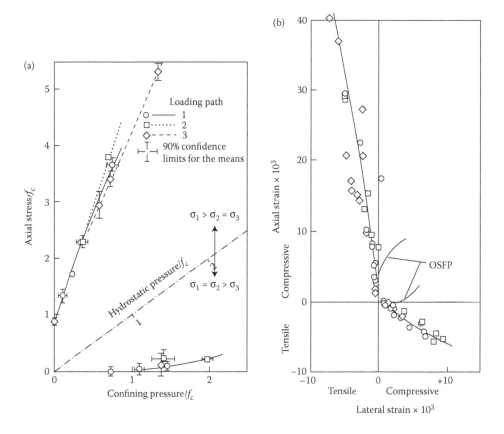

Figure 2.21 Typical OUFP envelopes ($f_c = 31.7$ MPa): (a) stress space; (b) strain space.

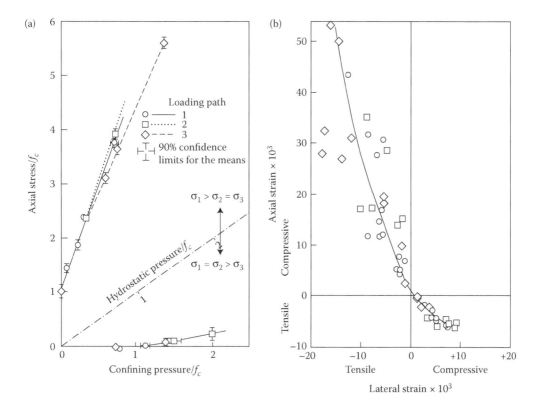

Figure 2.22 Typical US envelopes ($f_c = 31.7$ MPa): (a) stress space; (b) strain space.

both increasing and sustaining the maximum value of the hydrostatic stress of the test, and this causes the failure envelopes to expand in stress space.

It should be noted, however, that the above qualitative observations regarding the effect of the stress path on the various envelopes should not obscure the key fact that the variability of the data in the strain space is such that it is difficult to establish any path effects on the failure envelopes, that is the experimental scatter swamps any path-dependency effects even at high values of the hydrostatic-stress component. Moreover, the available results for the OUFP and US envelopes seem to indicate that, if any significant stress-path influence does occur, this can take place only at stress levels that are unlikely to be realised in ordinary structural-concrete applications. It can be concluded, therefore, that, for practical purposes, it is sufficiently accurate to neglect the effect of the stress path on: (*a*) the US envelope (as noted in Section 2.3.3.1, the latter can be taken to be practically identical to the corresponding OUFP envelope); (*b*) the stress–strain relations (since, allowing for data scatter, the strains are – like the stresses – effectively independent of the stress path).

The above conclusions are both confirmed and extended by some tests on cylinders in which the stress path consisted of the following three stages: increasing the hydrostatic pressure to a given value; decreasing the hydrostatic pressure to zero and applying an axial compressive stress to failure (Kotsovos and Newman 1981b). Such an approach provides some information on the changes in the material caused by previous histories of hydrostatic loading. Figure 2.23 shows the variation, with the maximum value of the hydrostatic stress of the test, of the stress levels at OSFP, OUFP and US obtained under subsequent uniaxial compression. This figure indicates that the above stress levels vary slightly up to a value

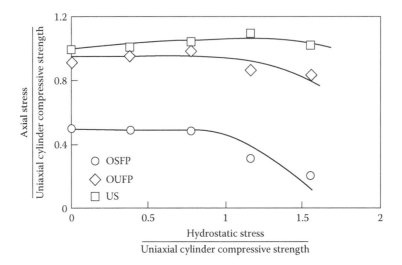

Figure 2.23 Variation of stress level at OSFP, OUFP and US with hydrostatic stress obtained from typical uniaxial-compression tests on concrete (with f_c = 47 MPa) which has been subjected to previous hydrostatic stress.

of the hydrostatic stress of approximately $0.8f_c$ beyond which they begin decreasing with increasing hydrostatic stress. If it is assumed that the fracture processes that occur under the hydrostatic portion of the stress path are related to the maximum value of the applied hydrostatic stress, the above variation reflects the effect of these fracture processes on the fracture processes that occur under the subsequently applied uniaxial compression.

As already mentioned, in a heterogeneous material such as concrete, the principal directions of the local stress fields that develop within the material when subjected to hydrostatic stress are indefinable and the cracking paths are of random orientation. Such cracking tends to create a more uniform stress and strain distribution within the material, because high stress and strain concentrations are relieved, and when situated in the path along which a crack is likely to propagate, tends also to inhibit such crack propagation. These effects may, in fact, cause the slight increase of the stress levels at OUFP and US with increase of hydrostatic stress up to about $0.8f_c$, as well as the increase in stress difference between OUFP and US which may be attributable to the delaying effect of the above fracture process on the unstable material behaviour above OUFP – see Figure 2.23. However, the figure suggests that, for values of hydrostatic stress higher than $0.8f_c$, the fracture processes under hydrostatic stress reduce the energy capacity of the material and thus the stress levels at OSFP, OUFP and US decrease with increasing hydrostatic stress. Nevertheless, any decrease in OUFP and, especially, US, appear to be quite small even up to the limit of available results (i.e., hydrostatic stress of around $1.5f_c$), particularly in view of the ever-present material-data scatter.

2.3.3.3 Fracture processes under generalised stress

Since any generalised state of stress can be decomposed into a hydrostatic and a deviatoric component, it would be expected that, under such a stress state, both types of fracture process discussed above would occur concurrently. The crack extension and propagation processes caused by the deviatoric component are considered to be those fracture processes which will eventually lead to ultimate collapse of the material. On the other hand, the fracture processes caused by the hydrostatic component will act as 'inhibitors' of the previous type of fracture process. Such an effect will tend to reduce the rate of the crack propagation processes caused by the deviatoric component, and thus increase the stress level required to cause ultimate failure, as is evident, for example, by reference to the OUFP and US envelopes for paths 1 and 2, which lie outside those for path 3 (the latter having a smaller hydrostatic-stress component since, in Figures 2.20 through 2.22, the hydrostatic stress corresponding to any given data point is obtained by a projection onto the hydrostatic axis that is either perpendicular to it (stress paths 1 and 2) or parallel to the axial-stress axis (stress path 3)). However, as mentioned earlier, such path dependency is small and, for practical purposes, can be ignored.

2.4 FAILURE MECHANISM IN CONCRETE STRUCTURES

2.4.1 A fundamental explanation of failure initiation based on triaxial material behaviour

The experimental information on concrete behaviour presented in the preceding section can be used to anticipate the typical mode of failure initiation in plain and RC structures by means of a simple reasoning based on two key features of concrete materials. These two fundamental characteristics are shown in Figure 2.24, which summarises much of the earlier discussion on material behaviour.

As is well known, concrete is weak in tension and strong in compression. Its primary purpose in an RC structural member is to sustain compressive forces, while steel reinforcement is used to cater for tensile actions with concrete providing protection to it. Therefore, since the structural role of concrete is concerned primarily with compressive stress states, the present discussion relates to its strength and deformational response under such conditions. Now, information on the strength and deformational properties of concrete is usually obtained by testing cylinder or prism specimens *under uniaxial compression*. Although typical stress–strain curves stemming from such tests have been shown in the preceding chapter, it is useful to refer to Figure 2.24a which depicts, in a generic sense, one further set of such curves. The figure serves as a reminder that, in addition to the strain in the direction of the loading (which usually constitutes the main – if not the sole – item of interest in current design thinking), the uniaxial test also provides information on the strain perpendicular to this direction. Furthermore, a typical plot of volumetric-strain variation appears in the figure. A characteristic feature of the curves in Figure 2.24a is that they comprise ascending and gradually descending branches. However, despite the prominence given to the latter in design, it was explained in Section 2.2.1.2 how experimental evidence shows quite conclusively that, unlike the ascending branch, the descending branch does not represent actual material behaviour: rather, it merely describes secondary testing-procedure effects resulting from the interaction between testing machine and specimen. This is an important observation concerning the behaviour of concrete at the material level, as the lack of strain softening, that is post-ultimate branch, justifies its being referred to as a brittle material. On the other hand, it turns out that considerations of the behaviour of concrete at the structural level make the actual post-ultimate response of the material irrelevant for, even if the latter were to exist, failure of concrete in a structure invariably occurs before the attainment of its ultimate compressive stress. The case for such a statement may be argued along the following lines.

Perhaps the most significant feature of concrete behaviour is the abrupt increase of the rate of lateral expansion a uniaxial-test specimen undergoes when the load exceeds a level

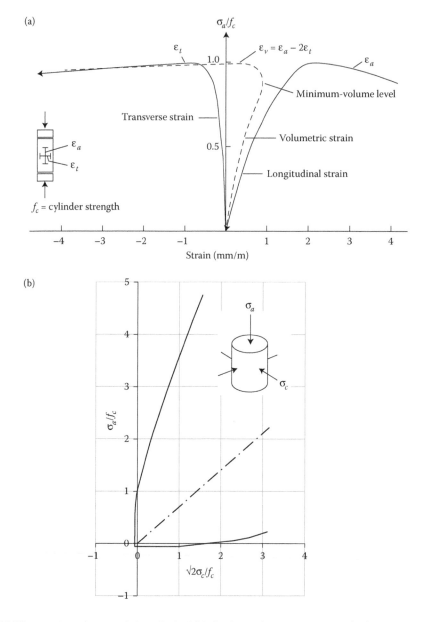

Figure 2.24 The two key characteristics of triaxial behaviour of concrete materials that govern the fundamental failure mechanism of structural concrete: (a) typical stress–strain curves obtained from
cylinders under uniaxial compression (note the rapid increase of tensile strains orthogonal to
the direction of principal stress once the OUFP level is exceeded); (b) typical failure envelope
of concrete under axisymmetric triaxial stress (note the large effect of even relatively small
secondary stresses σ_c on the load-carrying capacity σ_a).

close to, but not beyond, the peak stress. Such a feature was noted already in Section 2.2,
and the relevant stress level may be identified as the minimum-volume level (see Figure
2.24a) which marks the beginning of a dramatic volume dilation that, in the absence of any
frictional restraint at the interface between the ends of the specimen and the steel platens, is
considered to lead rapidly to failure even if the load remains constant. This is why the stress

at which concrete begins to expand is associated with a process governed essentially by void formation, and is termed the OUFP level which, for all practical purposes, may be equated to the failure load, as explained in Section 2.3.3.1. It is important to emphasise here that the rapid expansion at the minimum-volume level, with the tensile strain at right angles to the direction of maximum compressive stress soon exceeding the magnitude of the compressive strain, is a feature of both uniaxial and the more general triaxial compressive behaviour (see Figures 1.14 and 1.15, respectively).

The other key feature of concrete behaviour relates to the major role played by even relatively small (secondary) stresses when the true bearing strength of the material is being assessed. This may be illustrated by reference to Figure 2.24b, in which the variation of the peak axial compressive stress sustained by cylinders under various levels of confining pressure is indicated schematically. It is interesting to note that a small confining pressure of about 10% of the uniaxial cylinder compressive strength f_c, is sufficient to increase the load-carrying capacity of the specimen by as much as 50%. On the other hand, a small lateral tensile stress of about 5% of f_c is sufficient to reduce the cylinder strength by the same amount. Such behaviour implies that the presence of small secondary stresses that develop within a structural member in the region of the path along which compressive forces are transmitted to the supports should have a significant effect on the load-carrying capacity of the member: compressive stresses should increase it considerably while tensile stresses should – dramatically – have the opposite effect.

While well known, the above two fundamental characteristics of concrete at a material level (summarised in Figure 2.24) are rarely (if ever) mentioned; more important still from a design viewpoint, their implications for the behaviour of concrete in a structure do not appear yet to have been fully appreciated in terms of failure mechanisms resulting from the interaction of concrete elements in RC structures. In order to appreciate that such interactive behaviour is unavoidable irrespective of the type of structure and/or loading conditions, it is useful to recall that, owing to the heterogeneous nature of concrete, the stress conditions within a concrete structure or member can never be uniform even under uniform boundary conditions. As a result, even for the case of a cylinder subjected to uniform uniaxial compression, the development of triaxial stress conditions is inevitable on account of the setting up of secondary stresses that are essential for maintaining compatibility of deformation within the structure (see Figure 2.25). Under service loading conditions, the secondary stresses are negligible and can be ignored for design purposes. However, as the load increases, volume dilation occurs in a localised region where the stress conditions are the first to reach the minimum-volume level. Concrete dilation is restrained by the surrounding concrete and this is equivalent to the application of a confining pressure which, as Figure 2.24b indicates, should increase the strength of the dilating region. At the same time, the dilating region induces tensile stresses in the adjacent concrete and, on the basis of the information shown in Figure 2.24b, these should reduce the strength of the concrete.

The preceding reasoning for the failure of a specimen under nominally uniform stress conditions is even more evident in the general case of arbitrary structural systems in which there is always a localised region in compression where the OUFP level is exceeded before it is exceeded in surrounding regions (which are also in compression). As a result, the rate of tensile strain will increase abruptly in this region, thus inducing tensile stresses in the adjacent concrete. Concurrently, equilibrium requires that the surrounding concrete should restrain the expansion of the localised region. While this extra restraint further increases the strength of the localised region, the tensile stresses eventually turn the state of stress in the surrounding concrete into a state of stress with at least one of the principal stress components tensile and thus reduce the strength. Therefore, it is always the concrete surrounding

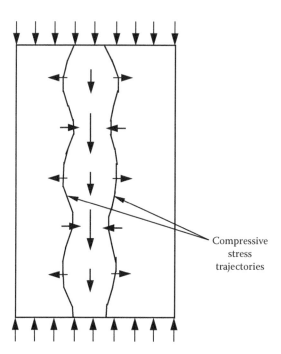

Figure 2.25 Schematic representation of the non-uniform stress distribution due to material heterogeneity within a concrete cylinder under uniform compressive stress.

the localised region of wholly compressive stresses that fails first, since its state of stress now has at least one tensile principal stress component.

It appears, therefore, that, owing to the interaction of the concrete elements within a structure, failure is unlikely to occur in regions where the compressive stress is largest. Instead, failure should occur in adjacent regions, where the compressive stresses may be significantly smaller, owing to the presence of small secondary tensile stresses that develop as discussed above. Such a failure mechanism indicates that concrete invariably fails in tension, and that a concrete structure collapses before the (usually triaxial) US of concrete in compression is exceeded anywhere within the structure. This notion that the concrete in the 'critical' zones of compression always fails by 'splitting' – never by 'crushing' – contrasts with widely held views that form the basis of current analysis and design methods for RC structures. Thus, most design procedures have been developed on the assumption that it is sufficient to rely almost entirely on uniaxial stress–strain characteristics for the description of concrete behaviour. This assumption may be justified by the fact that structural members are usually designed to carry stresses mainly in one particular direction, with the stresses that develop in the orthogonal directions being small enough to be assumed negligible for any practical purpose. However, such reasoning underestimates the considerable effect that small stresses have on the load-carrying capacity and on the deformational response of concrete beyond the in-service conditions. The ignoring of these small stresses in design necessarily means that their actual effect on structural behaviour is normally attributed to other causes that are expressed in the form of – as it turns out, erroneous – design assumptions. The following example will suffice to illustrate this. It is often pointed out that the strains recorded in the compressive zone of beams indicate that these are well in excess of that value that corresponds to the peak stress in a uniaxial cylinder or prism test. As a result, the argument is put forward that strain softening must be present since such large strains are observed only in the region of the descending branch of the uniaxial test. However, the true explanation lies in the fact that such regions are always subject to a state of triaxial compression, and this means that, although the peak stress has not been exceeded, the associated triaxial strains are much larger than their uniaxial-test counterparts around the OUFP level. Thus, for instance, an axial strain of around 7 mm/m marks practically the end of the descending branch for a concrete of $f_c \approx 46$ MPa in accordance with an ordinary (i.e., uniaxial) cylinder test (Figure 1.4); when subjected to a hydrostatic stress $\sigma_0 = 17$ MPa, the axial strain for the same concrete prior to the attainment of the descending branch is about three times this value, and becomes much higher still with increasing confinement σ_0 (Figure 1.14).

2.4.2 Triaxiality and failure initiation by macro-cracking: Some experimental and analytical evidence

The behaviour of structural concrete outlined in the preceding section has been predicted by analysis and verified by experiment (Kotsovos and Newman 1981a; Kotsovos 1982, 1987; Kotsovos and Pavlovic 1986). The inevitable triaxiality conditions in zones usually (misguidedly) deemed to be critical on account of large compressive action, and the associated failure initiation by tensile stresses adjacent to such zones will become evident throughout the various problems of later chapters tackled by means of FE modelling. Nevertheless, it is instructive to devote the present section to a preliminary illustration of the basic mechanism that governs the ultimate-load conditions in a concrete member.

Consider an RC beam designed in accordance with typical current regulations based on the ultimate-strength philosophy. The stress–strain characteristics of concrete in compression are considered to be adequately described by the deformational response of concrete specimens such as prisms or cylinders under uniaxial compression; thus, the ensuing stress distribution in the compressive zone of a cross section at the ultimate limit state, as proposed, for example, by BS 8110 (British Standards Institution 1985a,b) exhibits a shape similar to that shown in Figure 2.26a. The figure indicates that the longitudinal stress increases with the distance from the neutral axis up to a maximum value and then remains constant. Such a shape of stress distribution has been arrived at on the basis of both safety considerations and the widely held view that the stress–strain relationship of concrete in compression consists of both an ascending and a gradually descending portion, as illustrated in Figure 2.26b. In fact, the stress block in Figure 2.26a is based on the simplification that, beyond the peak stress, perfect plasticity may be assumed up to a strain of 0.0035; however, alternative stress blocks may also be used, either involving further simplification such as full plasticity leading to a rectangular stress block or derived by allowing for strain softening between peak stress and a strain of 0.0035 so that the shape of the stress block is curved throughout (British Standards Institution 1985b). The portion beyond the ultimate (i.e., peak) stress in Figure 2.26b defines the post-ultimate stress capacity of the material which, as indicated in Figure 2.26a, is generally considered to make a major contribution to the maximum load-carrying capacity of the beam. It will be noticed that the principal reasoning behind the stress block adopted for design purposes is based on the large compressive strains (in excess of 0.0035) measured on the top surface of an RC beam at its ultimate limit state, such strains being almost twice the value of the compressive strain ε_u at the peak-stress level under uniaxial compression. (Typically, ε_u is of the order of 0.002 – see Figure 2.26b.)

That the above design procedure is not borne out by experimental evidence can be shown by reference to the results obtained from a test series of three simply supported rectangular RC beams subjected to flexure under two-point loading (Kotsovos 1982). The details of a typical beam are shown in Figure 2.27, with the central portion under pure flexure constituting one-third of the span. The tension reinforcement consisted of two 6 mm diameter bars with a yield load of 11.8 kN. The bars were bent back at the ends of the beams so as to

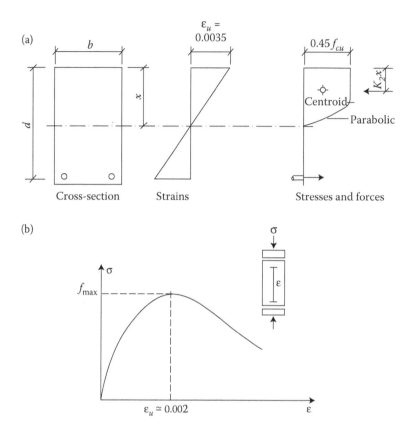

Figure 2.26 Characteristic design of a beam cross section for ultimate-load conditions: stress and strain distribution proposed by BS8110 (Part I) for a critical section at failure (f_{cu} = characteristic cube strength); (b) typical stress–strain relationship for concrete under uniaxial compression used to derive (a) ($f_{max} = f_c$ for cylinders; $f_{max} = f_{cu}$ for cubes.

provide compression reinforcement along the whole length of the shear spans. Compression and tension reinforcement along each shear span were linked by seven 3.2 mm diameter stirrups. Neither compression reinforcement nor stirrups were provided in the central portion of the beams. As a result of the above reinforcement arrangement all beams failed in flexure rather than shear, although the shear span-to-effective depth ratio was around 3. The beams, together with control specimens, were cured under damp hessian at 20°C for 7 days and then stored in the laboratory atmosphere (20°C and 40% relative humidity [RH]) for

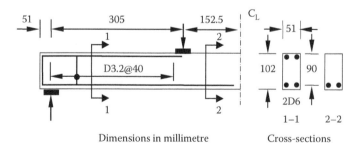

Dimensions in millimetre   Cross-sections

Figure 2.27 RC beam under two-point loading: beam details.

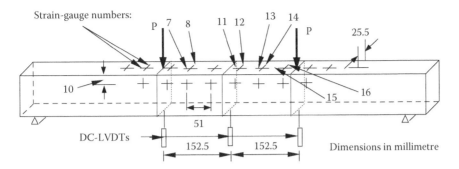

Figure 2.28 RC beam under two-point loading: beam instrumentation.

about 2 months, until tested. The cube and cylinder strengths at the time of testing were $f_{cu} = 43.4$ MPa and $f_c = 37.8$ MPa, respectively. Besides the load measurement, the deformational response was recorded by using both 20 mm long electrical resistance strain gauges and LVDTs. The strain gauges were placed on the top and side surfaces of the beams in the longitudinal and the transverse directions as shown in Figure 2.28. The figure also indicates the position of the LVDTs which were used to measure deflection at mid-span and at the loaded cross sections. Finally, the stress–strain characteristics in uniaxial compression for the concrete used in the investigation are depicted in Figure 2.29.

In presenting the salient results of the test series of beams, it is convenient to begin by showing the relationships between longitudinal (i.e., along the beam axis) and transverse (i.e., across the beam width) strains, as measured on the top surface of the girders. The relevant information is summarised in Figures 2.30a and b which refer to the strains recorded at the critical sections (i.e., throughout the middle third of the beam span) and within the shear spans respectively. Also plotted on these figures is the relationship between longitudinal and transverse strains derived on the basis of the uniaxial material characteristic of Figure 2.29 (or, equivalently, on the basis of the cylinder strength f_c, through the ensuing mathematical constitutive relations). Now, if the uniaxial-compression stress–strain characteristics of Figure 2.29 were to provide a realistic prediction of concrete behaviour in the

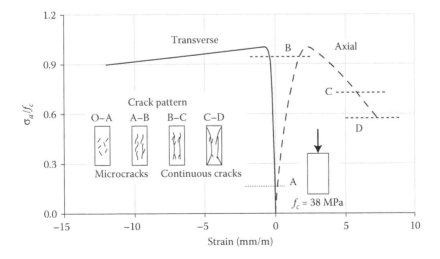

Figure 2.29 RC beam under two-point loading: stress–strain relationships under uniaxial compression for the concrete mix used.

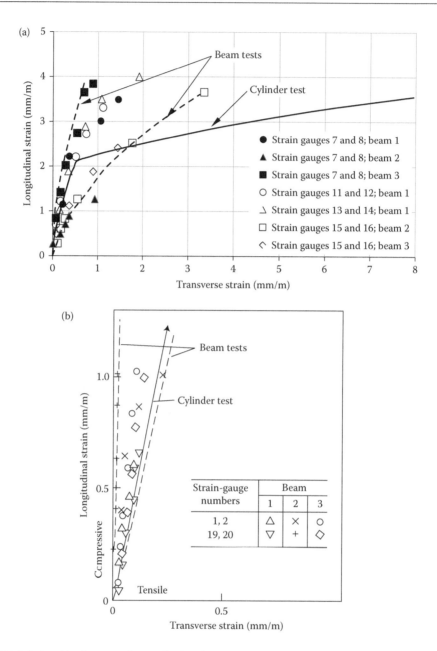

Figure 2.30 Relationships between longitudinal and transverse strains measured on the top surface of the RC beams under two-point loading (for strain-gauge locations, see Figure 2.28): (a) at critical sections; (b) within the shear spans.

compressive zone of the beams tested in flexure, then the relationships between longitudinal and transverse strains measured on the top surface of the beams would be expected to be compatible with their counterparts derived on the basis of the cylinder test; furthermore, longitudinal macro-cracks ought to appear on the top surface of the beams, as indicated in Figure 2.29, where typical crack patterns of axially compressed concrete cylinders around (B-C) and beyond (C-D) US are depicted schematically. It is apparent from Figure 2.30a, however, that, for the region of cross sections including a primary flexural crack, only the portion of the deformational relationship based on the uniaxial cylinder test up to the OUFP level can provide a realistic description of the beam behaviour. Beyond this minimum-volume level, there is a dramatic deviation of the cylinder strains from the beam relationships. Not only does such behaviour support the view that the post-peak branch of the deformational response of a cylinder in compression does not describe material response but, more importantly for present purposes, clearly proves that, while uniaxial stress–strain data may be useful prior to the attainment of the peak stress, they are insufficient to describe the behaviour once this maximum-stress level is approached. On the other hand, while Figure 2.30a demonstrates the striking incompatibility between cylinder specimen and structural member beyond compressive strains larger than about 0.002 (which, as noted earlier, corresponds to ε_u, the strain at the f_c (f_{cu}) level – see Figures 2.26 and 2.29), Figure 2.30b shows that the relationships between longitudinal and transverse strains measured on the top surface within the shear span of the beams are adequately described by the longitudinal strain–transverse strain relationship of concrete under uniaxial compression. It should be noted, however, that the relationships of Figure 2.30b correspond to stress levels well below US.

An indication of the causes of behaviour described by the relationships of Figures 2.30a and b may be seen by reference to Figure 2.31, which shows the change in shape of the transverse deformation profile of the top surface of beam 1 (but typical of all beams) with load increasing to failure. The characteristic feature of these profiles is that, within the 'critical' central portion of the beam, they all exhibit large local tensile strain concentrations which develop in the compressive regions of the cross sections where the primary flexural cracks, that eventually cause collapse, occur.

Although small strain concentrations may develop in these regions at early load stages before the occurrence of any visible cracking, they become large only when the ultimate limit state is approached and visible flexural cracks appear in the tension zones of the beams. Such a large and sudden increase in transverse expansion near the ultimate load is indicative of volume expansion and shows quite clearly that, even in the absence of stirrups, a triaxial state of stress can be developed in localised regions within the compressive zone. The local transverse expansion is restrained by concrete in adjacent regions (as indicated by the resultant compression forces F in Figure 2.31), a restraint equivalent to a confining pressure that will later be shown as being equivalent to at least 10% of f_c; hence, as Figure 2.24b indicates, the compressive region in the plane of a main flexural crack is afforded a considerable increase in strength so that failure is not initiated there. Concurrently, the expanding concrete induces tensile stresses in adjacent regions (these are indicated by the resultant tension forces F and $F/2$ in Figure 2.31), and this gives rise to a compression/tension state of stress. Such a stress state reduces the strength of concrete in the longitudinal direction and collapse occurs as a result of horizontal splitting of the compressive zone in regions between primary flexural cracks, as illustrated schematically in Figure 2.32a. Concrete crushing, which is widely considered to be the cause of flexural failure, thus appears to be a *post-failure* phenomenon that occurs in the compression zone of cross sections containing a primary flexural crack resulting from a loss of restraint provided previously by the adjacent concrete.

It may be concluded from the above, therefore, that the large compressive and tensile strains measured on the top surface of the central portion of the beams should be attributed to a *multiaxial* rather than uniaxial state of stress. A further indication that these large strains cannot be due to post-ultimate stress-strain characteristics is the lack of any *visible* longitudinal cracking on the top surface for load levels even near the maximum load-carrying capacity of the beams. As shown in Figure 2.29, such cracks characterise the post-US behaviour of concrete under compressive states of stress. Visible cracks occur predominantly on planes parallel to the top surface *at the moment* of final collapse. The typical view of the beam once the collapse of a member has taken place is depicted in Figure 2.32b, where the

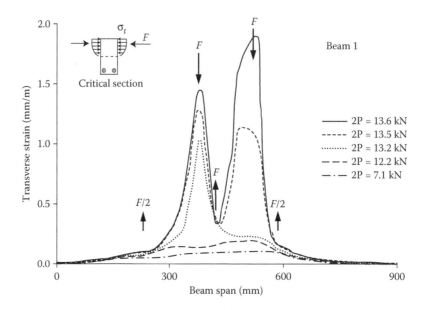

Figure 2.31 Typical variation of transverse deformation profile of loaded face of RC beams under two-point loading with increasing total load (2P) and schematic representation of resulting forces (F) and stresses (σ_t).

pair of main flexural cracks observed correspond to the peak tensile strain concentrations recorded experimentally in beam 1 (see Figure 2.31).It is interesting to note that the results described so far do not contradict the view expressed throughout this chapter that concrete in compression suffers a complete and immediate loss of load-carrying capacity when US is exceeded. The implication of the results of the beam tests is that, in the absence of a post-ultimate gradually falling branch of the stress–strain relationships of concrete in compression, the large compressive strains which characterise RC structures exhibiting 'ductile'

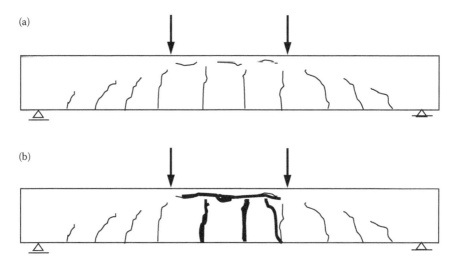

Figure 2.32 Typical failure mode of RC beam in flexure: (a) schematic representation of failure mechanism at collapse; (b) observed failure of beams following collapse. (From Kotsovos M. D., 1982, *Materials & Structures, RILEM*, 15(90), 529–537.)

behaviour under increasing load (i.e., behaviour characterised by load–deflection relationships exhibiting trends similar to those shown in Figure 2.33 for the under-reinforced members tested) are due to a complex multiaxial compressive state of stress which exists in any real structure at its ultimate limit state. Such stress states may be caused by secondary restraints imposed on concrete by steel reinforcement, boundary conditions, surrounding concrete, and so on. The significance of these restraints is, in most cases, not understood or simply ignored. It may thus be concluded that the US of concrete in localised regions exhibits significant variations dependent on the local *multiaxial* compressive state of stress within the compressive zone of an RC structure or member. The higher the multiaxial US of concrete at a critical cross section, the larger the corresponding compressive and tensile strains. The 'ductility' of the structure, therefore, seems to be dependent on the true (i.e., triaxial) US of concrete at critical cross sections rather than on stress redistributions attributable to post-ultimate material stress–strain characteristics, even if the latter were assumed to exist.

The state of compressive triaxial stresses compatible with the deformations and strains measured in the beams tested remains to be explored. In addition to the main longitudinal (σ_l) and the secondary transverse (σ_t) stresses, another set of secondary actions also exists, namely the radial stresses (σ_r) acting vertically. Clearly, vertical stresses must exist at, and in the vicinity of, the point loads, but the radial stresses referred to are additional to these and are more relevant for present purposes. These radial stresses are associated with the radial stress resultant (R) which develops within the deformed beam as a result of the inclination of the compressive (C) and tensile (T) stress resultants acting in the longitudinal direction. The above stress resultants are shown schematically in Figure 2.34, which indicates that, even the loaded face (which is generally assumed to be under plane-stress conditions), is subjected to a radial stress resultant. As long as the beam exhibits near-elastic behaviour, the radial stresses corresponding to the radial stress resultant are small in magnitude since they are distributed over the whole length of the central portion of the beam. However, when the central portion of the member starts to develop large deflections (see Figure 2.33) as a

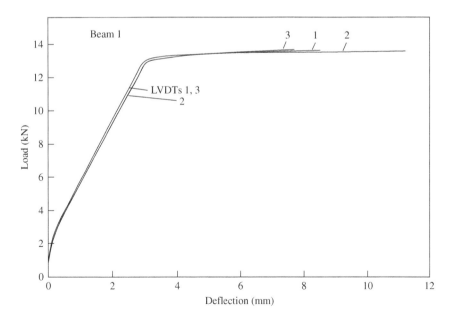

Figure 2.33 Typical load-deflection curves of RC beam under two-point loading (for LVDT's locations, see Figure 2.28).

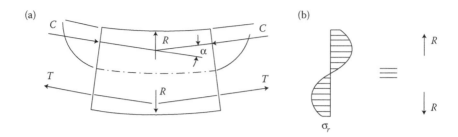

Figure 2.34 Schematic representation of radial stresses due to deflected shape of RC beams: (a) forces (R); (b) stresses (σ_r).

result of the formation of a 'plastic' zone caused by a critical flexure crack, the radial stresses become significant in magnitude since they tend to become localised and to concentrate over the plastic zone (Taylor and Al-Najmi 1980). For load levels close to the maximum load-carrying capacity of the beam, the mean value of the above radial stresses may be estimated – albeit roughly – as follows. If the inclination of the longitudinal compressive and tensile stress resultants is defined by the angle of discontinuity a resulting from the inelastic deformation of the 'plastic' zone (Figure 2.34), then

$$R = C \sin a = T \sin a \qquad (2.1)$$

Now, T is approximately equal to the total yield force of the reinforcement, that is, $T = 2 \times 11{,}800 = 23{,}600$ N (see earlier details), whereas an approximate value for a may be obtained by the ratio

$$a/2 = (\text{maximum mid} - \text{point deflection})/(\text{half} - \text{span of beam}) \qquad (2.2)$$

For a value of the maximum mid-point deflection approximately equal to 10 mm (see Figure 2.33) Equation 2.2 gives $a \approx 4.4 \times 10^{-2}$ rad which, when substituted in Equation 2.1, results in $R \approx 1000$ N. Finally, assuming that the length of the 'plastic' zone is 5 mm, a nominal value for the radial stresses (approximating the section width to approximately 50 mm) is $\sigma_r \approx 1000/(5 \times 50) \approx 4$ MPa. Hence, $\sigma_r \approx 0.1 f_c$ since, as noted previously, $f_c \approx 38$ MPa.

The order of magnitude of the transverse stresses σ_t may be assessed by reference to the estimate obtained for σ_r. Consider Figure 2.35, which shows the variation on the critical section of the average strains measured in the loading direction on the side faces of the beams with the transverse strains measured on the loaded surface. It is interesting to note from the figure that the strains measured on the side faces are slightly larger than those measured on the loaded face. This is considered as an indication that the average value of the stresses restraining the transverse expansion of the critical section should be at least as large as that of the radial stresses, that is, $\sigma_t > 0.1 f_c$. The transverse and radial stresses, therefore, combined with the longitudinal stresses give rise to a complex multiaxial compressive state of stress in the regions of the large tensile strain concentrations within the compressive zone of the beams. Under such a three-dimensional stress state, concrete can sustain both stresses and strains which can be considerably larger than those obtained in uniaxial material tests that form the basis of most current structural design.

How large are the main stresses (σ_l)? These would be expected to be at least 50% in excess of f_c because, as pointed out in Section 2.4.1, Figure 2.24b suggests that an axisymmetric confining pressure of about 10% of f_c boosts the actual strength by about one-half of its

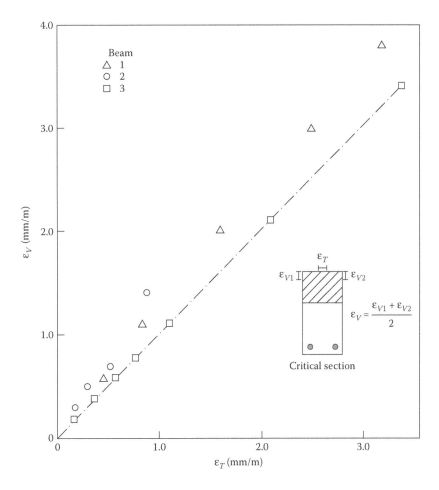

Figure 2.35 Relationships, with increasing load, between transverse and vertical strains at critical sections of RC beams under two-point loading.

original value. That this is indeed the case may be seen by reference to Figure 2.36a, which shows the tension (T) and compression (C) force resultants at a critical section of a beam. Since only an order-of-magnitude estimate of σ_l is required, average stress values may be used and hence it is sufficiently accurate to adopt a rectangular stress block. Now, earlier calculations for beam 1, which were presented in Chapter 1 (see Figure 1.13), gave $T(F_s) = C$ $(F_c) = 23,600$ N (i.e., ductile failure), while the ultimate load $P(V_f) = 6800$ N combined with a rounded-off value of the shear span of about 300 mm leads to the maximum-sustained bending moment (M_f) of approximately $6800 \times 300 \approx 2,040,000$ Nmm. The lever arm then follows at $z = 2,040,000/23,600 \approx 86.5$ mm, enabling the depth of the stress block to be estimated at $x = 2 \times (90-86.5) \approx 7$ mm. As before, the beam width may be approximated to 50 mm so that the compressive-zone stresses $\sigma_l = 23,600/(50 \times 7) \approx 67$ MPa, that is, the average value of the longitudinal stress at a critical section is 75% above f_c and, clearly, some of the actual local stresses will be even higher than this figure.

On the basis of the assumed distribution of secondary (i.e., 'confining') stresses σ_t and σ_r, (see Figures 2.31 and 2.34), it could be argued that the degree of triaxiality varies throughout the depth of the compressive zone in the manner shown in Figure 2.36b, with the longitudinal stresses σ_l increasing from the neutral plane up to a maximum value

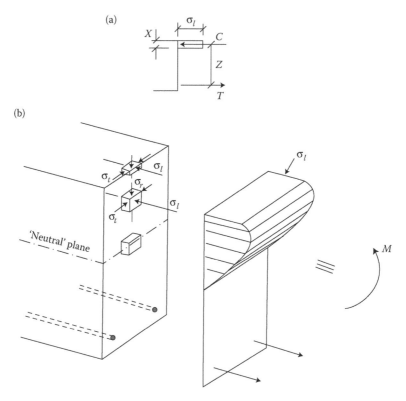

Figure 2.36 Longitudinal stresses σ_l in the critical compressive zone of the RC beams under two-point loading at failure: (a) assessment of average σ_l based on measured values of ultimate tensile force resultant and bending moment; (b) likely shape of σ_l distribution predicted on the basis of triaxial behaviour.

(where the confinement is greatest) and then gradually decreasing to a smaller value at the loaded face. If so, it might be suggested that, neglecting the inevitable stress variations across the beam width, which only a proper three-dimensional analysis could reveal, the shape of the σ_l distribution is not unlike that of the generally accepted stress-block shape derived on the basis of a uniaxial stress–strain relationship possessing a gradually descending post-ultimate branch which, as discussed earlier, is used by current design procedures recommended by codes of practice. However, the preceding study shows beyond doubt that, while both the large strains required for ductility and the shape of the stress block might appear as admitting the postulate that uniaxial material properties are applicable at a structural level, such a postulate does not accord with the actual mechanism of failure in a structure and, furthermore, leads to massive underestimates of the true stresses and transverse tensile strains under ultimate conditions. (In view of the latter, it is obvious that the various refinements in the shapes of the stress blocks – see the discussion at the start of the present section – are totally unjustified, so that the simplest stress-block shape, i.e., rectangular, might as well be used in ordinary design calculations.) Therefore, the main conclusion to be drawn from the preceding study is that the importance of triaxiality in elucidating what triggers the collapse of a structure and the sensitivity of triaxial failure envelopes to even small degrees of confinement make it mandatory to incorporate multiaxial material descriptions in any FE model aimed at accurate predictions of ultimate behaviour at the structural level.

The triaxiality in the previous study was provided by concrete itself: namely the compressive region of concrete where volume dilation was reached first and was restrained by the surrounding concrete. As already mentioned previously, such 'secondary' (but extremely significant) restraints on concrete could also be imposed by other agents such as, for example, steel reinforcement. As an illustration of the latter case, a simplified description of the effect of hoop reinforcement on the strength and deformation of what, in the absence of such reinforcement, is essentially a PC cylinder under uniform compression will now be presented. The results stem from an FE analysis (Kotsovos and Newman 1981a) of the type that will be described fully in later chapters. The hoop reinforcement is simulated by means of a spring support having a stiffness equivalent to the stiffness of the actual reinforcement. The results obtained from the FE analysis of this structural configuration (shown in Figure 2.37a) are depicted in Figure 2.37b–d. Figure 2.37b illustrates how the load-carrying capacity of concrete increases with decreasing spacing of the hoop reinforcement. The cause of such behaviour may be seen by reference to Figure 2.37c, which shows the effect of hoop-reinforcement spacing on the stress path to which concrete under increasing load is subjected. Figure 2.37c also includes the ultimate-strength envelope for PC and indicates that the confining pressure (σ_c) induced by the restraint which the hoop reinforcement imposes on the transverse expansion of concrete with increasing applied axial load increases with decreasing spacing of the reinforcement so that a higher axial stress (σ_a) is required for the ultimate-strength level to be exceeded. Finally, Figure 2.37d shows the variations of axial and lateral strains with increasing axial stress for various values of reinforcement spacing. The stress–strain relationships obtained from tests on PC under uniaxial compression are also shown in the figure for purposes of comparison. It should be noted that, for values of the reinforcement spacing up to about 50% of the diameter of the concrete member, the axial compressive strains corresponding at US are comparable to, or higher than, the maximum axial strain exhibited by PC under uniaxial compression.

It was pointed out in a previous discussion that, while a proper allowance for triaxial compression in regions usually considered as critical was a prerequisite in any formal modelling of a structure, it was, in fact, the surrounding concrete which initiates failure, as triaxial compression in the region being restrained is followed eventually by a state of stress in the adjacent zone (which does the restraining) with at least one principal-stress component being tensile. Therefore, it is this latter location, where failure is invariably triggered, which is truly critical. While this will become quite apparent in subsequent chapters describing various FE analyses, it is useful to conclude this section with an illustration of the typical failure mechanism in a concrete structure. This will be done by reference to the problem of a PC cylinder subjected to axisymmetric patch end-loading, as depicted in Figure 2.38a. Owing to symmetry, only one-eighth of the structure needs to be analysed, and the results (Kotsovos and Newman 1981a; Kotsovos 1981) appear in Figures 2.38b and c, with circumferential cracks denoted by short straight lines, while the regions of radial cracking are indicated by patterns of (mutually) orthogonal dashes. Characteristically, the analysis predicts cracking will occur in regions subjected to a state of stress with at least one of the principal stress components tensile. For the example under investigation, the most critical state of stress is that which initiates cracking in the region marked B in Figure 2.38b. With increasing load, this cracking propagates only into other regions subjected to similar states of stress, that is with at least one of the principal stress components tensile, up to the formation of the crack pattern corresponding to the collapse stage (see Figure 2.38c). Figure 2.38 also shows that region A, which is subjected to a wholly compressive state of stress, reduces progressively in size as the applied load increases above the level which causes crack initiation. This is due to the redistribution which transforms the state of stress at the periphery of region A from a wholly compressive state of stress (within the strength envelope) to a

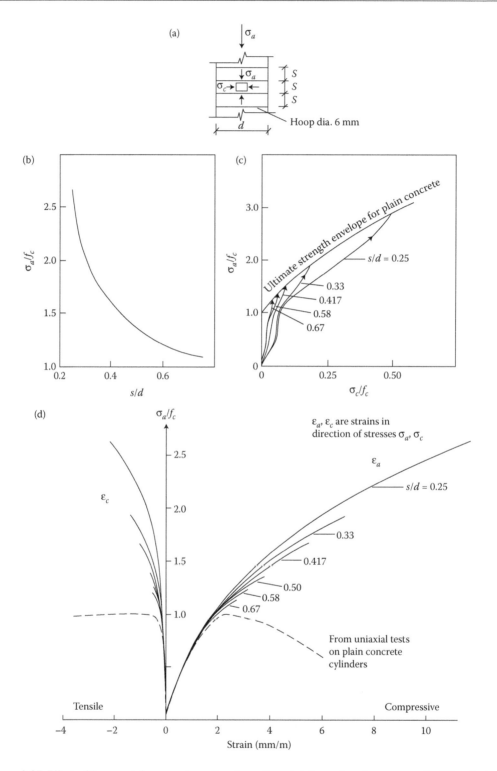

Figure 2.37 Effect of hoop reinforcement on the strength and deformational response of a PC cylinder: (a) structural form investigated; (b) relationship between strength and reinforcement spacing; (c) effect of reinforcement spacing on strength and stress path; (d) stress–strain relationships.

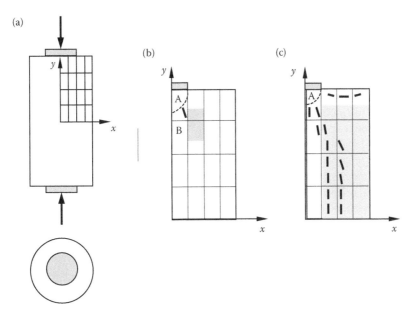

Figure 2.38 Characteristic failure mechanism of a PC cylinder under axisymmetric patch load: (a) elevation and plan views of the cylinder, showing portion actually analysed; (b) typical crack pattern at initiation of cracking; (c) typical crack pattern at collapse.

state of stress in which at least one of the principal stress components is tensile. When the strength of concrete under this latter state of stress is exceeded, cracking occurs and the size of region A (where failure still does not occur) is reduced further (see Figure 2.38c). Therefore, throughout the whole analysis, the strength of concrete in compression is not exceeded at any point in the member, so that collapse of the structure eventually occurs *before* the strength of concrete in region A is exceeded.

2.5 SUMMARY OF CHARACTERISTIC FEATURES OF CONCRETE RELEVANT TO MODELLING MATERIAL BEHAVIOUR

This chapter has been devoted to a detailed description of the various characteristics of concrete materials with reference to both phenomenological causes and the resulting effects which are of direct relevance to the behaviour of a structural-concrete member. Before embarking upon expressing the behaviour of concrete in an analytical form, it is useful to summarise very briefly the main findings of this chapter as regards the salient features of concrete.

Much of the experimental data on concrete properties may turn out to be unreliable on account of the interaction between test machine and tested specimen. A simple technique that is both sound and consistent is provided by the ordinary uniaxial-compression cylinder test. While the ascending branch of the stress–strain curve of a cylinder under uniaxial and/ or triaxial stress states provides a realistic description of material behaviour, the descending branch is a direct consequence of the frictional interaction between specimen and the loading device. If the latter frictional forces were to be completely removed, concrete would suffer an immediate loss of carrying capacity on reaching its peak stress. This suggests that, at the material level, concrete is brittle and that a full description of its response is embodied in the stress–strain characteristics up to the maximum sustained stress. Such characteristics represent the complex processes of fracture initiation and extension. Clearly, these processes are associated predominantly with volume decrease, and minimum volume is reached at the so-called OUFP level. From this point onwards, volume dilation begins as a result of void formation and the peak stress is quickly reached. Since the OUFP and the maximum-stress levels are very close to each other, the former may be considered to mark, for all practical purposes, failure of the brittle material, with the formation of (now visible) primary cracks. Just as the non-linear constitutive relations up to the OUFP level are associated with micro-cracking, the OUFP limit itself may be used to define the strength surface associated with macro-cracking.

The need is evident to employ triaxial strength envelopes which allow for the very considerable (local) triaxial strength (usually well in excess of f_c) in the 'critical' regions of a structural component (even when the latter might be considered to approximate quasi-uniaxial conditions). Although such triaxial effects have long been known (e.g., they were noted, among others, by Föppl [Timoshenko 1953]), their use in design has remained largely unexploited, despite the evidence that they govern the actual failure mechanism in structural concrete. In the case of non-linear FE analysis of concrete structures, due allowance of triaxiality is an essential prerequisite if the true failure mechanism is to be identified and an accurate ultimate-load estimate obtained. With regard to the mechanism that triggers failure, one constantly finds that this is governed by the combined effects of sudden volume expansion around the OUFP level and the very significant enhancement/deterioration in strength due to secondary compressive/tensile stresses. These cause the so-called critical regions, where maximum compressive-stress conditions are reached first, to gain further in strength so that the failure envelope (in compression) is never exceeded in those regions. Concurrently, adjacent zones are subjected eventually to a combination of compression-tension, and it is in these zones where failure actually begins by 'splitting', not 'crushing'. It would appear, therefore, that it is sufficient for a failure criterion to describe only those failure conditions in which the state of stress has at least one of the principal stress components tensile, thus bypassing altogether fully compressive stress states. This, of course, is not actually done in the FE model used, as it would clearly constitute a case of prejudging – if not imposing – the failure mechanism in a structure; rather, all possible failure criteria are allowed for, although, as will be seen in subsequent chapters, results invariably confirm that concrete in 'critical' compressive zones always fails in tension, never in compression.

REFERENCES

Ahmad S. H. and Shah S. P., 1979, Complete stress–strain curve of concrete and nonlinear design, *Proceedings CSCE-ASCE-ACI-CEB International Symposium on Nonlinear Design of Concrete Structures*, University of Waterloo, Ontario, pp. 61–81.

Barnard P. R., 1964, Researches into the complete stress–strain curve for concrete, *Magazine of Concrete Research*, 16(49), 203–210.

Brace W. F. and Bombolakis E. G., 1960, A note on the brittle crack growth in compression, *Journal of the Geophysics Research*, 68, 3709–3713.

British Standards Institution, 1985a, *British Standards, Structural use of concrete, BS8110 (Part 1, Code of practice for design and construction)*, British Standards Institution, London.

British Standards Institution, 1985b, *British Standards, Structural use of concrete, BS8110 (Part 2, Code of practice for special circumstances)*, British Standards Institution, London.

Cook J. and Gordon J. E., 1964, A mechanism for the control of crack propagation in all-brittle systems, *Proceedings of the Royal Society (London)*, Series A, 282, 508–520.

Evans R. H. and Marathe M. S., 1968, Microcracking and stress–strain curves for concrete in tension, *Materials and Structures, RILEM*, 1, 61–64.

Gerstle K. H., Aschl H., Bellotti R., Bertacchi P., Kotsovos M. D., Ko H.-Y., Linse D. H. et al., 1980, Behaviour of concrete under multiaxial stress states, *Journal of the Engineering Mechanics Division, ASCE*, 106(EM6), 1383–1403.

Gerstle K. H., Linse D. H., Bertacchi P., Kotsovos M. D., Ko H.-Y., Newman J. B., Rossi P. et al., 1978, Strength of concrete under multiaxial stress states, *Proceedings Douglass McHenry International Symposium on Concrete and Concrete Structures*, Mexico City, Mexico, ACI Publication SP-55, Detroit, USA, pp. 103–131.

Griffith A. A., 1921, The phenomena of rupture and flow in solids, *Philosophical Transactions of the Royal Society (London)*, Series A, 221, 163–198.

Heyman J., 1982, *The Masonry Arch*, Ellis Horwood, Chichester.

Hoek E. and Bieniawski Z. T., 1965, Brittle fracture propagation in rock under compression, *International Journal of Fracture Mechanics*, 1, 137–155.

Kotsovos M. D., 1974, Failure criteria for concrete under generalized stress, PhD thesis, University of London.

Kotsovos M. D., 1979a, Fracture processes of concrete under generalised stress states, *Materials & Structures, RILEM*, 12(72), 431–437.

Kotsovos M. D., 1979b, Effect of stress path on the behaviour of concrete under triaxial stress states, *ACI Journal, Proceedings*, 76(2), 213–223.

Kotsovos M. D., 1981, An analytical investigation of the behaviour of concrete under concentrations of load, *Materials & Structures, RILEM*, 14(83), 341–348.

Kotsovos M. D., 1982, A fundamental explanation of the behaviour of reinforced concrete beams in flexure based on the properties of concrete under multiaxial stress, *Materials & Structures, RILEM*, 15(90), 529–537.

Kotsovos M. D., 1983, Effect of testing techniques on the post-ultimate behavior of concrete in compression, *Materials & Structures, RILEM*, 16(91), 529–537.

Kotsovos M. D., 1987, Consideration of triaxial stress conditions in design: A necessity, *ACI Structural Journal, Proceedings* 84(3), 266–273.

Kotsovos M. D. and Cheong H. K., 1984, Applicability of test specimen results for the description of the behaviour of concrete in a structure, *ACI Journal, Proceedings*, 81(4), 358–363.

Kotsovos M. D. and Newman J. B., 1977, Behaviour of concrete under multiaxial stress, *ACI Journal, Proceedings*, 74(9), 453–456.

Kotsovos M. D. and Newman J. B., 1978, Generalised stress-strain relations for concrete, *Journal of the Engineering Mechanics Division, ASCE*, 104(EM4), 845–856.

Kotsovos M. D. and Newman J. B., 1981a, Plain concrete under load – A new interpretation, *IABSE Colloquium on Advanced Mechanics of Reinforced Concrete*, Delft, IABSE Final Report 34, pp. 143–158.

Kotsovos M. D. and Newman J. B., 1981b, Fracture mechanics and concrete behaviour, *Magazine of Concrete Research*, 33(115), 103–112.

Kotsovos M. D. and Pavlovic M. N., 1986, Non-linear finite element modelling of concrete structures: Basic analysis, phenomenological insight, and design implications, *Engineering Computations*, 3(3), 243–250.

Kotsovos M. D. and Pavlovic M. N., 1995, *Structural Concrete: Finite-Element Analysis for Limit-State Design*, Thomas Telford, London, UK, 550pp.

Newman J. B., 1973, Deformational behaviour, failure mechanisms and design criteria for concretes under combinations of stress, PhD thesis, University of London.

Newman J. B., 1974, Apparatus for Testing Concrete Under Multiaxial States of Stress, *Magazine of Concrete Research*, 26(89), 229–238.

Newman J. B., 1979, Concrete under complex stress, in *Developments in Concrete Technology* I, edited by Lyndon F. D., Applied Science Publishers, London, pp. 151–219.

Newman K. and Lachance L., 1964, The testing of brittle materials under uniaxial compressive stress, *Proceedings of ASTM*, 64, 1044–1067.

Shah S. P. and Sankar R., 1987, Internal cracking and strain-softening response of concrete under multiaxial compression, *ACI Materials Journal*, 84, 200–212.

Taylor R. and Al-Najmi A. Q. S., 1980, The strength of concrete in composite reinforced concrete beams in hogging bending, *Magazine of Concrete Research*, 32, 156–163.

Timoshenko S. P., 1953, *History of Strength of Materials*, McGraw-Hill, New York.

van Mier J. G. M., 1986, Multiaxial strain-softening of concrete, *Materials & Structures, RILEM*, 19(111), 179–200.

van Mier J. G. M., Shah S. P., Arnaud M., Balayssac J. P., Bassoul A., Choi S., Dasenbrock D. et al., 1997, Strain-softening of concrete in uniaxial compression (TC 148-SSC: Test methods for the strain-softening of concrete), *Materials & Structures, RILEM*, 30(198), 195–20.

Wang P. T., Shah S. P. and Naaman A. E., 1978, Stress-strain curves of normal and lightweight concrete in compression, *ACI Journal*, 75, 603–611.

Chapter 3

Modelling of concrete behaviour

The chapter presents an analytical description of the deformational and strength characteristics of concrete and steel. The analytical expressions proposed for concrete were derived from an analysis of data obtained from tests on specimens subjected to triaxial loading by using techniques capable of both inducing definable states of stress in the specimens and measuring reliably the deformational response of concrete. However, the important feature of the analysis was that it attributed the non-linear material behaviour to the *internal stresses* resulting from the fracture processes discussed in the preceding chapter. On the other hand, the analytical description of the properties of the steel reinforcement, for which there is little room for variation and disagreement, follows the recommendation of current codes for the design of concrete structures.

3.1 CONSTITUTIVE RELATIONS FOR CONCRETE

This section is concerned with deformational properties and presents an analytical description of the ascending portion of the stress–strain curve for concrete when subjected to short-term static loading conditions since, as discussed in Section 2.2, concrete is brittle in nature in that its post-peak behaviour is characterised by a complete and immediate loss of load-carrying capacity. The analytical description of the deformational behaviour is based on an analysis of triaxial stress–strain data obtained from tests on a wide range of concretes with uniaxial cylinder compressive strength f_c varying from about 15 to about 65 MPa (Kotsovos 1974; Kotsovos and Newman 1979). The tests were carried out at Imperial College London by following rules outlined in Section 2.1.1 (see also Newman 1974). In the analysis, use is made of the fact that the material non-linearities predominantly reflect the effect of fracture processes having two opposing effects on the material deformation.

> *Effect A*. Cracking causes a reduction of the high, predominantly tensile stress concentrations existing near the crack tips. This reduction in tensile stress can be assumed to be equivalent to the application of a compressive stress which tends to *reduce* the volume of concrete.
>
> *Effect B*. Cracking produces voids which tend to *increase* the volume of material.

It was shown in Section 2.3.3.1 that, depending on the combined effects of A and B on deformation, the cracking processes could be described qualitatively by reference to four stages. For present purposes, such a description may be simplified by dividing these fracture

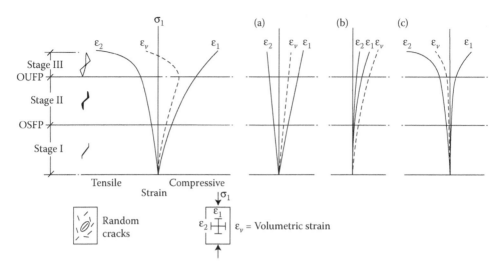

Figure 3.1 Stages of fracture processes and their effect on the stress–strain relationships of concrete. (a) Linear material properties, (b) effect of internal stresses, (c) effect of void formation.

processes into three stages, which are shown schematically in Figure 3.1. These can broadly be described as follows.

Stage I. Effect A is significant while effect B is insignificant since cracking is localised. As a result, the volume decreases.

Stage II. Effects A and B are significant, but effect A is greater than effect B and thus the volume continues to decrease. The beginning of this stage was previously termed the OSFP level, at which the rate of increase of strain ε_2 begins to exceed the rate of increase of strain ε_1, even though ε_1 still exceeds ε_2 (see Figure 3.1).

Stage III. In Stage III, both effects A and B are significant, but effect B is greater than effect A and this causes the volume to increase. The beginning of this stage was previously termed the OUFP level and can be defined easily since it corresponds to the level at which the volume of the material becomes a minimum (i.e., beyond the OUFP, ε_2 soon exceeds ε_1).

On the basis of the above considerations, therefore, it is evident that the deformational behaviour of concrete at all three stages may be decomposed into the following three components:

a. A 'linear' component throughout, dictated by the material characteristics and unaffected by the above fracture processes.
b. A non-linear component expressing the effect of the internal stresses caused by the fracture processes.
c. A non-linear component expressing the effect of void formation.

To quantify the above components using experimental stress–strain data appears to be an impossible task since the data describe overall material behaviour. Nevertheless, it is clear that their combined effects result in two broad regimes of (overall) non-linearity. The first of these, which forms the subject of the present section and relates to the constitutive relations up to the OUFP level, is governed primarily by a combination of (a) and (b). Therefore, although micro-cracking does cause some increase in the volume of the material owing to void formation, its predominant effect is to release the high tensile stresses at the crack tips which, as noted above, is statically equivalent to the application of an *internal* compressive state of stress that reduces the volume of concrete. Therefore, it is the latter effect (i.e., effect A) which is considered to be the underlying cause of the non-linear behaviour up to the minimum-volume level. This first regime of non-linearity (namely stages I and II in Figure 3.1), which is of a 'mild' nature, may be described analytically by means of the concept of the internal compressive stress state described above. Before this is done, however, it is necessary to discuss briefly in terms of suitable parameters the experimental data on the deformational behaviour of concrete during both loading and unloading.

3.1.1 Experimental data on observed behaviour

These can be summarised as follows.

3.1.1.1 Deformational behaviour during loading

The generalised stress–strain relationships for the experimental data corresponding to the ascending branch of concrete materials are expressed most conveniently by decomposing each state of stress and strain into hydrostatic and deviatoric components, that is, in the form of normal and shear octahedral stresses (σ_o, τ_o) and strains (ε_o, γ_o), respectively. (For a definition of these octahedral parameters, see Appendix A.) In this form of representation, the deformational behaviour of concrete up to the minimum-volume level under increasing stress (behaviour under decreasing stress is discussed later) can be described completely by reference to the strains produced upon the application, in turn, of hydrostatic and deviatoric components of stress.

The results of tests indicate that the deformational behaviour of concrete under hydrostatic stress σ_o can be fully described by the variation of the hydrostatic (volumetric) strain $\varepsilon_{o(h)}$ with σ_o, since the accompanying deviatoric strain $\gamma_{o(h)}$ has been found to be insignificant (Kotsovos and Newman 1978, Gerstle et al. 1980). (The subscript h serves as a reminder that the octahedral strains result from the application of a pure hydrostatic stress state.) These $\sigma_o - \varepsilon_{o(h)}$ relationships depend only on the uniaxial strength f_c, of the particular concrete, and typical results are shown as data points in Figure 3.2. It is evident that the variation of $\varepsilon_{o(h)}$ with σ_o is distinctly non-linear. Such non-linear behaviour is considered to reflect the dependence of volumetric strains on internal stresses resulting from redistributions that occur as a result of changes in the structure of the material when subjected to an externally applied hydrostatic stress. It would be expected that internal stresses can be decomposed into hydrostatic and deviatoric components, but the fact that the distortion of the material under a pure applied hydrostatic stress σ_o has been found to be insignificant (i.e., $\gamma_{o(h)} = 0$) indicates that the deviatoric component of the internal stress is negligible, and this may be attributed to the random orientation of the structural changes that concrete undergoes under external hydrostatic stress. Furthermore, it has also been found that specimens subjected to various levels of uniaxial compression below the failure level, when unloaded and then reloaded hydrostatically, exhibit a stress–strain relationship which is essentially the same as that exhibited by specimens under increasing hydrostatic stress without any previous (deviatoric) loading history (Newman 1973). In view of this experimental evidence, it appears realistic to consider that the volumetric strain $\varepsilon_{o(h)}$ of concrete due to the hydrostatic component σ_o of the applied state of stress is independent of the value of the applied deviatoric component τ_o, and hence $\varepsilon_{o(h)}$ may be expressed in terms of σ_o only. Finally, it should be pointed out that, since this variation of $\varepsilon_{o(h)}$ with σ_o indicates consolidation of the material

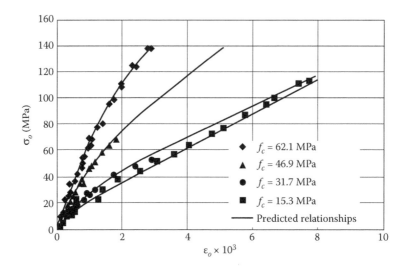

Figure 3.2 Typical experimental $\sigma_o - \varepsilon_{o(h)}$ relationships for various concretes.

that occurs at a progressively increasing rate, the hydrostatic component of internal stresses must be compressive and is considered to represent the reduction of high tensile-stress concentrations that occur near the crack tips as a result of crack extension; such a view is clearly consistent with the fracture mechanism of concrete under increasing stress discussed previously in the description of component (b) of the deformational behaviour, which governs non-linearity up to the OUFP level.

The application of an external deviatoric stress τ_o gives rise to both volumetric and deviatoric (shear) strains. Therefore, the deformational response of concrete under increasing deviatoric stress is defined by both $\tau_o - \gamma_{o(d)}$ and $\tau_o - \varepsilon_{o(d)}$ relationships. (The subscript *d* indicates that the octahedral strains are due to a pure deviatoric stress state.) Typical experimental results for the $\tau_o - \gamma_{o(d)}$ relations for various concretes are shown as data points in Figure 3.3. In addition, data points for both the $\tau_o - \gamma_{o(d)}$ and $\tau_o - \varepsilon_{o(d)}$ characteristics appear in Figures 3.4 and 3.5, respectively, for two possible stress paths ($\sigma_1 > \sigma_2 = \sigma_3$ [triaxial 'compression'] and ($\sigma_1 = \sigma_2 > \sigma_3$ [triaxial 'extension']). It is evident that both sets of relationships are essentially independent of the stress path, indicating that the influence of the direction of τ_o (i.e., rotational angle Θ – see Appendix A) on the octahedral (or deviatoric) planes (i.e., the planes orthogonal to the stress-space diagonal at various levels of σ_o) is negligible and that any stress-induced anisotropy is insignificant and can be ignored for practical purposes.

Additional confirmation of path independency (but concentrating now only on the $\tau_o - \gamma_{o(d)}$ relations) is provided by the results gathered in the international cooperative project referred to in Chapter 2, and which was based on the use of a common concrete mix by all participants (Gerstle et al. 1980). These findings appear in Figure 3.6 in the form of three curves that correspond to the different stress paths followed. (The stress paths denoted as 1 and 3 correspond to the triaxial 'compression' and 'extension', respectively, while 2 refers to the additional case of constant intermediate principal stress.) Mean curves were obtained for each participant's test series and for each path, and these, in turn, averaged as shown in the mean $\tau_o - \gamma_{o(d)}$ curves of Figure 3.6. It is obvious that these mean curves for the three paths coincide sufficiently closely so that the differences may be attributed to random effects, a conclusion confirmed by the fact that, among the various test series, there was no clear-cut ordering of stiffness for the different paths.

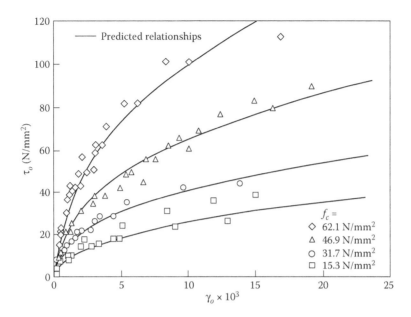

Figure 3.3 Typical experimental $\tau_o - \gamma_{o(d)}$ relationships for various concretes.

Such path independency of the relations $\tau_o - \gamma_{o(d)}$ and $\tau_o - \varepsilon_{o(d)}$ with respect to the direction of τ_o on the deviatoric plane defined for any given σ_o is an important and very useful property of concrete materials, as it considerably simplifies the analytical model to be subsequently derived. However, it leads next to the question of whether or not the two relations between τ_o and the strains it produces are in fact affected by the magnitude of σ_o itself. It turns out that, as in the case of the $\sigma_o - \varepsilon_{o(h)}$ relationships, the $\tau_o - \gamma_{o(d)}$ characteristic

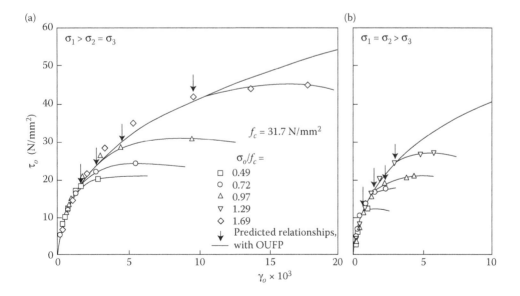

Figure 3.4 Typical experimental $\tau_o - \gamma_{o(d)}$ relationships for concrete (with $f_c = 31.7$ MPa) for two possible stress paths: (a) $\sigma_1 > \sigma_2 = \sigma_3$; (b) $\sigma_1 = \sigma_2 > \sigma_3$.

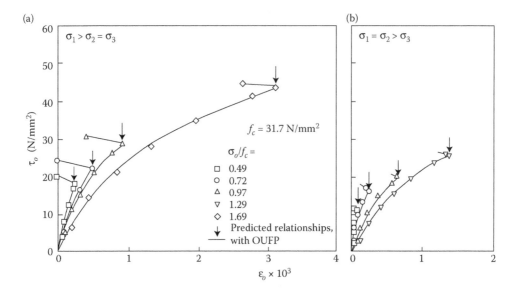

Figure 3.5 Typical experimental $\tau_o - \varepsilon_{o(d)}$ relationships for concrete (with $f_c = 31.7$ MPa) for two possible stress paths: (a) $\sigma_1 > \sigma_2 = \sigma_3$; (b) $\sigma_1 = \sigma_2 > \sigma_3$.

has also been found to be essentially unique, that is, for a given concrete (defined by f_c), γ_o is dependent on τ_o only and not on the level of σ_o. This is implicit in Figure 3.3 but may be seen more clearly by reference to Figure 3.4 where, for a given f_c, a single curve describes the $\tau_o - \gamma_{o(d)}$ relationship up to the OUFP level, that is, up to about the peak of the ascending branch that defines the range of material properties relevant for purposes of structural

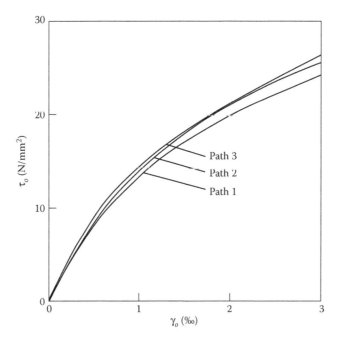

Figure 3.6 Deviatoric stress–strain curves obtained from triaxial tests using different loading paths.

applications. Therefore, the fact that, beyond the minimum-volume level, where the predominant effect on deformation is void formation (which is considered to cause the much faster rate of increase of $\gamma_{o(d)}$ with τ_o), the $\tau_o - \gamma_{o(d)}$ characteristic does become dependent on the value of σ_o has no bearing on the material model. Further evidence of the independence of the $\tau_o - \gamma_{o(d)}$ relations with respect to the location of octahedral planes is provided by some of the results of the various test series of the international cooperative project (Gerstle et al. 1980). The triaxial load history followed in a number of these tests allowed stress deviation within various octahedral planes of mean normal stress varying from about $0.5f_c$ to almost $2.0f_c$. Deviatoric stress–strain curves were plotted for all these octahedral planes and the results are shown in Figure 3.7. Here, deviations were imposed within different octahedral planes ranging from $\sigma_o = 0.55\ f_c\ (= 17.6\ \text{MPa})$ to $\sigma_o = 1.75\ f_c\ (= 56.3\ \text{MPa})$.

Each individual curve in Figure 3.7 represents the mean of several tests along each of these paths. It is clear that the curves are very closely bunched, with no systematic ordering according to σ_o. Other test series gave entirely similar results, especially as regards the very small random scatter of curves, thus confirming that the $\tau_o - \gamma_{o(d)}$ relations are independent of σ_o.

The $\tau_o - \varepsilon_{o(d)}$ characteristics, on the other hand, are a function of both f_c and σ_o, as is clear by reference to Figure 3.5. (It should be noted that the sharp deviations from the smooth curves on reaching the OUFP level are of no relevance to the analytical model.) Such characteristics, therefore, appear to represent the only form of interaction (i.e., coupling) between the hydrostatic and deviatoric components of the stress and strain states. As for the $\sigma_o - \varepsilon_{o(b)}$ relationships, the non-linearity of $\tau_o - \gamma_{o(d)}$ and $\tau_o - \varepsilon_{o(d)}$ is due primarily to the dependence of $\gamma_{o(d)}$ and $\varepsilon_{o(d)}$ on the internal stress redistributions that take place when an externally applied deviatoric stress is imposed. While the non-linear variation of $\gamma_{o(d)}$ with τ_o may be regarded as being caused by the deviatoric component of internal stresses, the non-linearity of $\varepsilon_{o(d)}$ with τ_o is considered to be dictated by the effect of the hydrostatic component of the internal stresses on the deformation of the material when subjected to applied deviatoric stress. The fact that $\varepsilon_{o(d)}$ increases (i.e., volume of specimen decreases) at an increasing rate implies that,

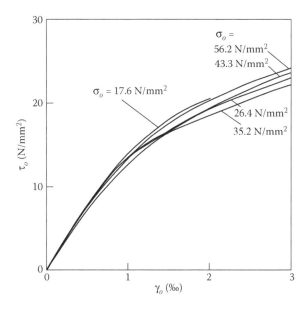

Figure 3.7 Deviatoric stress–strain curves for different octahedral planes.

as in the case of the $\sigma_o - \varepsilon_{o(h)}$ relationship, the hydrostatic component of internal stresses (under external deviatoric stress) is compressive and may also be considered to represent the reduction of the high tensile stress concentrations that occur near the crack tips as a result of crack extension.

3.1.1.2 Deformational behaviour during unloading

An indication of the stress–strain behaviour of concrete during unloading may be given by uniaxial and triaxial experimental data obtained from cyclic tests (Newman 1973, Kotsovos 1974, 1984, Spooner and Dougill 1975). The hysteresis loop exhibited by such data during the first cycle is so small that the same linear stress–strain relationship may be used to describe in a realistic way concrete behaviour during both unloading and subsequent reloading up to the maximum stress level experienced previously by the material (see Figure 3.8). Figure 3.8 indicates that this linear relationship has an essentially constant inclination and its distance, from the origin of the monotonically increasing stress–strain curve, increases with the maximum stress level of the cyclic load. For stress levels beyond this maximum stress, concrete response is described by the monotonically increasing stress–strain relationships discussed in the preceding section.

The above simple behaviour upon unloading may readily be understood by reference to the internal fracture processes that were used earlier to explain the causes of the observed behaviour under increasing stress. During unloading, these fracture processes cease and, as a result, concrete behaviour is then essentially elastic. Only when the maximum stress level previously experienced by the material is exceeded, the fracture can be resumed.

3.1.2 Mathematical description of deformational behaviour

The aim of this section is to incorporate the salient phenomenological features and related experimental data outlined in the preceding section into an analytical material model in a form suitable for use in the non-linear finite-element (FE) analysis of concrete structures. This can be done in several ways, and three possible approaches will be described in detail. The first of these will be most familiar to structural engineers used to dealing with a given number of material 'constants' that are required for defining the stress–strain relationships. The second approach is conceptually ideal in expressing the fracture processes associated with micro-cracking and the internal stresses which arise therefrom. For purposes of computer implementation, however, the third approach has been found to be most convenient (although such a standpoint is, arguably, a matter of personal taste); this is an eclectic procedure, combining the philosophies of the first two methods as needed. It should be evident that, whatever the formulation used to arrive at a suitable mathematical model, the key analytical ingredient is effectively a curve-fitting process based on reliably established experimental data of the kind discussed throughout this and the preceding chapters.

3.1.2.1 Three-moduli approach

As is well known, in the case of linearly elastic material behaviour in three dimensions, the six components of stress are related to their strain counterparts by the generalised law of Hooke involving a maximum of 21 independent constants. Material isotropy reduces the number of these constants to 2, which may be expressed in several (clearly interdependent) ways, such as by means of Lame's expressions or by the use of Young's modulus E and Poisson's ratio ν. For present purposes, where the octahedral representation of

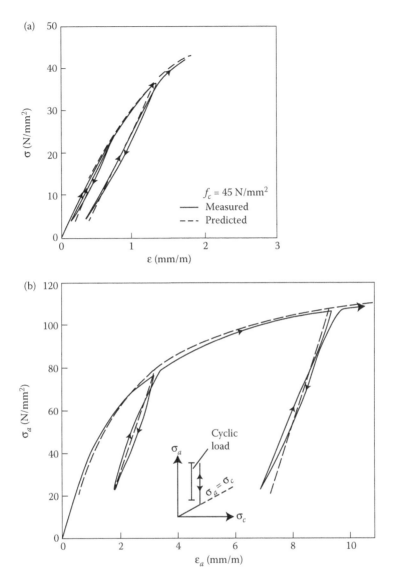

Figure 3.8 Measured and predicted stress–strain behaviour during loading and unloading/reloading for a typical concrete (with f_c = 45 MPa) under (a) uniaxial and (b) triaxial compression.

stresses and strains is adopted, perfect isotropy permits the decoupling of volume changes and distortional (i.e., shape) changes, with the two natural material constants becoming the bulk modulus K and the shear modulus G, as given by the following expressions (see Appendix A):

$$K = \sigma_o /(3\varepsilon_o) = E/[3(1 - 2v)]$$ (3.1)

$$G = \tau_o /(2\gamma_o) = E/[2(1 + v)]$$ (3.2)

Since these parameters are constant for linearly elastic materials, it follows that both the octahedral normal and shear stress–strain (straight line) curves are unique, as they are independent of stress magnitudes and stress ratios in the uniaxial, biaxial and triaxial ranges.

In the case of non-linear materials, parameters akin to relations (3.1) and (3.2) may still be employed, but then K and G are functions of the stress and strain levels in the material. For example, the following definitions could be used:

$$K_s = \sigma_o(\varepsilon_o)/(3\varepsilon_o) \tag{3.3}$$

$$G_s = \tau_o(\gamma_o)/(2\gamma_o) \tag{3.4}$$

where the subscript s indicates that the secant-modulus approach has been adopted. Now, it was argued in Section 2.3.3 that, up to about the OUFP level, concrete may be approximated by an isotropic model. Thus, expressions such as (3.3) and (3.4) can be used to describe the ascending branch of the stress–strain relations for concrete, or, more specifically, the $\sigma_o - \varepsilon_{o(h)}$ and $\tau_o - \gamma_{o(d)}$ characteristics, respectively. Such an approach, however, is not sufficient on account of the presence of the coupling between stress deviation and volume change. Owing to this $\tau_o - \varepsilon_{o(d)}$ characteristic, a third material modulus is needed, and a coupling modulus H has been proposed (Gerstle et al. 1980), defined as

$$H_s = \tau_o / \varepsilon_{o(d)} \tag{3.5}$$

Therefore, the total octahedral strains resulting from the application of an external stress state (σ_o, τ_o) may be written as

$$\varepsilon_o = \varepsilon_{o(h)} + \varepsilon_{o(d)} = \sigma_o/(3K_s) + \tau_o/H_s \tag{3.6}$$

$$\gamma_o = \gamma_{o(d)} = \tau_o/2G_s \tag{3.7}$$

It is evident that the presence of the second term on the right-hand side of expression (3.6) implies that the assumption of isotropy in concrete at the material macro-level may be used only in an approximate sense. Unlike a truly isotropic material, for which the principal stress and strain axes coincide, the coupling effect in concrete leads to the schematic representation of Figure 3.9, and the requirement for a third modulus for what may be described more accurately as quasi-isotropic behaviour. While data on the moduli K_s and G_s associated with true isotropy will be presented later, it may be of interest to include here some information on the additional modulus that has emerged from the international cooperative project in which various testing techniques were used (Gerstle et al. 1980). This cooperative work produced, among other data, a number of relations and, although these allow only a superficial estimate of H_s to be achieved, it is worth exhibiting here the main trend of results. If a straight-line relation between τ_o and $\varepsilon_{o(d)}$ is assumed (clearly an approximation – see e.g., Figure 3.5), H_s will be a parameter that varies only with hydrostatic pressure σ_o. In order to assess this variation, the mean $\tau_o - \varepsilon_{o(d)}$ curves of all triaxial series considered were plotted individually and their moduli determined. These were then plotted against σ_o and are shown as data points in Figure 3.10 (the solid curve refers to the resulting approximate mean $H_s - \sigma_o$ characteristic with the symbol names being the abbreviations of the names of laboratories which participated in the cooperative programme [Kotsovos and Pavlovic 1995]). It is evident that the coupling effect tends to become very slight at low values of σ_o, as indicated by the corresponding high values of H_s.

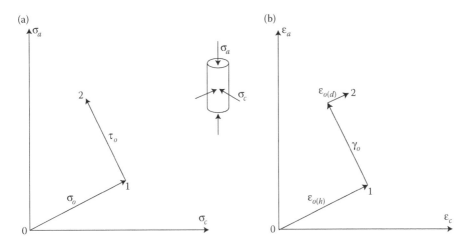

Figure 3.9 Schematic representation of the coupling effect between stress deviation τ_o and volume change $\varepsilon_{o(d)}$: (a) stress path followed; (b) resulting strains.

3.1.2.2 Internal-stress approach

Consider a typical ascending branch of the stress–strain relations of concrete, such as that represented by the curve of Figure 3.11. (The tensorial notation and convention adopted for the stresses and the strains are used here in a generic sense; the main specific application of Figure 3.11 and associated Equations 3.8 through 3.10 – see below – concerns, of course, octahedral stress–strain characteristics.) As explained already, the non-linearity of such a stress–strain relationship is governed essentially by the fracture processes identified with micro-cracking and which may be viewed as being equivalent to the application of an internal compressive state of stress. It follows, therefore, that, once the departure from linearity

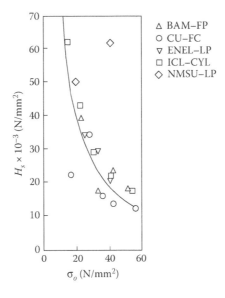

Figure 3.10 Coupling modulus versus octahedral normal stress.

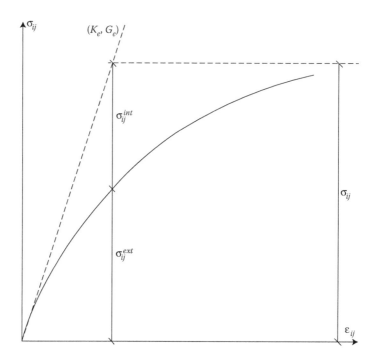

Figure 3.11 The internal-stress concept used to account for the non-linear constitutive relations of concrete materials.

is established and quantified, it can be thought of as the overall effect of these internal stresses at the structural level. Then, the use of this internal state of stress, in conjunction with the initial elastic moduli K_e, G_e, E_e, ν_e (tangent to the curve at the origin), is sufficient to describe fully the non-linear constitutive characteristic(s). In this way, the knowledge of the internal stress state throughout the ascending range of the stress–strain relations enables the non-linear response of concrete to be expressed analytically by using linear material properties, namely the two moduli associated with isotropic behaviour (Kotsovos 1984). Thus, when σ_{ij} refers to a stress related to the state of strain by constant material properties, the generalised form of Hooke's law may be written as

$$\varepsilon_{ij} = -(\nu_e / E_e)\sigma_{kk}\delta_{ij} + [(1 + \nu_e)/ E_e]\sigma_{ij} \tag{3.8}$$

or, equivalently

$$\varepsilon_{ij} = -[(3K_e - 2G_e)/(18K_eG_e)]\sigma_{kk}\delta_{ij} + (1/2G_e)\sigma_{ij} \tag{3.9}$$

in which the tensorial summation convention is implicit (as is the well-known meaning of Kronecker's delta), so that $\sigma_{kk} = 3\sigma_o$ and $2\,\varepsilon_{ij} = \gamma_{ij}$. By decomposing the total strains ε_{ij} and stresses σ_{ij} into their hydrostatic components, $\varepsilon_o = \varepsilon_{kk}/3$ and $\sigma_o = \sigma_{kk}/3$, and deviatoric components, $e_{ij} = \varepsilon_{ij} - \varepsilon_o\delta_{ij}$ and $s_{ij} = \sigma_{ij} - \sigma_o\delta_{ij}$, the following simple alternative expression for Hooke's law in its generalised form is obtained:

$$\varepsilon_{ij} = \varepsilon_o\delta_{ij} + e_{ij} = [\sigma_o /(3K_e)]\delta_{ij} + s_{ij} /(2G_e) \tag{3.10}$$

For any of the above expressions (3.8) through (3.10) to be applicable to concrete behaviour, it is evident that the following equation must hold:

$$\sigma_{ij} = \sigma_{ij}^{ext} + \sigma_{ij}^{int} \tag{3.11}$$

where the superscripts *ext* and *int* refer to the external (applied) and internal (microcracking) stress states, respectively. The strain component due to σ_{ij}^{ext} is recoverable during unloading whereas that due to σ_{ij}^{int} is permanent and equivalent to the strain state caused by the maximum level of σ_{ij}^{int} previously experienced by the material. The elastic recovery can be evaluated from any of the Equations 3.8 through 3.10 by setting $\sigma_{ij}^{int} = 0$ (Kotsovos 1984). However, the use of Equations 3.8 through 3.10 for the evaluation of the strain state corresponding to a given stress level requires a quantitative description of the elastic constants K_e and G_e; the internal state of stress σ_{ij}^{int}; a criterion defining loading and unloading; and a criterion defining the variation in stress space of the external state of stress causing failure. Qualitative descriptions of the first three of the above items are presented in the following, while the failure criterion will be given in Section 3.2.

The elastic moduli K_e and G_e (in MPa) corresponding to the initial properties (i.e., those for $\sigma_{ij}^{ext} = 0$) of the stress–strain curves for concrete may be expressed as follows (Kotsovos 1984):

$$K_e = 11000 + 3.2f_c^2 \tag{3.12}$$

$$G_e = 9224 + 136f_c + 3296 \times 10^{-15}f_c^{8.273} \tag{3.13}$$

where the uniaxial cylinder compressive strength f_c is also expressed in MPa.

Equations 3.12 and 3.13 are valid for values of f_c in the range 15–65 MPa. Outside this range, K_e and G_e remain constant and equal to their values for $f_c = 15$ and 65 MPa; this is also true for all the other parameters to be defined subsequently (namely *b*, *A*, *d*, *C*, *k*, *I*, *m*, *n*). (Such approximations are based on the observations that, beyond $f_c \approx 65$ MPa, any parameter variations flatten out to nearly constant values, while below $f_c \approx 15$ MPa, the relevant characteristics for the latter f_c value represent sufficiently accurate averages for the scatter of results in this low concrete-strength range.)

By decomposing each state of stress and strain into a hydrostatic and a deviatoric component, the internal stress state may be quantified by using experimental data similar to those shown in Figures 3.2 through 3.5. A regression analysis of $\sigma_o - \varepsilon_{o(h)}$ data (see Figure 3.2) has led to the following analytical expression for the relationship between external stress and resulting strain (Kotsovos 1984):

$$\varepsilon_{o(h)} = (\sigma_o + 3aK_e\sigma_o^b)/(3K_e) \quad \text{for } \sigma_o/f_c \leq 2 \tag{3.14}$$

$$\varepsilon_{o(h)} = [\sigma_o + 3abK_e(2f_c)^{b-1}\sigma_o + 3a(1-b)K_e(2f_c)^b]/(3K_e) \quad \text{for } \sigma_o/f_c > 2 \tag{3.15}$$

where K_e is given by Equation 3.12, while *a* and *b* are parameters that depend on the material properties and can be evaluated by regression analysis. The relation for *b* may be expressed in the following form:

$$b = 2.0 + 1.81 \times 10^{-8}f_c^{4.461} \tag{3.16}$$

A regression analysis for a is not actually carried out at this stage, since this parameter is later absorbed into another one (A), as will be seen below. In keeping with the notion that the non-linear variation of $\varepsilon_{o(h)}$ with σ_o, as expressed by Equation 3.14, is to be attributed to internal stresses caused by the fracture processes that occur under increasing hydrostatic stress, it is useful to recall the description in Section 2.3.3.2 of these processes, which take the form of micro-cracks that are randomly distributed and oriented. When the micro-cracks are situated in the path along which a potential crack is likely to propagate, they will tend to increase the energy required to start the process and therefore act as crack 'inhibitors'. Since the number of such crack-propagation inhibitors will increase with the level of hydrostatic stress, it is realistic to assume that this type of fracture process diminishes progressively with increasing stress and that, therefore, the cause of the non-linearity of the $\sigma_o - \varepsilon_o$ relationship should eventually cease to exist. Such considerations are supported by experimental evidence which shows that, for hydrostatic stress levels higher than $2f_c$, the $\sigma_o - \varepsilon_o$ relationship becomes linear (see e.g., the curve for $f_c = 15.3$ MPa in Figure 3.2). The validity of Equation 3.14, therefore, is assumed to extend up to $\sigma_o = 2f_c$. For higher values, a linear variation of σ_o with ε_o may be expressed by the tangent of Equation 3.14 at $\sigma_o = 2f_c$, and this is what Equation 3.15 refers to.

For Equations 3.14 and 3.15 to be compatible with the first term of the right-hand side of Equation 3.10, the hydrostatic component of the internal state of stress ($\sigma_{o(h)}^{int} \equiv \sigma_{ih}$) resulting from the external hydrostatic stress ($\sigma_o^{ext} \equiv \sigma_o$) must be

$$\sigma_{ih}/f_c = A(\sigma_o/f_c)^b \quad \text{for } \sigma_o/f_c \leq 2 \tag{3.17}$$

$$\sigma_{ih}/f_c = 2^{b-1}Ab(\sigma_o/f_c) + 2^b A(1-b) \quad \text{for } \sigma_o/f_c > 2 \tag{3.18}$$

where

$$A = 3aK_e f_c^{b-1} \tag{3.19}$$

with a regression analysis of experimental data yielding the following expressions for A (which now incorporates – and defines – a)

$$A = 0.516 \quad \text{for } f_c \leq 31.7 \tag{3.20}$$

$$A = 0.516/[1 + 0.0027(f_c - 31.7)^{2.397}] \quad \text{for } f_c > 31.7 \tag{3.21}$$

As discussed previously, the deviatoric component of the internal stress state under applied σ_o is insignificant since the measured deviatoric strain under external hydrostatic stress has been found to be negligible. It appears, therefore, that Equations 3.17 and 3.18 describe completely the internal stress state which develops within concrete when it is subjected to increasing external hydrostatic stress. On the other hand, while an *external hydrostatic stress* gives rise to an internal stress state which is also purely *hydrostatic*, the internal stress state which develops within concrete under *external deviatoric stress* consists of both a *hydrostatic* and a *deviatoric* component.

The deviatoric component ($\tau_{o(d)}^{int} \equiv \tau_{id}$) of the internal stress state that develops as a result of an externally applied deviatoric stress can be quantified on the basis of an analytical description of $\tau_o - \gamma_o$ data similar to those shown in Figure 3.3. Such an analytical description may take the form (Kotsovos 1984)

$$\gamma_o = (\tau_o + 2cG_e\tau_o^d)/(2G_e) \tag{3.22}$$

where G_e is given by Equation 3.13, while c and d are material parameters. A regression analysis for d yields the following expressions:

$$d = 2.12 + 0.0183f_c \quad \text{for } f_c \leq 31.7 \tag{3.23}$$

$$d = 2.7 \quad \text{for } f_c > 31.7 \tag{3.24}$$

As before, when the parameter a was discussed, there is no need at this stage to perform a regression analysis for c, as the latter is to be absorbed into another parameter (C). Now, for Equation 3.22 to be compatible with the second term of the right-hand side of Equation 3.10, τ_{id} must be

$$\tau_{id}/f_c = C(\tau_o/f_c)^d \tag{3.25}$$

where

$$C = 2cG_ef_c^{d-1} \tag{3.26}$$

With C both incorporating and defining the parameter c, all that remains to be done is to perform a regression analysis on the experimental data for the τ_o relations. This yields the following expressions:

$$C = 3.573 \quad \text{for } f_c \leq 31.7 \tag{3.27}$$

$$C = 3.573/[1 + 0.0134(f_c - 31.7)^{1.414}] \quad \text{for } f_c > 31.7 \tag{3.28}$$

The hydrostatic component ($\sigma_{o(d)}^{int} \equiv \sigma_{id}$) of the internal stress state resulting from an external deviatoric stress is considered to be the fundamental cause of the volumetric response $\varepsilon_{o(d)}$ of concrete under deviatoric stress. Nominal values of σ_{id} for a given concrete may easily be obtained by using $\tau_o - \varepsilon_{o(d)}$ and $\sigma_o - \varepsilon_{o(h)}$ relationships such as those shown in Figures 3.5 and 3.2, respectively. For a value of the volumetric strain $\varepsilon_{o(d)}$ corresponding to a given level of applied stress (σ_o, τ_o) (see Figure 3.5), a value of hydrostatic stress – which may be considered to represent a nominal value for σ_{id} – can be obtained from the $\sigma_o - \varepsilon_{o(h)}$ relationship of Figure 3.2. In this way, the $\tau_o - \varepsilon_{o(d)}$ relationships of Figure 3.5 can be transformed into $\sigma_{id} - \tau_o$ relationships such as those shown, for example, in Figure 3.12. This procedure (Kotsovos 1980) is illustrated schematically in Figure 3.13. A regression analysis of experimental data similar to those shown in Figure 3.12 has led to the following analytical expression for σ_{id} (Kotsovos 1984):

$$\sigma_{id}/f_c = M(\tau_o/f_c)^n \tag{3.29}$$

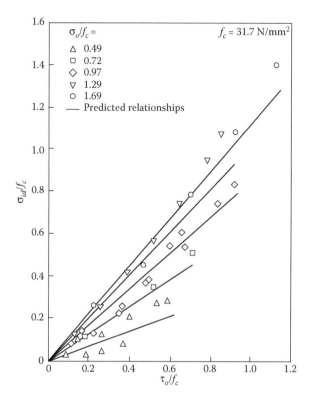

Figure 3.12 Variation of σ_{id} with τ_o for various σ_o, for a typical concrete with $f_c = 31.7$ MPa.

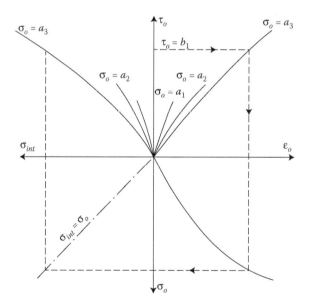

Figure 3.13 Schematic representation of the approach used to evaluate σ_{id} for a given combination of σ_o and τ_o.

where

$$M = k/[1 + l(\sigma_o/f_c)^m]$$ (3.30)

and k, l, m, n are material parameters which may be expressed in terms of f_c as follows:

$$k = 4/[1 + 1.087(f_c - 15)^{0.23}]$$ (3.31)

$$l = 0.222 + 0.01086 f_c - 0.000122 f_c^2$$ (3.32)

$$m = -2.415 \quad \text{for } f_c \leq 31.7$$ (3.33)

$$m = -3.531 + 0.0352 f_c \quad \text{for } f_c > 31.7$$ (3.34)

$$n = 1 \quad \text{for } f_c \leq 31.7$$ (3.35)

$$n = 0.3124 + 0.0217 f_c \quad \text{for } f_c > 31.7$$ (3.36)

It is evident that, up to a uniaxial cylinder strength of 31.7 MPa, the relationships between σ_{id} and τ_o are always linear (since $n = 1$), as is clear, for example, by reference to Figure 3.12. For higher values of f_c, $n \neq 1$, and hence such relationships become non-linear.

It should be noted that the coupling between stress deviation and volume change, expressed in terms of σ_{id}, has been carried out in a manner which, although referring to the latter parameter as an internal stress, in fact treats σ_{id} as an external stress. This is clear if it is recalled that σ_{id} is eventually obtained from the $\sigma_o - \varepsilon_{o(h)}$ relationship (see e.g., Figure 3.13), so that $\sigma_{id} = \sigma_o$ refers to the σ_{ij}^{ext} (and not the σ_{ij}^{int}) component (see Figure 3.11). Therefore, while σ_{ih} and τ_{id} are indeed internal stresses (compatible with the generic definition 3.11), σ_{id} is an *equivalent* external stress required to produce the actual $\varepsilon_{o(d)}$. While it would be an easy matter to express the coupling by means of a truly internal stress (as was done for σ_{ih} and τ_{id}), it turns out to be more convenient – from a computational viewpoint – to follow the present approach, as will be shown subsequently.

Finally, a criterion that defines loading and unloading may also be expressed in terms of the internal-stress concept. As discussed above, the internal stress state which develops within concrete under loading consists of three components, namely σ_{ih}, τ_{id} and σ_{id} (the latter coupling parameter is expressed as an equivalent external stress), defined by Equations 3.17 and 3.18, 3.25 and 3.29, respectively. These equations indicate that, while σ_{ih} and τ_{id} occur independently of each other under increasing external hydrostatic and deviatoric stresses, respectively, σ_{id} occurs under combined external hydrostatic and deviatoric stress and represents the fundamental cause of all observed interaction between the hydrostatic and deviatoric components of the external stresses and corresponding strains. (Although σ_{id} is caused primarily by the deviatoric component of the applied stress (τ_o) – see expression (3.29) – its actual value is also governed by the relevant octahedral plane defined by the hydrostatic component of the applied stress, as indicated by the 'constant' M, expression (3.30). In view of the above, loading, defined as any change of the external stress state that results in an increase of the level of the internal stresses, may be classified as follows.

    a. *Hydrostatic loading.* This occurs when the current external σ_o exceeds any previous external σ_o, thus resulting in an increase of σ_{ih}.

    b. *Deviatoric loading.* This occurs when the current external τ_o exceeds any previous external τ_o, thus resulting in an increase of σ_{id}.

    c. *Combined loading.* This occurs when the current combination of external σ_o and τ_o results in a σ_{id} larger than any previous σ_{id}. This loading may occur when at least one of (a) and (b) is true.

On the basis of the above, unloading occurs when any of (a), (b) or (c) is not true. Furthermore, it is implied that various combinations of loading and unloading may take place simultaneously.

3.1.2.3 Combined approach

The combined approach is based on the use of two variable mechanical properties – namely the moduli K and G – which account for the non-linearities in the $\sigma_o - \varepsilon_{o(h)}$ and $\tau_o - \gamma_{o(d)}$ relationships, respectively, combined with the use of the σ_{id} variable that accounts for the coupling effect $\tau_o - \varepsilon_{o(d)}$. It will be recalled that, although based on the internal-stress concept, the latter parameter was expressed earlier as an equivalent external stress: the reason for this is that the coupling effect may then be allowed for by simply adding σ_{id} to the externally applied hydrostatic stress. Such an approach enables the total octahedral strains (3.6) and (3.7) caused by an arbitrary, externally applied stress state (σ_o, τ_o) to be rewritten as follows:

$$\varepsilon_o = \varepsilon_{o(h)} + \varepsilon_{o(d)} = (\sigma_o + \sigma_{id})/(3K_s) \tag{3.37}$$

$$\gamma_o = \gamma_{o(d)} = \tau_o/2G_s \tag{3.7}$$

In this way, the three-moduli approach has now been modified by the replacement of the third modulus (H_s) by an equivalent superimposed stress state (σ_{id}) based on the internal-stress concept.

While $\sigma_{id}(\sigma_o, \tau_o, f_c)$ accounts for the coupling between τ_o and $\varepsilon_{o(d)}$, $K_s(\sigma_o, f_c)$ and $G_s(\tau_o, f_c)$ are secant bulk and shear moduli, respectively, should such a coupling not exist (i.e., they are obtained by ignoring σ_{id}). It was shown earlier how curve fitting of experimental uniaxial, biaxial and triaxial data has enabled expression (3.29) for σ_{id} to be derived. Similarly, a regression analysis of experimental information has led to expressions (3.14), (3.15) and (3.22), and it is from these that $K_s = (1/3)(\sigma_o/\varepsilon_o)$ and $G_s = (1/2)(\tau_o/\gamma_o)$, respectively, can easily be derived in the following form:

$$K_s/K_e = 1/[1 + A(\sigma_o/f_c)^{b-1}] \quad \text{for } \sigma_o/f_c \leq 2 \tag{3.38}$$

$$K_s/K_e = 1/[1 + 2^{b-1}Ab - 2^b(b-1)A(\sigma_o/f_c)^{-1}] \quad \text{for } \sigma_o/f_c \leq 2 \tag{3.39}$$

$$G_s/G_e = 1/[1 + C(\tau_o/f_c)^{d-1}] \tag{3.40}$$

In an analogous manner, the tangent bulk and shear moduli (again, σ_o neglecting σ_{id}) which relate stress and strain *increments*, $K_t = (1/3)(d\sigma_o/d\varepsilon_o)$ and $G_t = (1/2)(d\tau_o/d\gamma_o)$ may readily be obtained by differentiation of expressions (3.14), (3.15) and (3.22), respectively. The result is

$$K_t / K_e = 1/[1 + bA(\sigma_o/f_c)^{b-1}] \quad \text{for } \sigma_o/f_c \leq 2 \tag{3.41}$$

$$K_t / K_e = 1/[1 + 2^{b-1}Ab] \quad \text{for } \sigma_o/f_c > 2 \tag{3.42}$$

$$G_t / G_e = 1/[1 + dC(\tau_o/f_c)^{d-1}] \tag{3.43}$$

The above formulae for the secant and tangent moduli are illustrated in Figures 3.14 and 3.15 by reference to a particular concrete $f_c = 31.7$ MPa. (It should be noted that, in the case of G, the experimental data shown cover both triaxial-'compression' and triaxial-'extension' cases.)

Since σ_{id} is a pure hydrostatic correction, expressions (3.37) and (3.7) are equivalent to the following relations in global coordinate directions:

$$\varepsilon_{ij} = (\sigma_{ij} + \sigma_{id}\delta_{ij})/(2G_s) - (3\nu_s/E_s)(\sigma_o + \sigma_{id})\delta_{ij} \tag{3.44}$$

where $E_s(\sigma_o, \tau_o, f_c)$ and $\nu_s(\sigma_o, \tau_o, f_c)$ are the secant Young's modulus and Poisson's ratio, respectively, derived from K_s and G_s by the following standard formulae of linear elasticity (obtained through the use of Equations 3.1 and 3.2):

$$E = (9KG)/(3K + G) \tag{3.45}$$

$$\nu = (3K - 2G)/(6K + 2G) \tag{3.46}$$

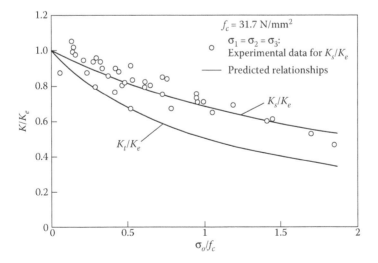

Figure 3.14 Typical variation of the bulk moduli (K_s, K_t) with σ_o for a given concrete (with $f_c = 31.7$ MPa).

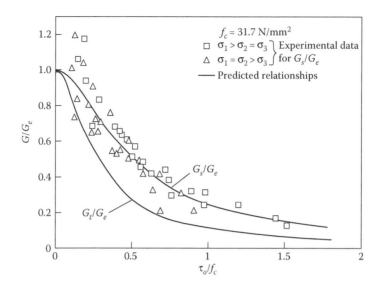

Figure 3.15 Typical variation of the shear moduli (G_s, G_t) with τ_o for a given concrete (with $f_c = 31.7$ MPa).

Expression (3.44) is obtained by noting that the two terms of the strain tensor $\varepsilon_{ij} = \varepsilon_o\delta_{ij} + e_{ij}$ (see [3.10]) must be dealt with separately when the constitutive relations are introduced in this tensor relation. Thus, while the model does not impose any deviation from non-linear elasticity with regard to the second term (i.e., $e_{ij} = (\sigma_{ij} - \sigma_o\delta_{ij})/2G_s$), the first term does involve the correction σ_{id} to the actual, applied σ_o (i.e., $\varepsilon_o\delta_{ij} = (\sigma_o + \sigma_{id})\,\delta_{ij}/(3K_s)$.

Expressions (3.44) form the basis for calculating global strains from global stresses. The actual procedure reduces to the following steps.

- The octahedral stresses (σ_o, τ_o) are calculated either from the principal stresses (σ_1, σ_2, σ_3) – computed previously on the basis of the global stresses σ_{ij}, that is, (σ_x, σ_y, σ_z, τ_{xy}, τ_{xz}, τ_{yz}) – or directly from the first and second stress invariants expressed in terms of σ_{ij}. (See Appendix A.)
- K_s, G_s and E_s, v_s are calculated.
- The hydrostatic correction (i.e., coupling stress) σ_{id} is calculated.
- Global strains ε_{ij} are calculated.

It is clear that the computation of global strain increments from global stress increments follows the same procedure, but with the material 'constants' now defined in terms of tangent values K_t, G_t, E_t and v_t.

3.1.3 Accuracy of the mathematical model for the constitutive relations

As explained in Section 3.1, the cracking processes up to failure are governed, broadly, by the following two (opposite) effects: a reduction in the predominantly tensile stress concentrations near the crack tips (leading to volume reduction), and a production of voids (leading to volume increase). The proposed constitutive model is based on the first of these effects, through the use of the mechanical properties (K_s, G_s) and the coupling parameter σ_{id}. On the other hand, the effect of void formation has been ignored in the stress–strain expressions thus derived.

The conceptual accuracy of the model may be judged by reference to Figure 3.16, which shows some triaxial data for a particular concrete. Three degrees of refinement in the constitutive modelling were applied in turn, and their relative success in mimicking the experimental information can be seen from the resulting curves. First, only the mechanical properties are taken into account, and it is evident that such limited means cannot provide an accurate description of the behaviour beyond the case of pure hydrostatic stress states. Once the stress path departs from this state (at $\sigma_o = 35$ MPa in Figure 3.16), the effect of σ_{ij} requires that σ_{id} (in addition, of course, to G_s) should also be accounted for. Therefore, for general stress states the inclusion of the coupling effect is necessary for a good correlation between model and experimental data to be achieved up to the OUFP level. At this point, the effect of void formation begins to play an increasingly important role and must be given due consideration if the stress–strain paths beyond the OUFP level are to be reproduced (as in Figure 3.16, where curve III is based on the inclusion of void formation in the constitutive model [Kotsovos and Newman 1979]). However, as noted in Section 2.3.3.1, the OUFP level may be taken to coincide – for all practical purposes – with the US level, so that a sufficiently accurate description of the constitutive relations up to failure is attained by the inclusion of the effect of the internal (i.e., mechanical properties and coupling characteristics) while the effect of void formation is disregarded altogether.

The actual accuracy of the model with respect to a large – and varied – body of available experimental data may be seen by reference to Figures 3.2 through 3.5, 3.10, 3.12, 3.14 and 3.15: in most of these the predictions of the adopted model appear as solid lines. It is evident that the proposed mathematical expressions derived on the basis of 'best fit' to experimental data provide a very satisfactory description of the deformation of concrete at a material level, whether individual tests or combined data from more than one test (e.g., Figure 3.15) are considered. Further comparisons between the model and experimental results stemming from specimens tested at Imperial College under triaxial stress conditions appear in Figures 3.17 and 3.18 which refer to triaxial-'compression'

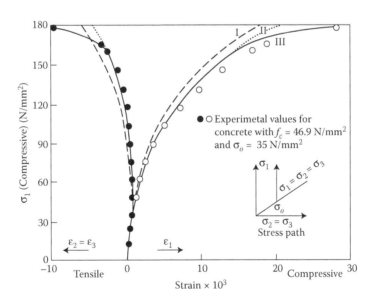

Figure 3.16 Typical stress–strain relations under triaxial stress predicted by the model for a given concrete (with $f_c = 31.7$ MPa). (I) including mechanical properties of the model only; (II) including mechanical properties and effects of internal stresses only; (III) including mechanical properties and effects of internal stresses and void formation.

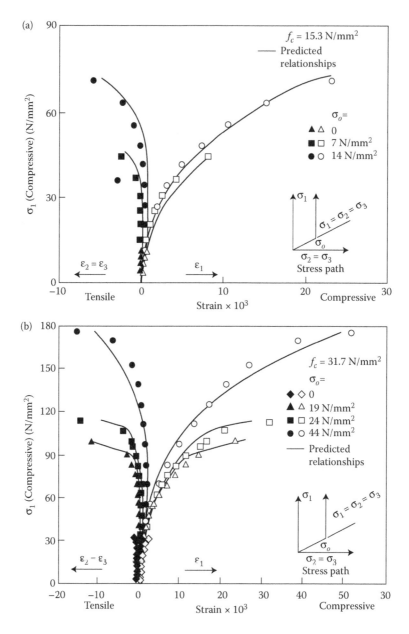

Figure 3.17 Stress–strain relationships for various concretes under stress states $\sigma_1 > \sigma_2 = \sigma_3$ (triaxial 'compression'): (a) f_c = 15.3 MPa; (b) f_c = 31.7 MPa; (c) f_c = 46.9 MPa; (d) f_c = 62.1 MPa. *(Continued)*

and triaxial-'extension' loading paths, respectively; once again, a close fit of the predicted relationships with most of the data is apparent, suggesting that the proposed mathematical model is adequate for all stress paths. The mathematical stress–strain relationships are also in good agreement with the results of biaxial (and uniaxial) stress conditions, as can be seen by reference to Figure 3.19, which is based on the experimental values obtained at the Technical University of Munich (Kupfer et al. 1969). (Although carried out well before the international cooperative project aimed at elucidating the effect of testing procedures

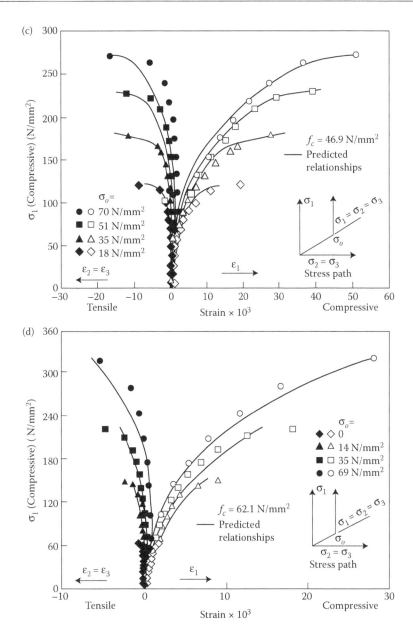

Figure 3.17 (Continued) Stress–strain relationships for various concretes under stress states $\sigma_1 > \sigma_2 = \sigma_3$ (triaxial 'compression'): (a) f_c = 15.3 MPa; (b) f_c = 31.7 MPa; (c) f_c = 46.9 MPa; (d) f_c = 62.1 MPa.

which was referred to in Chapter 2 [Gerstle et al. 1980], the Munich tests represent reliable experimental data in the same way as the tests performed subsequently at Imperial College: unlike the Imperial College results, however, which use sequential paths and a near-uniform stress state applied by the loading system through the adoption of cylinders with a height-to-width ratio of 2.5, the Munich tests were based on a proportional stress path and the application of a near-uniform state of strain on the specimens through brush-bearing platens; the close correlation between predicted and experimental relationships indicates that the effect of testing conditions on the material deformation in the Munich

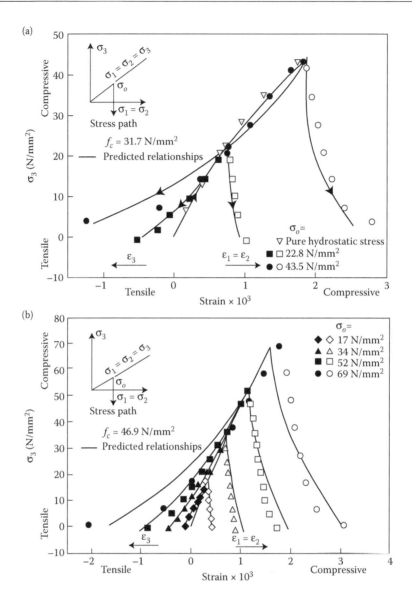

Figure 3.18 Stress–strain relationships for various concretes under stress states $\sigma_1 = \sigma_2 > \sigma_3$ (triaxial 'exten-sion'): (a) f_c = 31.7 MPa; (b) f_c = 46.9 MPa; (c) f_c = 62.1 MPa. *(Continued)*

experiments is also insignificant for practical purposes.) Finally, in addition to the special case $\sigma_2/\sigma_1 = 0$ forming part of the more general biaxial-compression data of Figure 3.19 for f_c = 31.7 MPa, Figure 3.20 shows the model's close fit to experimental values of uni-axial compression for a much wider range of concrete strengths (Kotsovos 1980). Up to f_c = 40 Mpa, these experimental values refer to given data points (Barnard 1964), while for f_c > 40 MPa they are seen to correlate very closely with empirical relationships between the applied stress and the corresponding strain in the direction of loading proposed else-where (Popovics 1973).

In assessing the model's accuracy, the sheer impossibility of fully reproducing experimen-tal data should also be borne in mind, even when comparing results of tests on concrete

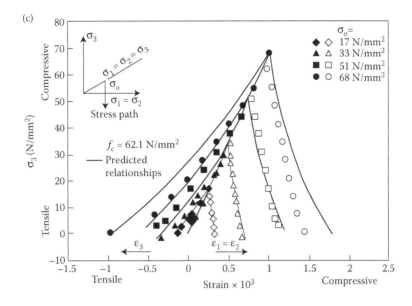

Figure 3.18 (*Continued*) Stress–strain relationships for various concretes under stress states $\sigma_1 = \sigma_2 > \sigma_3$ (triaxial 'extension'): (a) $f_c = 31.7$ MPa; (b) $f_c = 46.9$ MPa; (c) $f_c = 62.1$ MPa.

specimens stemming from a common mix and performed by means of the same apparatus. This scatter of results due to material variability (within a given mix) is illustrated in Figure 3.21a–d (Kotsovos 1974, Kotsovos and Newman 1981), which show the typical range of relevant values for a concrete mix tested triaxially at four values of confining pressure (three specimens were used for each of these confining-pressure values). Such degree of scatter, which is also representative of typical uniaxial and biaxial stress–strain relationships, indicates clearly that further refinement of the mathematical model and/or its component parameters would not be justified.

3.2 STRENGTH ENVELOPES FOR CONCRETE

3.2.1 Experimental data on, and mathematical description of, failure surfaces

The use of non-linear computer-based methods for the analysis of concrete structures subjected to complex stress states requires that both the strength and the deformational properties of concrete should be expressed in a suitable form. The deformational properties have been the subject of the previous section in which a mathematical description of the stress–strain behaviour of the material under generalised stress was outlined. The present section complements the above constitutive properties and is concerned with the mathematical description of the strength properties of concrete (Kotsovos 1979). Such a mathematical description is considered essential, since most of the strength criteria proposed to date for use in practical structural design (e.g., Hannant 1974, Kotsovos and Newman 1977, Hobbs et al. 1977, Lowe 1978, Newman and Newman 1978) have not been expressed in a suitable form for computer applications. Furthermore, certain criteria (Hannant 1974) have been formulated in such a way that the convexity principle (Drucker 1967) does not hold, whereas the formulation of others has been based on the over-simplified assumptions that the effect of the intermediate principal stress on the strength properties is negligible

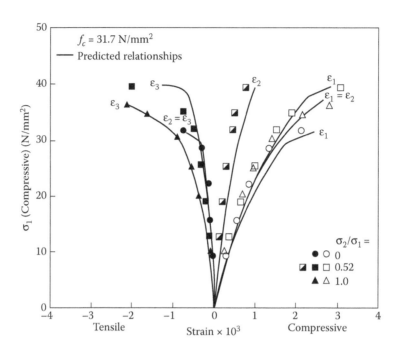

Figure 3.19 Stress–strain relationships for under biaxial (and uniaxial) compression for a typical concrete (with $f_c = 31.7$ MPa).

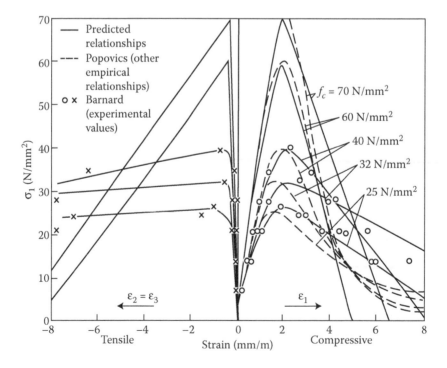

Figure 3.20 Stress–strain relationships for various concretes under uniaxial compression.

(a)

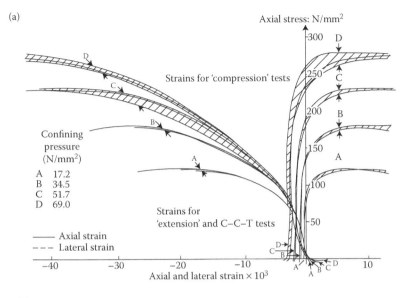

(b)

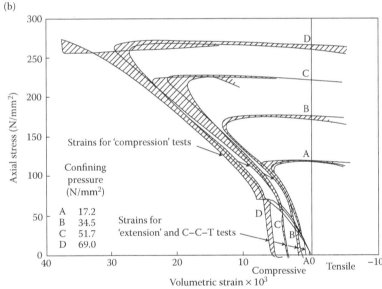

Figure 3.21 Typical scatter on (triaxial) constitutive material data for a given concrete (with $f_c = 46.9$ MPa) tested under various levels of maximum confining pressure. (a) Variation of axial and lateral strains with total axial stress for triaxial 'compression', triaxial 'extension' and C–C–T tests; (b) variation of volumetric strain with total axial stress for triaxial 'compression', triaxial 'extension' and C–C–T tests; (c) variation of lateral strain with axial strain for triaxial 'compression', triaxial 'extension' and C–C–T tests; (d) variation of lateral strain with axial strain for triaxial 'extension' and C–C–T tests.

(*Continued*)

(Hobbs et al. 1977) or that concrete behaves elastically up to a limiting principal tensile strain which defines ultimate strength and is regarded as a material constant (Lowe 1978).

The derivation of mathematical expressions given here is based on an analysis of strength data obtained in the course of investigations of the behaviour of concrete under multi-axial stress states carried out at Imperial College and described elsewhere (Newman 1973,

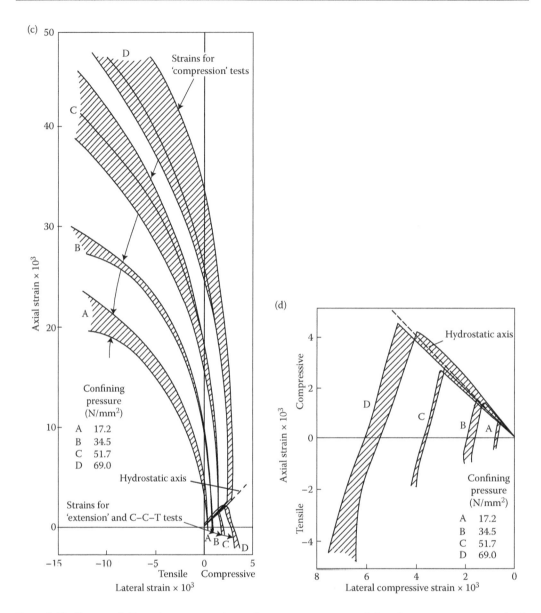

Figure 3.21 (Continued) Typical scatter on (triaxial) constitutive material data for a given concrete (with $f_c = 46.9$ MPa) tested under various levels of maximum confining pressure. (a) Variation of axial and lateral strains with total axial stress for triaxial 'compression', triaxial 'extension' and C–C–T tests; (b) variation of volumetric strain with total axial stress for triaxial 'compression', triaxial 'extension' and C–C–T tests; (c) variation of lateral strain with axial strain for triaxial 'compression', triaxial 'extension' and C–C–T tests; (d) variation of lateral strain with axial strain for triaxial 'extension' and C–C–T tests.

Kotsovos 1974, 1979, Newman and Newman 1978). The testing techniques used to obtain these data (see Section 2.1.1 and Newman 1974) have been validated by comparing them with those obtained in the international cooperative programme of research into the effect of testing techniques and apparatus on the behaviour of concrete under biaxial and triaxial stress states referred to previously (Gerstle et al. 1978, 1980).

As in the case of the constitutive relations, a mathematical description of the strength envelope of concrete, which is governed by combinations of maximum stresses that define a given failure criterion, is most readily formulated in terms of hydrostatic and deviatoric components acting on the octahedral plane. Therefore, it is convenient to define the stress space by the orthogonal coordinate system $(\sigma_1, \sigma_2, \sigma_3)$ of principal stresses. (The convention that compressive stresses are positive will be adopted.) Then, viewing the coordinate system and the octahedral (or deviatoric) plane from the hydrostatic axis, which intersects this plane at right angles, it is easy to see that the stress space can be divided into six regions, within which the following conditions are satisfied:

region 1 : $\sigma_1 \geq \sigma_2 \geq \sigma_3$ (3.47)

region 2 : $\sigma_1 \geq \sigma_3 \geq \sigma_2$ (3.48)

region 3 : $\sigma_2 \geq \sigma_1 \geq \sigma_3$ (3.49)

region 4 : $\sigma_2 \geq \sigma_3 \geq \sigma_1$ (3.50)

region 5 : $\sigma_3 \geq \sigma_1 \geq \sigma_2$ (3.51)

region 6 : $\sigma_3 \geq \sigma_2 \geq \sigma_1$ (3.52)

These regions are shown clearly in Figure 3.22.

The transformation of the orthogonal coordinate system $(\sigma_1, \sigma_2, \sigma_3)$ into the cylindrical coordinate system $(z = (3)^{1/2}\sigma_o, r = (3)^{1/2}\tau_o, \vartheta)$ has been outlined in Appendix A. Accordingly, z is related to the hydrostatic stress that coincides with the space diagonal $\sigma_1 = \sigma_2 = \sigma_3$, while the radius r is similarly related to the magnitude of the deviatoric stress component, the rotational variable ϑ defining the latter's orientation on the octahedral plane. (Clearly, the hydrostatic and deviatoric stresses are obtained by contracting the (z, r) coordinates by a constant factor of $(3)^{1/2}$.) With these preliminaries, the strength envelope may be described by reference to both coordinate systems, and this is shown in Figure 3.22. The resulting ultimate-strength variation obeys the convexity principle usually associated with failure surfaces (Drucker 1967), and is open in compression since concrete can sustain increasing values of deviatoric stress for increasing hydrostatic compressive stress levels, that is, cross sections of the strength envelope (perpendicular to the z axis) become larger as σ_o increases.

If isotropic material behaviour is assumed, the ultimate-strength surface possesses a six-fold symmetry about the space diagonal $\sigma_1 = \sigma_2 = \sigma_3$. Therefore, it follows that only one-sixth of the closed curve defining the failure boundary on a deviatoric plane (Figure 3.22b) is required for its definition. Now, experimental data are readily obtainable for τ_{oe} and τ_{oc} (the factor $(3)^{1/2}$ will henceforth be dropped, that is, the deviatoric plane will be shrunk to the curve τ_{ou} rather than $(3)^{1/2}\tau_{ou}$). These values correspond to axisymmetric stress states easily imposed in a tri-axial test. Thus, τ_{oe} (for $\vartheta = 0°$) is obtained by setting $\sigma_1 = \sigma_2 > \sigma_3$ (triaxial 'extension') while τ_{oc} (for $\vartheta = 60°$) follows by setting $\sigma_1 > \sigma_2 = \sigma_3$ (triaxial 'compression'). In this way, τ_{oe} and τ_{oc} values can be determined for various levels of hydrostatic stress σ_o. For each σ_o, the value of τ_{ou} for any ϑ intermediate between 0° and 60° may be interpolated between the values of τ_{ou} at 0° and 60° by means of the following expression (Willam and Warnke 1974):

$$\tau_{ou} = \{2\tau_{oc}(\tau_{oc}^2 - \tau_{oe}^2)\cos\vartheta + \tau_{oc}(2\tau_{oe} - \tau_{oc})$$
$$[4(\tau_{oc}^2 - \tau_{oe}^2)\cos^2\vartheta + 5\tau_{oe}^2 - 4\tau_{oc}\tau_{oe}]^{1/2}\}/[4(\tau_{oc}^2 - \tau_{oe}^2)\cos^2\vartheta + (2\tau_{oe} - \tau_{oc})^2] \quad (3.53)$$

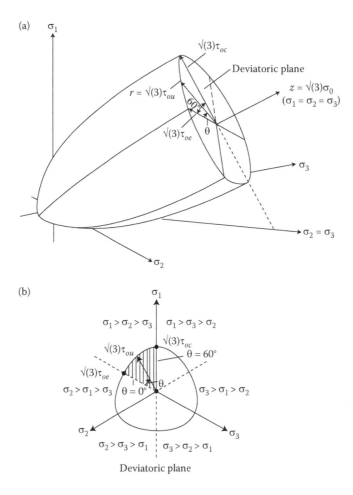

Figure 3.22 Schematic representation of the ultimate-strength surface: (a) general view in stress space; (b) typical cross section of the strength envelope with a deviatoric plane (i.e., a plane of constant σ_o, viewed along the axis $\sigma_1 = \sigma_2 = \sigma_3$).

This expression describes a smooth convex curve with tangents perpendicular to the directions of τ_{oe} and τ_{oc} at $\vartheta = 0°$ and $60°$, respectively (see Figure 3.22). Therefore, it follows that a full description of the strength surface may be established once the variations of τ_{oe} and τ_{oc} with σ_o are determined.

Figure 3.23 shows such variations of τ_{oe} and τ_{oc}. These combinations of octahedral stresses (σ_o, τ_o) at the ultimate-strength level, which appear normalised with respect to the uniaxial cylinder compressive strength f_c were obtained from triaxial tests carried out at Imperial College on a wide range of concretes (with f_c varying approximately between 15 and 65 MPa) subjected to the axisymmetric stress states $\sigma_1 > \sigma_2 = \sigma_3 > 0$ (triaxial 'compression'), $\sigma_1 = \sigma_2 > \sigma_3 > 0$ (triaxial 'extension') and $\sigma_1 = \sigma_2 > 0 > \sigma_3$ (triaxial 'tension' C–C–T). Full details of these tests can be found elsewhere (Kotsovos and Newman 1977, Gerstle et al. 1978). Figure 3.23 indicates that, for the portion of the stress space investigated, the ultimate-strength envelopes are essentially independent of f_c, that is, the type of concrete. Furthermore, since the stress-path effects on ultimate strength have been found small enough to be regarded as insignificant for practical purposes (see Section 2.3.3.3), the two envelopes

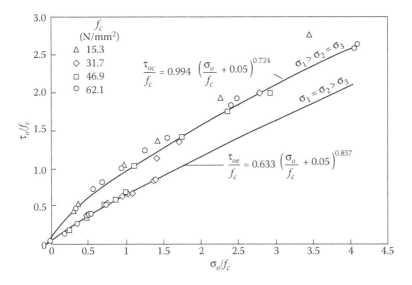

Figure 3.23 Combinations of octahedral stresses at ultimate strength for concrete under triaxial 'compression' and triaxial 'extension'.

of Figure 3.23 is considered to describe adequately the strength of most ordinary concretes likely to be encountered in practice when these are subjected to axisymmetric stress states. This lack of influence of loading history for both stress–strain relations and the ultimate-strength envelope was discussed in Section 2.3.3.2, where it was argued that the unsystematic variability of the relevant data is larger than the scatter due to path dependency. Figure 3.24 shows the justification for such an argument in the case of failure data: it is evident that the scatter of ultimate stresses for concrete of a given f_c for different loading paths is smaller than the scatter of ultimate values for concretes of different f_c, but following a given loading path (Kotsovos 1984). A similar justification for adopting the OUFP level as the failure limit (as opposed to the slightly higher maximum sustained stress level – see Section 2.3.3.1) may be seen by reference to Figure 3.25, which shows that the unsystematic variation of the maximum stress level for various concretes far exceeds the deviation between OUFP and US level for a given concrete (Kotsovos 1984).

A mathematical description of the two strength envelopes in Figure 3.23 may be obtained by fitting curves to the experimental data. Such an approach leads to the following expressions:

$$\tau_{oc}/f_c = 0.944(\sigma_o/f_c + 0.05)^{0.724} \tag{3.54}$$

$$\tau_{oe}/f_c = 0.633(\sigma_o/f_c + 0.05)^{0.857} \tag{3.55}$$

Equations 3.54 and 3.55 represent two open-ended convex envelopes the slopes of which tend to become equal to that of the space diagonal as σ_o tends to infinity. These expressions, together with Equation 3.53, define an ultimate-strength surface which conforms with generally accepted shape requirements such as six-fold symmetry and convexity with respect to the space diagonal, open-ended shape which tends to become cylindrical as σ_o tends to infinity and so forth (Franklin 1970). A three-dimensional (3-D) view of this ultimate-strength surface is shown in Figure 3.26.

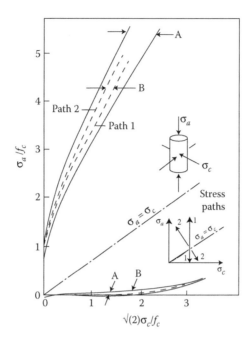

Figure 3.24 A, unsystematic variability of 'failure' data obtained from tests using stress path 1 for concretes with f_c varying between approximately 15 and 65 MPa; B, stress-path effect on 'failure' data for a typical concrete (with $f_c = 31.7$ MPa).

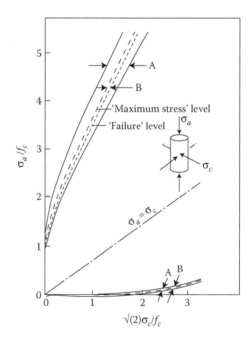

Figure 3.25 A, unsystematic variability of 'maximum stress' level exhibited by concretes with f_c varying between approximately 15 and 65 MPa; B, deviation of 'failure' level from 'maximum stress' level for a typical concrete (with $f_c = 47$ MPa).

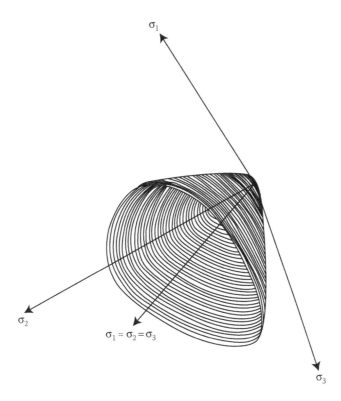

Figure 3.26 Three-dimensional view of the predicted ultimate-strength surface.

It will be noticed that the validity of expressions (3.54) and (3.55) is limited by the constraint that tensile hydrostatic stress states cannot exceed 5% of the uniaxial cylinder compressive strength f_c. This leads to consideration of the question of what experimental data there are for states of stress in which at least two of the principal stresses are tensile, and how the model describes the failure envelopes under such conditions. Experimental evidence of this type is very scarce and, moreover, is invariably associated with large scatter. Under such circumstances, the model smoothly extrapolates the C–C–T portion of the failure surface into regions where more than one principal stress is tensile. A typical cross section of the failure envelope is shown in Figure 3.27 (corresponding to the axisymmetric case) (Kotsovos and Pavlovic 1986), and the result is a smooth surface in the 'tension' region which provides a conservative estimate to a parameter that is subject to a very large degree of unsystematic variability, and which, further-more, represents a small absolute order of magnitude (relative to other stress values) in the stress space. (It is also important to recall the well-known fact that the testing of brittle materials in tension is usually more problematic than the determination of their compressive properties; this was stressed already by Föppl [Timoshenko 1953] and is still largely relevant today.)

On the basis of expressions (3.53) through (3.55), checks may be carried out to ascertain whether a state of stress lies inside or outside the failure envelope. The actual procedure consists of the following steps.

- The octahedral stresses and the rotational variable (σ_o, τ_o, ϑ) are calculated either from the principal stresses (σ_1, σ_2, σ_3) – computed previously on the basis of the global stresses σ_{ij}, that is, (σ_x, σ_y, σ_z, τ_{xy}, τ_{xz}, τ_{yz}) – or directly from the first, second and third stress invariants expressed in terms of σ_{ij} (see Appendix A).

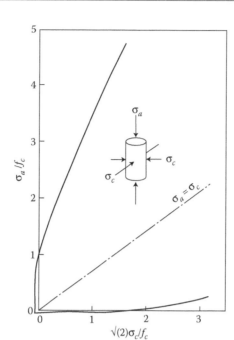

Figure 3.27 Complete ultimate-strength envelope for concrete under axisymmetric stress.

- The ultimate deviatoric stresses at $\vartheta = 0°$ and $60°$ (i.e., τ_{oe} and τ_{oc}, respectively) are calculated for the existing state of hydrostatic stress σ_o.
- The ultimate deviatoric stress τ_{ou} for the existing rotational angle ϑ is calculated on the basis of the interpolation formula defined by (τ_{oe}, τ_{oc} and ϑ).
- The values of τ_o and τ_{ou} are compared; if $\tau_o > \tau_{ou}$, the state of stress lies outside the failure envelope.

3.2.2 Accuracy of the mathematical model for the failure surfaces

It is important to assess the accuracy of the proposed model by comparing the predictions with a wider body of experimental data than that which provided the basis for the derivation of the mathematical expressions. This was done in Section 3.1.3 for the stress–strain relations, and a similar exercise is carried out in the present section for the strength envelopes of concrete.

The intersections of the predicted ultimate-strength surface with the planes $\sigma_1 = 0$, $\sigma_2 = 0$, $\sigma_3 = 0$ represent the ultimate-strength envelopes for concrete under biaxial stress. Figure 3.28a–c show the biaxial strength envelopes (normalised with respect to f_c) for concrete under compression–compression (C–C), compression–tension (C–T) and tension–tension (T–T), respectively, as predicted by the proposed general expressions. The figures also include most of the experimental data published up to the end of 1973 (Hannant 1974). Figure 3.29 shows the biaxial strength envelope predicted for concrete under C–C, together with experimental data obtained in the international cooperative project referred to earlier in this chapter (Gerstle et al. 1978).

The larger scatter of the data shown in all of the above figures is evident. In the case of C–C and C–T, the mathematical model is seen to provide a fair average to such scatter. On

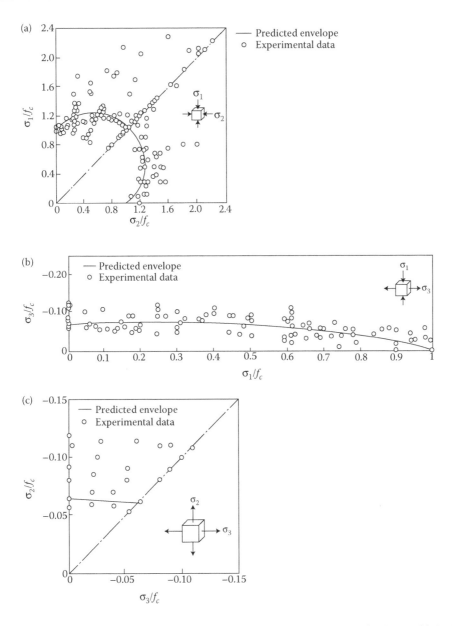

Figure 3.28 Predicted biaxial strength envelopes for concrete with experimental values published up to 1973 (Hannant 1974) (it should be noted that the envelope $\sigma_1 > \sigma_2 > \sigma_3$, where compression is positive, has been adopted [Kotsovos 1979]): (a) compression–compression; (b) compression–tension; (c) tension–tension.

the other hand, the model gives a lower bound for the T–T case, in an attempt to ensure a conservative estimate of a parameter which is especially sensitive to the method of testing used; hence, even though the uniaxial tensile strength of concrete is often taken to be of the order of 10% of f_c, such a figure is halved in the present model.

As explained in Chapter 2, the wide variation of the experimental data shown in all the above figures has been attributed mainly to the different testing techniques used and, in

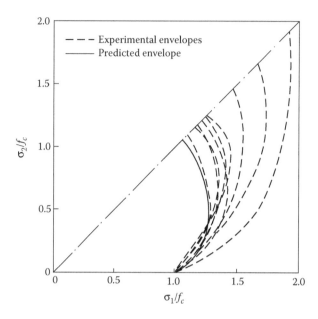

Figure 3.29 Predicted biaxial strength envelopes for concrete under compression–compression with experimental values published between 1973 and 1978.

particular, to the degree of frictional restraint at the platen–specimen interfaces (Gerstle et al. 1978). Now, it is generally accepted that one of the most efficient ways of minimising this frictional restraint is by loading through 'brush' platens, and it is by using such a loading system that the biaxial strength envelope shown in Figure 3.30 has been obtained (Kupfer et al. 1969). Figure 3.30 also includes the predicted envelope and it is significant that a close correlation between predicted and experimental envelopes is now attained.

So far, the accuracy of the proposed strength envelope has been assessed by comparison of the mathematical prediction with available experimental data obtained under biaxial-loading conditions. Next, results from triaxial tests are considered. Figure 3.31a–c show the strength envelopes, expressed in terms of normalised principal stresses, predicted by the proposed expressions for concrete under various triaxial axisymmetric stress states, together with the experimental values corresponding to the data published up to the end of 1973 (Hannant 1974). Figure 3.32a and b show the same envelopes corresponding to Figure 3.31a and b, respectively, but now expressed in terms of octahedral stresses (and for a given concrete), together with the experimental values obtained in the international cooperative project (Gerstle et al. 1978).

The octahedral strength envelope predicted by the proposed expressions for $\vartheta = 30°$ is shown in Figure 3.33, while Figure 3.34 depicts the intersection of the ultimate-strength surface with the deviatoric plane $\sigma_o = 34.5$ MPa. Both these figures include experimental values obtained in the international cooperative project (Gerstle et al. 1978). As for the other triaxial data, the proposed analytical expressions are seen to give a satisfactory fit to experimental results. In conclusion, the suggested ultimate-strength surface appears to provide a simple generalised mathematical representation of failure in concretes under any type of short-term loading condition. The surface conforms to generally accepted shape requirements and has been found to produce a close fit for most biaxial and triaxial strength data published to date.

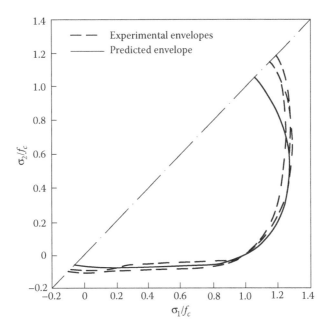

Figure 3.30 Predicted biaxial strength envelopes for concrete with experimental envelopes obtained in tests for which 'brush' platens were used.

3.3 DEFORMATIONAL AND YIELD CHARACTERISTICS OF REINFORCING STEEL

The deformation and strength of reinforcing steel bars are adequately described by reference to their uniaxial properties. (Therefore, as will be seen in later chapters, the reinforcement is almost invariably modelled by means of line FEs possessing stiffness along their longitudinal axes only.) Such material modelling is compatible with the negligible importance attached to the effect of the so-called dowel action in shear-transfer mechanisms, as explained elsewhere (Kotsovos and Pavlovic 1986, Kotsovos et al. 1987).

Figures 3.35 and 3.36 show the tri-linear diagrams adopted for the stress–strain characteristic of steel bars up to and including their plastification. The first of these was used in the two-dimensional (2-D) version of the FE model, which was developed when the 1972 *Code of practice for the structural use of concrete* (CP110) was in force (British Standards Institution 1972). Such a diagram, intended primarily for mild-steel reinforcement, broadly follows this code's recommendations but the third branch of the characteristic is given a small slope (instead of being horizontal, as in the code) in order to avoid the numerical difficulties that would result if an abrupt stiffness change to $E = 0$ were to occur (Bedard 1983).

The diagram corresponding to Figure 3.36, on the other hand, was adopted in the 3-D version of the FE model, this having been developed after the 1985 *Code of practice for structural concrete* (BS 8110) became operative (British Standards Institution 1985). The influence of the latter version of the code can be seen, for example, in the fact that high-strength steels are equally catered for and, also, in the lack of differentiation between tension and compression reinforcement.

As one would expect, the choice between Figures 3.35 and 3.36 usually has negligible effect on structural behaviour predicted by the FE model. In both cases, the first two branches of the characteristic are completely defined once f_y (the yield stress in MPa) is

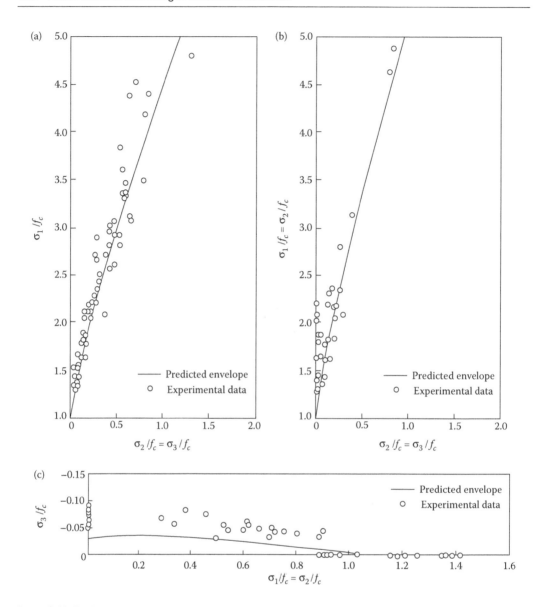

Figure 3.31 Predicted triaxial strength envelopes for concrete under axisymmetric stress states with experimental envelopes published up to 1973: (a) $\sigma_1 > \sigma_2 = \sigma_3 > 0$ (triaxial 'compression'); (b) $\sigma_1 = \sigma_2 > \sigma_3 > 0$ (triaxial 'extension'); (c) $\sigma_1 = \sigma_2 > 0 > \sigma_3$ (triaxial 'tension').

specified. The third branch has a much smaller effect on the behaviour of a structure and hence is either given a small nominal slope (approximately 1% of the initial slope) and a cut-off ultimate strain $\varepsilon_u = 0.12$ (2-D model), or it requires the specification of the maximum, that is, ultimate, stress f_u, and the corresponding ultimate strain ε_u (for the 3-D model). (Since f_u and ε_u are rarely reported for mild steel, the 3-D version of the model automatically assumes $f_u = 1.15 f_y$ and $\varepsilon_u = 0.12$ for such steels [Gonzalez Vidosa 1989].)

It should be stressed that, although a bilinear diagram might appear to be more suitable in the case of mild steels (and is, in fact, recommended by the 1985 code), it turns out to be very convenient to predict earlier yielding of the steel (i.e., the second branch) in the course

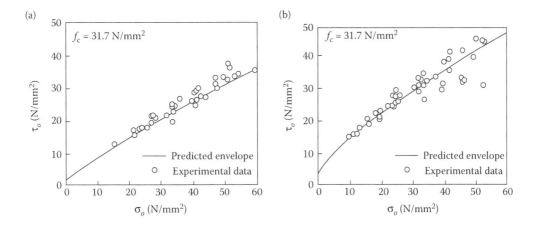

Figure 3.32 Predicted triaxial combinations of octahedral stresses at ultimate strength for a given concrete (with $f_c = 31.7$ MPa) together with experimental values published between 1973 and 1978: (a) $\vartheta = 60°$ (triaxial 'compression'); (b) $\vartheta = 0°$ (tiaxial 'extension').

of numerical analysis. This avoids the uncertainty of (single) yielding detected in the last and non-converged load step of the analysis and, furthermore, clearly helps to differentiate ductile from brittle predictions.

The tri-linear stress–strain diagram adopted by Gonzalez Vidosa was complemented with an additional linear branch (Cotsovos 2004) describing steel behaviour under cyclic loading (see Figure 3.37). Figure 3.37 indicates that this additional branch has an inclination equal to that of the first branch of the monotonic diagram, and, after yielding, its distance from the origin of the latter is linked to the maximum stress of the cyclic load.

3.4 A SUMMARY OF CHARACTERISTIC FEATURES OF CONCRETE RELEVANT TO MODELLING OF MATERIAL BEHAVIOUR

This chapter has been devoted to a detailed description of the mathematical expressions derived for the deformational and strength characteristics of concrete under triaxial states

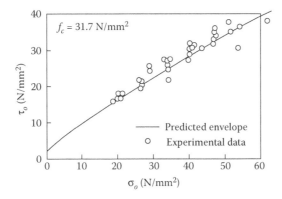

Figure 3.33 Predicted triaxial combinations of octahedral stresses at ultimate strength for a given concrete (with $f_c = 31.7$ MPa) under triaxial stress states with $\vartheta = 30°$ together with experimental values published between 1973 and 1978.

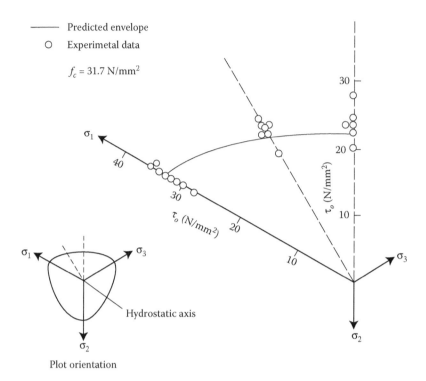

Figure 3.34 Predicted triaxial strength envelope on the deviatoric plane $\sigma_o = 34.5$ MPa for a given concrete (with $f_c = 31.7$ MPa) together with experimental values published between 1973 and 1978.

of stress. These expressions have been derived by regression analysis of valid experimental data obtained from tests carried out at Imperial College London.

The deformational characteristics of concrete have been expressed in the form of bulk K and shear G moduli and the internal stress σ_{id}, the latter describing the coupling between stress deviation τ_o and volume change $\varepsilon_{o(d)}$. Both K and G and σ_{id} have been found to be independent of the rotational variable ϑ, since K and G are functions of the hydrostatic σ_o and deviatoric τ_o

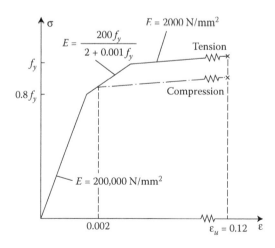

Figure 3.35 Constitutive and strength relationships for steel (2-D FE package).

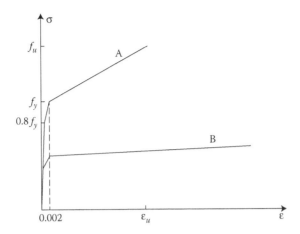

Figure 3.36 Constitutive and strength relationships for steel (3-D FE package): A, high-yield steel; B, mild steel. (Note: for purposes of clarity, only the salient features for the A characteristics are marked on the vertical axis.)

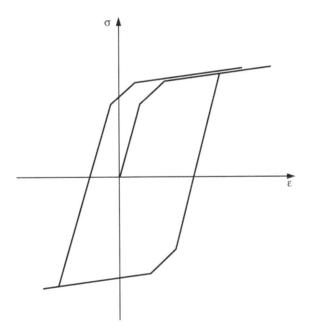

Figure 3.37 Constitutive and strength relationships for steel under cyclic loading.

stresses, respectively, only, whereas σ_{id} is a function of σ_o and τ_o. The strength characteristics of concrete are represented by a surface in stress space which possesses six-fold symmetry about the space diagonal $\sigma_1 = \sigma_2 = \sigma_3$ and is fully defined by its meridians for $\vartheta = 0°$ and $60°$ (i.e., the variations of τ_o with σ_o for $\sigma_1 > \sigma_2 = \sigma_3$ and $\sigma_1 = \sigma_2 > \sigma_3$, respectively) and an interpolation function which describes the variation of τ_o with ϑ ($0° \le \vartheta \le 60°$) for any given σ_o.

It is important to note that the single input variable required by the proposed expressions for defining the strength and deformational properties of concrete under arbitrary multiaxial stress conditions is f_c – the uniaxial cylinder compressive strength of concrete.

Thus, both the constitutive relationships and the strength envelopes for concrete under uni-axial, biaxial or triaxial stress states are fully and unequivocally determined (irrespective of the stress path followed) upon specification of f_c. Such a parameter is readily available in everyday engineering practice, and this is in marked contrast with the parameters required by other, more complex models of concrete for which the number of variables may be large, their determination difficult and the ensuing values of doubtful validity. Many such 'theories' are no more than curve-fitting exercises for a particular set of experimental data and, hence, it is not surprising that, besides the inherent awkwardness, their accuracy is usually limited to the particular circumstances and/or structural form for which they were originally derived, more general problems necessitating the adjustment of parameters or the use of altogether new models for concrete behaviour. The model being proposed here, on the other hand, is applicable to all structural forms under any (short-term static) loading and/or boundary conditions, the generality of its predictions being consistent with the fact that a comprehensive description of concrete response – all the way to failure – is provided by the simple uniaxial-strength parameter f_c which could not possibly be the subject of controversy.

It should also be noted that, at the material level, concrete is brittle and that a full descrip-tion of its response is embodied in the stress–strain characteristics up to the maximum sustained stress. Such characteristics represent the complex processes of fracture initiation and extension. The overall effect of these have been approximated in a simple manner by describing the ascending portion of the stress–strain relations as one due to progressive micro-cracking that can be explained with the aid of the 'internal compressive stress' con-cept, the latter being equivalent to the release of tensile-stress concentrations at the tips of (usually barely – if at all – visible) cracks. Clearly, this process is associated predominantly with volume decrease, and minimum volume is reached at the so-called OUFP level. From this point onwards, volume dilation begins as a result of void formation and the peak stress is quickly reached. Since the OUFP and the maximum-stress levels are very close to each other, the former may be considered to mark, for all practical purposes, the failure of the brittle material, with the formation of (now visible) primary cracks. Just as the non-linear constitutive relations up to the OUFP level are associated with micro-cracking, the OUFP limit itself may be used to define the strength surface associated with macro-cracking. The quasi-isotropy of concrete permits the description of the stress–strain relations in terms of the two isotropic material parameters as well as a third 'constant' which accounts for cou-pling between deviatoric stress and ensuing hydrostatic strain. Strength envelopes indicate that failure is essentially a function of deviatoric stresses and that permissible values of these increase with the level of hydrostatic stress.

REFERENCES

Barnard P. R., 1964, Researches into the complete stress–strain curve for concrete, *Magazine of Concrete Research*, 16(49), 203–210.

Bedard C., 1983, Non-linear finite element analysis of concrete structures, PhD thesis, University of London.

British Standards Institution, 1972, Code of practice for the structural use of concrete, CP110 (Part 1. Design, materials and workmanship), British Standards Institution, London.

British Standards Institution, 1985, British Standard. Structural use of concrete, BS8110 (Part 1. Code of Practice for design and construction), British Standards Institution, London.

Cotsovos D. M., 2004, Numerical investigation of structural concrete under dynamic (earthquake and impact) loading, PhD thesis, University of London, UK.

Drucker D. C., 1967, *Introduction to Mechanics of Deformable Solids*, McGraw-Hill, New York.

Franklin J. A., 1970, Classification of Rock According to its Mechanical Properties, *Rock Mechanics Report No. T.1*, Imperial College, London.

Gerstle K. H., Linse D. H., Bertacchi P., Kotsovos M. D., Ko H-Y., Newman J. B., Rossi P. et al., 1978, Strength of concrete under multiaxial stress states, *Proceedings Douglass McHenry International Symposium on Concrete and Concrete Structures*, Mexico City, Mexico, ACI Publication SP-55, Detroit, USA, pp. 103–131.

Gerstle K. H., Aschl H., Bellotti R., Bertacchi P., Kotsovos M. D., Ko H-Y., Linse D. H. et al., 1980, Behaviour of concrete under multiaxial stress states, *Journal of the Engineering Mechanics Division, ASCE*, 106(EM6), 1383–1403.

Gonzalez Vidosa F., 1989, Three-dimensional finite element analysis of structural concrete under static loading, PhD thesis, University of London.

Hannant D. J., 1974, Nomograms for the failure of plain concrete subjected to short-term multi-axial stresses, *The Structural Engineer*, 52, 151–165.

Hobbs D. W., Pomeroy C. D. and Newman J. B., 1977, Design stresses for concrete structures under combined states of stress, *The Structural Engineer*, 55, 151–164.

Kotsovos M. D., 1974, Failure criteria for concrete under generalized stress, PhD thesis, University of London.

Kotsovos M. D., 1979, Mathematical description of the strength properties of concrete under generalised stress, *Magazine of Concrete Research*, 31(108), 151–158.

Kotsovos M. D., 1980, A mathematical model of the deformational behaviour of concrete under generalised stress based on fundamental material properties, *Materials & Structures, RILEM*, 13(76), 289–298.

Kotsovos M. D., 1984, Concrete – A brittle fracturing material, *Materials & Structures, RILEM*, 17(98), 107–115.

Kotsovos M. D., Bobrowski J. and Eibl J., 1987, Behaviour of RC T-beams in shear, *The Structural Engineer*, 65B(1), 1–9.

Kotsovos M. D. and Newman J. B., 1977, Behavior of concrete under multiaxial stress, *ACI Journal, Proceedings* 749, 453–456.

Kotsovos M. D. and Newman J. B., 1978, Generalised stress-strain relations for concrete, *Journal of the Engineering Mechanics Division, ASCE*, 104(EM4), 845–856.

Kotsovos M. D. and Newman J. B., 1979, A mathematical description of the deformational behavior of loading under complex loading, *Magazine of Concrete Research*, 31, 77–90.

Kotsovos M. D. and Newman J. B., 1981, Fracture mechanics and concrete behaviour, *Magazine of Concrete Research*, 33(115), 103–112.

Kotsovos M. D. and Pavlovic M. N., 1986, Non-linear finite element modelling of concrete structures: Basic analysis, phenomenological insight, and design implications, *Engineering Computations*, 3(3), 243–250.

Kupfer H., Hilsdorf H. K. and Rusch H., 1969, Behavior of concrete under triaxial stresses, *ACI Journal*, 66, 656–666.

Lowe P. G., 1978, Deformation and fracture of plain concrete, *Magazine of Concrete Research*, 30, 200–204.

Newman J. B., 1973, Deformational Behaviour, Failure Mechanisms and Design Criteria for Concretes Under Combinations of Stress, PhD thesis, University of London.

Newman J. B., 1974, Apparatus for testing concrete under multiaxial states of stress, *Magazine of Concrete Research*, 26(89), 229–238.

Newman J. B. and Newman K., 1978, *Development of design Criteria for Concrete Under Combined States of Stress*, Technical Note 93, CIRIA, London.

Popovics S., 1973, A numerical approach to the complete stress–strain curve of concrete, *Cement and Concrete Research*, 3, 583–599.

Spooner D. C. and Dougill J. W., 1975, A quantitative assessment of damage sustained in concrete during compressive loading, *Magazine of Concrete Research*, 27, 151–160.

Timoshenko S. P., 1953, *History of Strength of Materials*, McGraw-Hill, New York.

Willam K. J. and Warnke E. P., 1974, *Constitutive model for the triaxial behaviour of concrete*, Seminar on concrete structures subjected to triaxial stresses, Instituto Sperimentale Modeli e Strutture, Bergamo, Paper III-1.

Chapter 4

Structure modelling for static problems

The structure modelling is based on the finite-element (FE) approach which is so widely used nowadays – both in practice and for research – that a detailed treatment of it in the present work – where the technique is employed merely as a numerical tool – would be superfluous. Instead, only the briefest of general outlines will be given here, mainly in order to fix ideas and notation, and the ensuing condensed summary will form the starting point to the subject of interest, namely the non-linear finite-element analysis (NLFEA) of concrete structures. On the other hand, emphasis will be placed on the description of the manner in which widely used numerical techniques have been modified and used for the development of a non-linear strategy that forms the basis of the numerical scheme proposed in the present book as the most suitable one for the analysis of concrete structures.

4.1 FINITE-ELEMENT METHOD

As is often remarked, much of the popularity of the FE method (FEM) stems from its ready visualisation and obvious physical interpretation. Thus, the modelling of a continuum by discretising it into a finite number of components or elements has both a mathematical and a physical counterpart in the solution of actual discrete problems such as, for example, frameworks, where nodes connect standard units (i.e., simple elements, the behaviour of which may be deduced once the relevant parameters – usually displacements – at their end nodes have been determined). In the case of a continuum, however, the implicit assumption is usually made that, in order to attain the exact solution, the discretisation process should be extended *ad infinitum*, although, for engineering purposes, a finite degree of subdivision will eventually be sufficiently accurate. In this respect, the FEM may well be compared with its analytical analogue, the Ritz technique, where the assumed fields must be admissible and, also, complete.

4.1.1 Direct formulation of FE characteristics

By far the most popular FE approach in structural problems is based on assumed displacement fields. Consider an individual FE, having a given number of nodal points along its boundaries. Then the (continuous) displacement field within the element, u (where the components of u depend on the dimensions of the problem), is deemed to be obtainable by interpolating between (i.e., by operating on) the relevant nodal parameter d (such as displacements and, possibly, their derivatives at these locations). In matrix form

$$u = [N]d \tag{4.1}$$

where $[N]$ is the matrix of shape functions relating the continuous field u to the discrete set d. The shape functions represent the approximating interpolation to the actual function within the element (i.e., between nodes), with the known nodal values d providing the basis for the interpolation even though the vector d itself will usually constitute also an approximation to the true values at the nodes. The number of parameters that define a given interpolating function must match the number of nodal degrees of freedom (DOF). As an example, consider a straight beam FE of length l which spans between nodes 1 (coordinate $x = 0$) and 2 ($x = l$). Under the assumption of a predominantly flexural response, the displacement field consists of a single variable, namely the transverse displacement w. It is usual to represent the interpolating functions by polynomials. For the beam FE under consideration, the cubic polynomial

$$w = c_1 + c_2 x + c_3 x^2 + c_4 x^3 \tag{4.2}$$

consists of four parametric constants, and hence four nodal DOF are required, that is, two at each of the end nodes 1 and 2. It is customary to take displacements (w_1, w_2) and slopes (dw_1/dx, dw_2/dx) as the four nodal DOF. Then the vectors u and d are given by

$$u = w \tag{4.3}$$

$$d = \begin{bmatrix} w_1 \\ \theta_1 \\ w_2 \\ \theta_2 \end{bmatrix} \tag{4.4}$$

Clearly, the constants $c_1 - c_4$ must be adjusted so that the relevant nodal parameter(s) in d are obtained on specification of the coordinates of the given node. This is achieved by the following 1×4 shape-function matrix:

$$[N] = [N_1 N_2 N_3 N_4] \tag{4.5}$$

where

$$
\begin{aligned}
N_1 &= 1 - \frac{3x^2}{l^2} + \frac{2x^3}{l^3}; \quad & N_2 &= x - \frac{2x^2}{l} + \frac{x^3}{l^2}; \\
N_3 &= \frac{3x^2}{l^2} - \frac{2x^3}{l^3}; \quad & N_4 &= -\frac{x^2}{l} + \frac{x^3}{l^2}
\end{aligned}
\tag{4.6}
$$

It should be evident that, since Equation 4.1 is valid for all d, the N corresponding to a given node is unity at that node and zero at all the other nodes. (It should be noted that, in the case of N_2 and N_4, these must be differentiated before the preceding check, which refers, obviously, to dimensionally compatible variables, is applied; however, in all subsequent FE formulations, relevant to the modelling of concrete structures, no such mixture of nodal variable types will arise as these consist exclusively of displacements.)

Once the displacement field u has been obtained throughout a given FE, the vector of strains, ε, follows upon operating on u by means of a suitable linear operator $[L]$, that is

$$\varepsilon = [L]u \tag{4.7}$$

and, through the use of Equation 4.1

$$\varepsilon = [B]d \tag{4.8}$$

where

$$[B] = [L][N] \tag{4.9}$$

that is, $[B]$ is made up of differentials of the shape functions contained in $[N]$. For instance, in the previous beam example, the generalised strain is simply the curvature, that is, $\varepsilon \approx -d^2w/dx^2$. Alternatively, if two-dimensional (2-D) problems of plane elasticity are considered, where u consists of two variables, so that

$$u = \begin{bmatrix} u \\ v \end{bmatrix} \tag{4.10}$$

the strain vector would be given by

$$\varepsilon = \begin{bmatrix} \varepsilon_x \\ \varepsilon_y \\ \gamma_{xy} \end{bmatrix} = \begin{bmatrix} \partial u/\partial x \\ \partial v/\partial y \\ \partial u/\partial y + \partial v/\partial x \end{bmatrix} = \begin{bmatrix} \partial/\partial x & 0 \\ 0 & \partial/\partial y \\ \partial/\partial y & \partial/\partial x \end{bmatrix} \begin{bmatrix} u \\ v \end{bmatrix} \tag{4.11}$$

with the 3×2 matrix in Equation 4.11 defining $[L]$ for this particular problem type.

Finally, the stress state σ in the FE may be obtained upon specification of the matrix of constitutive relations, $[D]$, which links stresses and strains. In general form

$$\sigma = [D](\varepsilon - \varepsilon_o) + \sigma_o \tag{4.12}$$

where, for the sake of completeness, the vectors of initial strains (ε_o) and stresses (σ_o) have been included, although these are seldom considered. For the beam example, $[D] = EI$, that is, the flexural rigidity, while in the instance of the plane-stress case in 2-D elasticity, Equation 4.12 reduces (ignoring ε_o and σ_o) to

$$\begin{bmatrix} \sigma_x \\ \sigma_y \\ \tau_{xy} \end{bmatrix} = \frac{E}{1 - v^2} \begin{bmatrix} 1 & v & 0 \\ v & 1 & 0 \\ 0 & 0 & (1 - v)/2 \end{bmatrix} \begin{bmatrix} \varepsilon_x \\ \varepsilon_y \\ \gamma_{xy} \end{bmatrix} \tag{4.13}$$

The computation of the stresses directly from the nodal parameters d may be written as

$$\sigma = [D][B]d \tag{4.14}$$

where, for numerical purposes, it is worth noting that $[D]([B]d)$ requires fewer operations than $([D][B])d$.

The equilibrium of an FE subject to nodal actions p_n as well as loads p_e which are distributed throughout the element, may be tackled by means of virtual-work considerations or through the principle of minimum potential energy. By adopting the former approach, the application of a set of virtual displacements δd at the nodes will produce element displacements

$$\delta u = [N]\delta d \tag{4.15}$$

and internal strains

$$\delta\varepsilon = [B]\delta d \tag{4.16}$$

W_i, the internal work done by the stresses through the volume V of the element, is then

$$W_i = \int_v \delta\varepsilon^T\sigma \, dV \tag{4.17}$$

while W_e, the external work done by the nodal actions and distributed forces, amounts to

$$W_e = \delta d^T p_n + \int_v \delta u^T p_e \, dV \tag{4.18}$$

On equating W_i and W_e, and recalling that the result must hold for all values of δd^T, the following is obtained

$$\left(\int_v [B]^T[D][B]dV\right)d = p_n + \int_v [N]^T p_e \, dV \tag{4.19}$$

where relations (4.15) and (4.16) have been used. The last term in Equation 4.19 shows that any distributed – and/or concentrated (if present) – loading that is not acting directly at the nodes must be converted to nodal actions in a 'work-equivalent' manner through the use of the shape-function matrix; such 'consistent' nodal forces may then be added to p_n and the combined result, that is, the total nodal actions, will be denoted simply by p. As a final preliminary, the definition of the stiffness matrix for the element is

$$[k] = \int_v [B]^T[D][B] \, dV \tag{4.20}$$

so that the equilibrium statement sought becomes

$$[k]d = p \tag{4.21}$$

Evidently, since there is a one-to-one correspondence between the respective DOF that make up d and p, $[k]$ is always a square matrix.

4.1.2 Generalisation to the whole structure

So far, it has been assumed implicitly that d is known so that the computation of u, ε and σ can proceed. The determination of the kinematic nodal DOF requires the analysis of the whole structure, modelled by the full assemblage of FEs. Such an analysis is based on the so-called stiffness or displacement method, which requires the solution of the system of linear equations

$$[K]d = f \qquad\qquad\qquad (4.22)$$

where d is now understood to represent all the generalised nodal displacements to be determined, while f consists of the vector of generalised forces acting on these nodes, which is obtained by summing the contributions of all elements at every node. Similarly, the structure stiffness matrix $[K]$ is assembled by adding the contributions of the various element stiffness matrices $[k]$; this process is based on the node numbering at the structural level (rather than that of the individual elements) and on the structure or global coordinate system (hence element stiffness matrices derived in a local coordinate system must be transformed to the structure axes – this is achieved by the relevant transformation matrices made up of the direction cosines defining the relative orientation between the two coordinate systems or, as in the case of the isoparametric and Lagrangian elements adopted for modelling reinforced concrete (RC) structures discussed in the following section, through the use of the Jacobian matrix linking the derivatives of $[N]$ with respect to the two sets of coordinates). Once $[K]$ has been assembled, it constitutes a singular matrix since, at that stage, the structure being modelled is free to undergo any combination of rigid-body movements; the singularity disappears on specification of the relevant boundary conditions, and this is equivalent to the removal of the rows and columns corresponding to the unknown reactions and the associated known (usually zero) displacements, respectively.

Since the superposition of the k-matrices results in the overlapping of the stiffnesses of only those elements meeting at the common node(s), the resulting $[K]$ is sparse, and, with suitable nodal numbering, also 'banded', that is, the non-zero coefficients cluster around the diagonal. Moreover, on account of the linearity of the problems being considered, stiffness matrices are symmetric. All these properties are used to advantage in achieving an efficient method of solution to the set of equations defined in Equation 4.22.

On the basis of the foregoing, it is evident that the approximation inherent in the discretisation process described by Equation 4.22 is subject to automatic enforcement of both equilibrium and compatibility at the nodes, and that, in addition, compatibility is also satisfied within each element. On the other hand, compatibility may or may not be satisfied at the non-nodal element junctions, depending on the type of FE adopted. Furthermore, exact equilibrium is usually not attained within elements, nor across their common boundaries. However, provided that certain basic convergence criteria are met, mesh refinement will gradually reduce these local equilibrium violations and, if incompatible elements are used, the effects of such inter-element continuity breakdowns. For a full description of such convergence requirements and, indeed, for further background to the FE method in general, the reader is referred to standard texts on the subject (e.g., Zienkiewicz 1977, Cook 1981, Cook et al. 1989).

4.1.3 Finite elements selected

Throughout this book, only three-dimensional (3-D) analysis will be implemented. Problems of 2-D and axisymmetric analysis of structural concrete are fully discussed in Kotsovos and Pavlovic (1995), where cases of 3-D analysis of concrete structures under static monotonic loading are also presented. The FE elements selected for this type of analysis are shown in Figure 4.1. Since all the elements used are of the curved, isoparametric type with three nodes along the various edges – arguably, the most widely employed FEs in current practice – much of the essential discussion pertaining to their salient features is common to all of

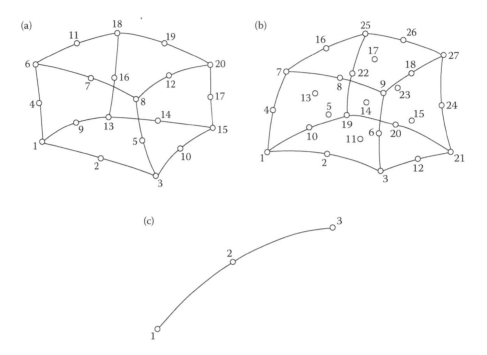

Figure 4.1 Finite elements used in 3-D analysis: (a) 20-node serendipity brick element; (b) 27-node Lagrangian brick element; (c) 3-node isoparametric line element.

them, as will become evident in the brief outline that follows. (For a more detailed general background, see e.g., Zienkiewicz 1977; with regard to the advantages of the isoparametric formulation in respect to non-linear concrete modelling in particular, a good summary may be found in Philips and Zienkiewicz 1976.)

Solid brick elements are the basis for modelling concrete, with three DOFs (displacements) at each node, the full stress and strain tensors being given in the output. The 20-node serendipity unit shown in Figure 4.1a has been primarily used for the analysis of structural members under static monotonic loading (Kotsovos and Pavlovic 1995), whereas the 27-node Lagrangian element shown in Figure 4.1b, despite the additional 21 DOFs required by the insertion of one central and six mid-face nodes, has been exclusively used for the analysis of RC members under static cyclic and dynamic (seismic and impact) loading. The steel reinforcement is assumed to consist of line elements, which are modelled by means of the three-node FE bars shown in Figure 4.1c, with a specified cross-sectional area and a total of nine displacement DOFs, although its stiffness is still limited to axial properties only, so that the bar admits solely direct stresses/strains along its length on account of its total lack of bending rigidity.

A natural choice for the shape, or interpolating, functions in the case of elements with three-node edges – a feature common to all of the above FEs – is of the polynomial type that yields a quadratic variation of displacement(s) along these edges. Such shapes usually take the form of the well-established serendipity or Lagrangian formulations on the basis of which the elements in Figure 4.1 are derived. The adoption of second-order polynomials ensures displacement continuity between adjacent elements since a parabola is uniquely defined by three points, in this case the three (common) nodes along the inter-element boundary. A direct consequence of this is that the strains, which, for all the elements under consideration are functions of the first derivatives of displacements (as e.g., in the 2-D case given by

Equation 4.11), always remain finite at the interface between elements, thus satisfying the usual convergence criteria. In addition to this consideration of C_o-continuity, the quadratic family of elements provides a good balance between accuracy (with an obvious improvement over linear-element formulations) and complexity (as the number of DOFs per element is well below those of higher-order models, especially in 3-D problems). Furthermore, the quadratic interpolation functions provide the basis for the curved shapes of the elements in Figure 4.1 that enables the ready modelling of arbitrary geometric contours through the isoparametric formulation, some of the relevant features of which will be summarised subsequently. Before these are discussed, however, it is useful to illustrate the type of shape functions used for the elements adopted. In so doing, it may be helpful to imagine that the various one-dimensional (1-D) and 3-D units have been temporarily 'straightened' (thus consisting of lines and cubes, respectively) although, as implied above, the same shape functions will also be applied later to their curved counterparts.

The interpolation functions [N] of the various elements are given in terms of local, element coordinates, and it is usual practice to make the latter dimensionless. These natural or intrinsic coordinates have origins at the centre of the element, attaining values of ±1 at its ends (end nodes in 1-D and mid-facets in 3-D). Such local coordinates are to be denoted by ξ, η, ζ (or ξ_i) in contrast to the global or structure coordinates x, y, z (or x_i). The various Ns may be checked – and, in fact, are often actually constructed – by recalling that a given N_I that refers to node I is equal to unity at that node and zero at all other nodes. The following three cases (in increasing dimensional and number of nodes order) will help to illustrate the general approach usually adopted.

Consider first the bar (now line) element with its origin at the central node bisecting the element. The intrinsic coordinate ξ runs along its axis and the Lagrangian interpolation functions may be written in terms of it as follows:

Corner node ($\xi_I = \pm 1$): $N_I = (1/2)\xi(\xi_I + \xi)$ (4.23)

Mid-side node ($\xi_I = 0$): $N_I = 1 - \xi^2$ (4.24)

where, as stated already, the subscript I defines the node in question. Turning now to a 3-D example, the shape functions for the 20-node serendipity element (now a cube) are described in terms of the intrinsic coordinate system ξ, η, ζ (or ξ_i) emanating from the centre of the unit.

Corner nodes ($\xi_I = \pm 1, \eta_I = \pm 1, \zeta_I = \pm 1$):

$$N_I = (1/8)(1 + \xi\xi_I)(1 + \eta\eta_I)(1 + \zeta\zeta_I)(\xi\xi_I + \eta\eta_I + \zeta\zeta_I - 2)$$ (4.25)

Mid-side nodes ($\xi_I = 0, \eta_I = \pm 1, \zeta_I = \pm 1; \xi_I = \pm 1, \eta_I = 0, \zeta_I = \pm 1; \xi_I = \pm 1, \eta_I = \pm 1, \zeta_I = 0$):

$$N_I = (1/4)(1 - \xi^2)(1 + \eta\eta_I)(1 + \zeta\zeta_I) \quad \text{for } \xi_I = 0$$ (4.26)

$$N_I = (1/4)(1 - \eta^2)(1 + \xi\xi_I)(1 + \zeta\zeta_I) \quad \text{for } \eta_I = 0$$ (4.27)

$$N_I = (1/4)(1 - \zeta^2)(1 + \xi\xi_I)(1 + \eta\eta_I) \quad \text{for } \zeta_I = 0$$ (4.28)

The shape functions for each of the 27 nodes of the Lagrangian element in Figure 4.1b are also described in terms of the intrinsic coordinate system ξ, η, ζ (or ξ_i) emanating from the centre of the unit. In what follows, these are expressed in an explicit form for clarity purposes.

Internal node:

$$N_{14} = (1 - \xi^2)(1 - \eta^2)(1 - \zeta^2) \qquad (4.29)$$

Mid-face nodes:

$$N_{11} = (1 - \xi^2)(1 - \eta^2)(1 - \zeta)/2 - N_{14}/8 \qquad (4.30)$$

$$N_{17} = (1 - \xi^2)(1 - \eta^2)(1 + \zeta)/2 - N_{14}/8 \qquad (4.31)$$

$$N_5 = (1 - \xi^2)(1 - \eta)(1 - \zeta^2)/2 - N_{14}/8 \qquad (4.32)$$

$$N_{15} = (1 + \xi)(1 - \eta^2)(1 - \zeta^2)/2 - N_{14}/8 \qquad (4.33)$$

$$N_{23} = (1 - \xi^2)(1 + \eta)(1 - \zeta^2)/2 - N_{14}/8 \qquad (4.34)$$

$$N_{13} = (1 - \xi)(1 - \eta^2)(1 - \zeta^2)/2 - N_{14}/8 \qquad (4.35)$$

Mid-edge nodes:

$$N_2 = (1 - \xi^2)(1 - \eta)(1 - \zeta)/4 - (N_5 + N_{11})/2 - N_{14}/8 \qquad (4.36)$$

$$N_{12} = (1 + \xi)(1 - \eta^2)(1 - \zeta)/4 - (N_{11} + N_{15})/2 - N_{14}/8 \qquad (4.37)$$

$$N_{20} = (1 - \xi^2)(1 + \eta)(1 - \zeta)/4 - (N_{11} + N_{23})/2 - N_{14}/8 \qquad (4.38)$$

$$N_{10} = (1 - \xi)(1 - \eta^2)(1 - \zeta)/4 - (N_{11} + N_{13})/2 - N_{14}/8 \qquad (4.39)$$

$$N_8 = (1 - \xi^2)(1 - \eta)(1 + \zeta)/4 - (N_5 + N_{17})/2 - N_{14}/8 \qquad (4.40)$$

$$N_{18} = (1 + \xi)(1 - \eta^2)(1 + \zeta)/4 - (N_{15} + N_{17})/2 - N_{14}/8 \qquad (4.41)$$

$$N_{26} = (1 - \xi^2)(1 + \eta)(1 + \zeta)/4 - (N_{17} + N_{23})/2 - N_{14}/8 \qquad (4.42)$$

$$N_{16} = (1 - \xi)(1 - \eta^2)(1 + \zeta)/4 - (N_{13} + N_{17})/2 - N_{14}/8 \qquad (4.43)$$

$$N_4 = (1 - \xi)(1 - \eta)(1 - \zeta^2)/4 - (N_5 + N_{13})/2 - N_{14}/8 \qquad (4.44)$$

$$N_6 = (1 + \xi)(1 - \eta)(1 - \zeta^2)/4 - (N_5 + N_{15})/2 - N_{14}/8 \qquad (4.45)$$

$$N_{24} = (1 + \xi)(1 + \eta)(1 - \zeta^2)/4 - (N_{15} + N_{23})/2 - N_{14}/8 \tag{4.46}$$

$$N_{22} = (1 - \xi)(1 + \eta)(1 - \zeta^2)/4 - (N_{13} + N_{23})/2 - N_{14}/8 \tag{4.47}$$

Corner nodes:

$$N_1 = (1 - \xi)(1 - \eta)(1 - \zeta)/8 - (N_2 + N_4 + N_{10})/2 + (N_5 + N_{11} + N_{13})/4 - N_{14}/8 \tag{4.48}$$

$$N_3 = (1 + \xi)(1 - \eta)(1 - \zeta)/8 - (N_2 + N_6 + N_{12})/2 + (N_5 + N_{11} + N_{15})/4 - N_{14}/8 \tag{4.49}$$

$$N_{21} = (1 + \xi)(1 + \eta)(1 - \zeta)/8 - (N_{12} + N_{20} + N_{24})/2 + (N_{11} + N_{15} + N_{23})/4 - N_{14}/8 \tag{4.50}$$

$$N_{19} = (1 - \xi)(1 + \eta)(1 - \zeta)/8 - (N_{10} + N_{20} + N_{22})/2 + (N_{11} + N_{13} + N_{23})/4 - N_{14}/8 \tag{4.51}$$

$$N_7 = (1 - \xi)(1 - \eta)(1 + \zeta)/8 - (N_4 + N_8 + N_{16})/2 + (N_5 + N_{13} + N_{17})/4 - N_{14}/8 \tag{4.52}$$

$$N_9 = (1 + \xi)(1 - \eta)(1 + \zeta)/8 - (N_6 + N_8 + N_{18})/2 + (N_5 + N_{15} + N_{17})/4 - N_{14}/8 \tag{4.53}$$

$$N_{27} = (1 + \xi)(1 + \eta)(1 + \zeta)/8 - (N_{18} + N_{24} + N_{26})/2 + (N_{15} + N_{17} + N_{23})/4 - N_{14}/8 \tag{4.54}$$

$$N_{25} = (1 - \xi)(1 + \eta)(1 + \zeta)/8 - (N_{16} + N_{22} + N_{26})/2 + (N_{13} + N_{17} + N_{23})/4 - N_{14}/8 \tag{4.55}$$

The quadratic shape functions for the adopted elements with three-node edges have now been presented, and the way in which those straight-edged units can be distorted so as to produce the types of FEs sketched in Figure 4.1 may next be considered. Such a distortion of straight-edged elements through mapping to produce FEs with curved boundaries that will fit more easily complex problem geometries – and thus allow coarser meshes for the same degree of accuracy – may be formulated in several ways. In general, the local (straight) coordinates become a set of curvilinear coordinates when mapped onto a global (Cartesian) set of coordinates, the type and degree of distortion depending on the shape functions chosen to establish such a coordinate transformation. Of particular interest is the case when the shape functions [N] adopted as the interpolation functions for the relevant parameter in the FE analysis are also used as the shape functions for the mapping transformation of coordinate geometries. Such an isoparametric formulation is especially advantageous for the elements with three-node sides based on quadratic interpolation functions, since the resulting parabolic distortions of the edges (as in Figure 4.1) are usually sufficient for practical purposes. In addition, displacement continuity at inter-element boundaries of the parent units also ensures the continuity of the distorted elements, while the choice of the mid-side locations for the non-corner edge nodes for the FEs of Figure 4.1 automatically guarantees uniqueness, that is, one-to-one mapping.

At this stage, it must be borne in mind that, in the course of FE computations, element stiffness matrices and related load vectors (as defined by the first and last terms of expression (4.19) must be evaluated in the form of integrals involving quantities that are to be operated upon with respect to global axes x_i: e.g., $[B]$ involves first derivatives of $[N]$ with respect to $x/y/z$ (in the present work $[L]$ – see Equation 4.9 – always consists of first derivatives of the type illustrated earlier for the 2-D case – see Equation 4.11 – and readily extendible not only to the 3-D unit but also to the 1-D element, as the latter possesses no bending stiffness); similarly, the infinitesimal volume dV throughout which the integration is performed is given by $dxdydz$. On the other hand, the N_Is are defined in terms of local (now curvilinear) coordinates ξ_i and hence a transformation is required between global and local derivatives; and, similarly, the integral itself must be computed through the use of local coordinates, so that a further transformation is needed to obtain the correct scale factor, with the concomitant change of integration limits.

The transformation involving first derivatives of N_I can be derived by invoking the chain rule for partial differentiation. This will be illustrated by reference to the general 3-D case which, clearly, also encompasses its simpler 2-D and 1-D counterparts. In matrix form, the required expression is

$$
\begin{bmatrix} \partial N_I/\partial\xi \\ \partial N_I/\partial\eta \\ \partial N_I/\partial\zeta \end{bmatrix} = \begin{bmatrix} \partial x/\partial\xi & \partial y/\partial\xi & \partial z/\partial\xi \\ \partial x/\partial\eta & \partial y/\partial\eta & \partial z/\partial\eta \\ \partial x/\partial\zeta & \partial y/\partial\zeta & \partial z/\partial\zeta \end{bmatrix} \begin{bmatrix} \partial N_I/\partial x \\ \partial N_I/\partial y \\ \partial N_I/\partial z \end{bmatrix}
\tag{4.56}
$$

To determine the constituent elements of the square Jacobian matrix linking the vectors $\partial N_I/\partial\xi_i$ and $\partial N_I/\partial x_i$, use is made of the shape functions relating the two sets of coordinates x_i and ξ_i. Since the isoparametric formulation has been adopted, in which both displacements and coordinates are

$$
[J] = \begin{bmatrix} \partial N_1/\partial\xi & \partial N_2/\partial\xi & \partial N_3/\partial\xi & - & - \\ \partial N_1/\partial\eta & \partial N_2/\partial\eta & \partial N_3/\partial\eta & - & - \\ \partial N_1/\partial\zeta & \partial N_2/\partial\zeta & \partial N_3/\partial\zeta & - & - \end{bmatrix} \begin{bmatrix} x_1 & y_1 & z_1 \\ x_2 & y_2 & z_2 \\ x_3 & y_3 & z_3 \\ - & - & - \\ - & - & - \end{bmatrix}
$$

$$
= \begin{bmatrix} \Sigma(\partial N_I/\partial\xi)x_I & \Sigma(\partial N_I/\partial\xi)y_I & \Sigma(\partial N_I/\partial\zeta)z_I \\ \Sigma(N_I/\partial\eta)x_I & \Sigma(N_I/\partial\eta)y_I & \Sigma(\partial N_I/\partial\eta)z_I \\ \Sigma(\partial N_I/\partial\zeta)x_I & \Sigma(\partial N_I/\partial\zeta)y_I & \Sigma(\partial N_I/\partial\zeta)z_I \end{bmatrix}
\tag{4.57}
$$

where, in the last expression, the Σ may be left out provided that the repeated index is still understood to imply summation. The vector $\partial N_I/\partial\xi_i$ on the left-hand side in Equation 4.56 being also readily obtainable, the required vector $\partial N_I/\partial x_i$ of derivatives with respect to global coordinates is finally obtained by inverting $[J]$ and multiplying the result by the array $\partial N_I/\partial\xi_i$.

The second coordinate transformation, pertaining to the region of integration, involves a scaling factor between the two coordinate systems which may readily be shown to be simply the determinant of $[J]$ – usually referred to plainly as the 'Jacobian' (see e.g., Zienkiewicz 1977, Irons and Ahmad 1980). Thus, for the general 3-D case

$$
dV = dxdydz = (\det[J])d\xi d\eta d\zeta
\tag{4.58}
$$

and, considering, for example, the computation of its element stiffness matrix, the following can be written as

$$[k] = \int_{-1}^{1}\int_{-1}^{1}\int_{-1}^{1} [B]^T[D][B](\det[J]d\xi d\eta d\zeta \tag{4.59}$$

Clearly, 2-D and 1-D elements will be, essentially, of the same form while, obviously, consisting of double and single integrals, respectively.

Integrals such as those forming the basis of the element-stiffness computations are often too complex to be calculated exactly by analytical means. This is particularly true in the case of the more sophisticated curved elements stemming from the isoparametric formulation, so that resort must be made to numerical-integration techniques (just as the inverse of $[J]$ is also calculated numerically in such problems). In these methods, the function being integrated (F) is evaluated at a finite number of points within the element, and each of these values is multiplied by the length of the interval associated with that F value, which length may be thought of as an adequate weighting factor W; the sum of all the products WF constitutes an approximation to, and in certain instances actually coincides with, the true value of the integral. With specific reference to the intrinsic coordinates and integration limits associated with the isoparametric formulation, 1-D, 2-D and 3-D integrals may therefore be expressed as follows:

$$I_1 = \int_{-1}^{1} F(\xi)d\xi = \sum_i W_i F(\xi_i) \tag{4.60}$$

$$I_2 = \int_{-1}^{1}\int_{-1}^{1} F(\xi,\eta)d\xi d\eta = \sum_i \sum_j W_i W_j F(\xi_i \eta_j) \tag{4.61}$$

$$I_3 = \int_{-1}^{1}\int_{-1}^{1}\int_{-1}^{1} F(\xi,\eta,\zeta)d\xi d\eta d\zeta = \sum_i \sum_j \sum_k W_i W_j W_k F(\xi_i, \eta_j, \zeta_k) \tag{4.62}$$

where the number of points (i.e., range(s) of i, j and k) are usually the same in all directions (and are symmetrically located with respect to the origin). When the above are applied to an integrand matrix – as, for example, in Equation 4.59 – it should be borne in mind that each coefficient of $[B]^T[D][B](\det[J])$ must be integrated in turn through the relevant formula given in Equations 4.60 through 4.62; the resultant I defines the single coefficient k_{ij} in $[k]$.

For a given number of points at which F is evaluated, Gauss's method of numerical integration or quadrature provides the best degree of accuracy. These optimal sampling-point locations and their respective weighting coefficients are readily available in tabulated form for various integration orders n, all data being specified to a substantial number of significant figures as required for precision purposes (see, for instance, Zienkiewicz 1977 and Cook 1981).

Numerically integrated matrices obviously depend on the order of integration adopted. Accordingly, different numbers of sampling points will produce different stiffness matrices, and the question thus arises as to the optimum – or at least an adequate – quadrature rule. Although such a choice varies from problem to problem – and can be ascertained only through numerical experimentation – a general guideline is that as low an order of integration is usually desirable provided that it does not lead to numerical instability. Several arguments may be adduced in favour of low-order integration. First, there is the computation time, which is quite considerable as its order of magnitude can be comparable to that of equation solving. The consideration is of special significance in non-linear work, where element stiffness matrices are usually constantly updated. Bearing in mind that numerical integration is proportional to the square of the number of DOF in the element times the number of sampling points, it is clear that the change from a two-point to a three-point rule will more than approximately double – in 2-D cases – and triple – in 3-D problems – the integration cost. Another advantage of low-order quadrature algorithms is that these result generally in a softening of the element, thus counteracting the overly stiff nature inherent in FEs derived on the basis of displacement or stiffness formulations. A third favourable characteristic associated with fewer sampling points refers to the locations at which stresses are computed. It is well known that calculations of stresses along edges or faces of the element tend to be inaccurate, especially at the nodes. This is particularly true of the quadratic C_o elements used in the present work. It turns out that the optimal locations at which stresses should be calculated coincide with the Gauss points in the interior of the element (if required, stresses at edges/faces or nodes may then be extrapolated from the values obtained at Gauss points). Now, the use of low-order or 'reduced' integration has often been found to provide not only superior answers to those attained by means of 'full' (i.e., higher-order) integration but also the optimum locations for stress computation. For example, such behaviour is exhibited by serendipity elements where 2×2 integration tends to improve the predictions of the 3×3 rule and, in addition, the stress-sampling locations corresponding to the four-point quadrature happen to be best irrespective of which of the two integration rules is used.

The above advocacy of low-order integration algorithms must be tempered with the requirement that it should not give rise to numerical-instability problems. Such difficulties stem from the possible presence of kinematic mechanisms which, as is well known, lead to, and are detectable by, singular systems of equations or, in the present context, result in stiffness-matrix singularity. The cause of these singularities is traceable to zero-energy deformation modes, associated with certain patterns of nodal displacements compatible with a strain field which is zero at all sampling points. Now, it is obvious that, as the number of such quadrature points decreases, the likelihood of occurrence of zero-energy modes becomes greater. In fact, it can be argued that, if the number of independent linear relations, between the nodal unknowns, that are available at all the integrating points is below the number of these nodal parameters, the stiffness matrix will be singular (Zienkiewicz 1977). (The independent relations may be obtained by multiplying the number of strain components per quadrature point times the number of such points; the unknowns are simply the number of DOF per node times the number of nodes, minus those DOF which have been eliminated through the imposition of kinematic restraints.) To show that singularities will definitely not occur is not so easy and, therefore, while the condition that the total number of strains should be larger than the total number of DOF is a requirement for an element/ system to be well behaved, it is always advisable to conduct the eigenvalue test so as to determine the actual number of zero-energy modes (Cook 1981). As the eigenvalue test is usually performed on the full stiffness matrix of a given element, that is, without imposing the necessary nodal restraints necessary to prevent rigid-body motions, a proper element formulation should yield the correct number of mechanisms associated with this type of translation and rotation in space (i.e., three and six mechanisms for 2-D and 3-D elements, respectively); any additional mechanisms reveal the presence of zero-energy modes connected with low-order numerical-integration rules.

In view of the above, care must always be exercised when choosing, at the outset of an analysis, reduced-integration techniques, and this is particularly relevant in non-linear problems of the type considered here, where gradual cracking may give rise to drastic stiffness changes within the element(s) concerned. Not surprisingly, therefore, decisions to under-integrate should be explored with due caution and this is why such questions have been given appropriate prominence in Kotsovos and Pavlovic (1995), where evidence is provided which shows that the use of 20-node serendipity and 27-node Lagrangian brick elements with $2 \times 2 \times 2$ and $3 \times 3 \times 3$, respectively, integration rules provide similar predictions of the behaviour of RC members under short-term static monotonic loading.

4.2 NON-LINEAR ANALYSIS

The preceding section has dealt with the general problem of the FE discretisation process for an arbitrary linear system. Thus, linearity was assumed to apply at the three levels of statics (equilibrium equations written in the initial, undeformed geometry, as displacements are assumed to be small), kinematics (linearised strain–displacement compatibility equations, as strains are also taken to be small) and constitutive relations (Hooke's law or its generalised version deemed to be applicable). Such linearisation assumptions lead directly to a mathematical model possessing the following desirable features: uniqueness of solution, use of superposition, and, most important perhaps, the ready availability of efficient programs for the solving of linear systems of equations based on well-established mathematical tools and algorithms. Although nature tends to be distinctly non-linear – often highly so – it turns out that a large number of structural-engineering problems may be tackled on the basis of linearly elastic concepts, at least for purposes of achieving an adequate level of performance under ordinary 'working' or 'serviceability' conditions. On the other hand, the understanding of ultimate-load conditions or, indeed, the more rational and/or economic design of structures necessitates the consideration of non-linear effects: this is especially true in the case of structural-concrete problems. Unlike linear analyses, non-linear systems cannot be solved directly but rely on various iterative or 'search' techniques; all of these, however, are based invariably on repeated solutions of linear systems until convergence is achieved. Owing to the possibility of non-uniqueness, care must be exercised – often by appeal to physical reasoning – in order to ensure that the converged solution attained is actually the correct one.

The analysis of a non-linear structural system, which has been discretised in accordance with the stiffness formulation, still proceeds through the solution of the set of Equations 4.22, but now the stiffness matrix is a function of the load/displacement level. For convenience, this will be denoted by the statement $[K] = [K(d)]$. In what follows, only the briefest of outlines on the main iterative procedures for non-linear problems will be presented, with priority of choice given eventually to incremental methods. Discussion will be restricted to 'softening' structures, where this term is now used to denote systems for which the $f - d$ path is 'convex', in the sense that the stiffness decreases with increasing f, as in the case of structural concrete where a steady degradation of stiffness occurs as the load is augmented; 'hardening' or 'concave' systems (i.e., exhibiting a steady increase of stiffness with loading) – which may require different iteration strategies – need not be considered for present purposes.

4.2.1 Direct iteration method

It is useful to begin with a concise description of the main numerical devices for attaining the solution at a given load level in a single series of iterations, without regard to the previous $f - d$ path, that is non-incrementally; subsequently, it will become evident that the incremental techniques are based essentially on identical principles. Perhaps the most basic

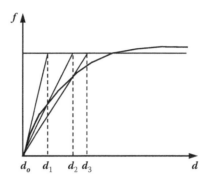

Figure 4.2 Direct iteration method.

solution type is that of 'direct iteration' (also known as 'functional iteration' or 'successive substitution/approximation'). In this method, successive solutions are performed, each iteration making use of the previous solution for the unknown(s) d to predict the improved, current value of $[K(d)]$

$$d_{n+1} = [K(d_n)]^{-1}f; \quad n = 0, 1, 2, 3, \ldots \tag{4.63}$$

The initial guess is usually taken to be $d_o = 0$, and the process is deemed to have converged when $d_{n+1} - d_n \to 0$. It will be found convenient to illustrate this method (and, also, subsequent ones) graphically by reference to the 1-D case (i.e., f, d constitute a single-DOF system), although, clearly, the same type of iterative behaviour extends to multi-DOF systems. Figure 4.2 shows the implementation of direct iteration in the search for the solution corresponding to the load level f in the given non-linear response $f - d$, with the initial guess taken as $d_o = 0$. It can be seen that the unknowns are the displacements, the secant slope being used in each iteration. This 'secant modulus' or 'variable stiffness' approach tends to be expensive since $[K]$ must be revised and a new set of linear equations solved for each iteration. Furthermore, symmetry in the matrix of coefficients need not necessarily result when direct iteration is employed (Zienkiewicz 1977), and this means that the more efficient algorithms, based on the fact that $[K]$ for linear problems is symmetrical, may not always be applicable. Another drawback of the scheme is that its convergence is not guaranteed and cannot be predicted *a priori*. In addition, as the number of DOF increases, coupling of stiffness terms might lead to instability of the iterative technique (Owen and Hinton 1980).

4.2.2 Newton–Raphson method

A more sophisticated process of iteration is the well-known Newton–Raphson (NR) method. This can be outlined as follows. Unless convergence has occurred, $[K]d = f$ will not be satisfied at any stage of the iteration, and hence a system of residual forces $\Delta \bar{f}$ can be assumed to exist, so that

$$\Delta \bar{f} = f - [K]d \tag{4.64}$$

that is $\Delta \bar{f}$ may be viewed as a measure of the system's current departure from the required state of equilibrium. Now, a better approximation exists at

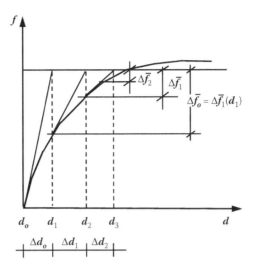

Figure 4.3 Newton–Raphson method.

$$d_{n+1} = d_n + \Delta d_n; \quad n = 0, 1, 2, 3, \ldots \tag{4.65}$$

where the NR approximation for the increment or correction Δd_n may be written as

$$\Delta d_n = [K_t(d_n)]^{-1}\Delta \bar{f}(d_n); \quad n = 0, 1, 2, 3, \ldots \tag{4.66}$$

and where the subscript in $[K]$ indicates that the latter is the tangential stiffness matrix. With increasing number of iterations n, convergence is achieved as $\Delta \bar{f}$ and/or $\Delta d_n \to 0$. The process is shown schematically in Figure 4.3 for a single-DOF system, with initial guess d_o ($= 0$ here) leading to $\Delta \bar{f}_o$ which, in turn, yields the correction so that $d_1 = d_o + \Delta d_o$ becomes an improved approximation, and so on. In this figure, the slopes of the $f-d$ characteristic at the locations corresponding to the various d_n are the 1-D counterparts of the tangential stiffness matrix $[K_t]$ used in multiple-DOF problems, as can readily be seen by invoking the well-known argument whereby Taylor's series are curtailed beyond the first derivative (Cook 1981). (For a version of the NR technique in which the matrix of coefficients linking $\Delta \bar{f}$ and Δd is not symmetric, see Owen and Hinton 1980; $[K_t]$, on the other hand, is always symmetric, and hence conforms to those equation-solving algorithms that make use of this property.) In general, the NR or 'tangential stiffness' method converges more rapidly and exhibits superior stability than the direct-iteration scheme. Again, however, there is no guarantee of convergence, especially if the initial guess is not close to the actual solution and/or combinations of 'convex' and 'concave' characteristics are encountered throughout the region of iteration. As for the direct-iteration process, the NR technique is demanding computationally, since each iteration requires the assembly of the updated matrix $[K_t]$ and the concomitant linear-equation solving.

4.2.3 Modified

NR method instead of tackling a new system of equations for each iteration, the following approximation could be made:

$$[K_t(d_n)] \approx [K_t(d_o)] \tag{4.67}$$

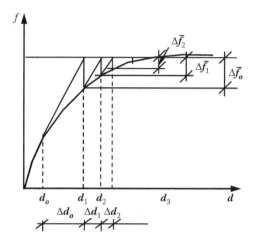

Figure 4.4 Modified NR method.

so that

$$\Delta d_n = [K_t(d_o)]^{-1} \Delta \bar{f}(d_n); \quad n = 0, 1, 2, 3, \ldots \tag{4.68}$$

throughout the entire search process. This algorithm is known as the 'initial/constant stiffness' method, and also as the 'modified Newton–Raphson' approach; its schematic illustration is depicted in Figure 4.4 for the 1-D case (with $d_o \neq 0$). The tangential stiffness matrix corresponding to the initial guess is assembled only once and, on reduction or factorisation of the set of equations and the storing of the result, solutions required in subsequent iterative cycles can be obtained at a much reduced computational effort. This significant saving in computing cost per iteration, however, is countered by a lower rate of convergence when compared with the formal NR algorithm. Once again, although convergence is usually achieved, this cannot be guaranteed for all cases, and sometimes divergence may be encountered in situations where the more rapidly converging NR technique is successful. The relative economics and convergence rates of the initial and tangential stiffness methods depend on the degree and type of non-linearity of the system considered. The optimum algorithm is usually obtained by combining both methods so that $[K]$ is updated to $[K_t]$ only occasionally during iterations.

4.2.4 Generalised Newton–Raphson method

As stressed repeatedly in the preceding paragraphs, none of the previous methods, in which the unknowns were the total displacements d, converge in all cases. Only *incremental* procedures, where the unknowns are the changes Δd due to increments in loading Δf, can provide some assurance on convergence. Furthermore, such methods enable a full study of the load deformation behaviour of a structure to be made; besides its obvious usefulness, the complete knowledge of the $f - d$ characteristic followed at the structural level becomes mandatory when the solution is dependent, not only on the current displacements, but also on the previous loading history. Evidently, with sufficiently small increments, convergence may be ensured and the local linearisation at each iterative step becomes fully justified. Then

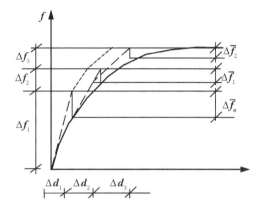

Figure 4.5 Incremental with one-step NR correction.

the NR method suggests itself as a natural iterative technique (its incremental version being termed the 'generalised Newton–Raphson' method), with the initial value now always taken as $d_o = 0$. A possible general algorithm might be

$$[K_t(d_n)]\Delta d_n = \Delta f_n + \Delta \bar{f}_{n-1}; \quad n = 0, 1, 2, 3,\ldots \tag{4.69}$$

where Δf_n represents the increment in the actual load, while $\Delta \bar{f}_{n-1}$ stands for the residual out-of-balance forces from the previous load step. This is often referred to as the 'incremental with one-step NR correction', and is illustrated for the 1-D case in Figure 4.5 by means of the dashed lines. It is evident that, despite the fact that the single residual-force corrections result in effective load increments ($\Delta f_o; \Delta f_1 + \Delta \bar{f}_o; \Delta f_2 + \Delta \bar{f}_1$; etc.), a small (usually cumulative) drift from the true solution path occurs. On the other hand, when $\Delta \bar{f} = 0$ in Equation 4.69, the algorithm becomes in essence the matrix counterpart of Euler's numerical method for the solution of a differential equation; as can be seen by the path indicated by the dotted lines in Figure 4.5, the cumulative drift now becomes larger in this purely incremental algorithm without corrections. At the other extreme, by setting $\Delta f = 0$ in Equation 4.69, the formal NR method (in its incremental form) is recovered, namely sufficient iterations are performed for each load increment in order to converge to the actual solution before the next external-load increment is applied. Obviously, any degree of transition between the extremes $\Delta \bar{f} = 0$ and $\Delta f = 0$ could be specified: for example, many small external-load increments with few iterations in each, or fewer but larger external-load steps coupled with a substantial number of corrective iterations for each of them. For overall economy, any such gradations may be combined with the constant-stiffness iteration algorithm.

4.2.5 Concluding remarks

Reliable algorithms for the non-linear analysis of concrete structures by means of the FEM should employ incremental techniques. On the basis of the preceding outline, a summary of the three most widely used versions associated with the incremental NR method (INRM) is contained in Figure 4.6. (It should be noted that all plots refer to a given load step or increment.) Whether on their own, in combination, or slightly amended, they will be found to constitute the backbone of the iterative-search process in the present work.

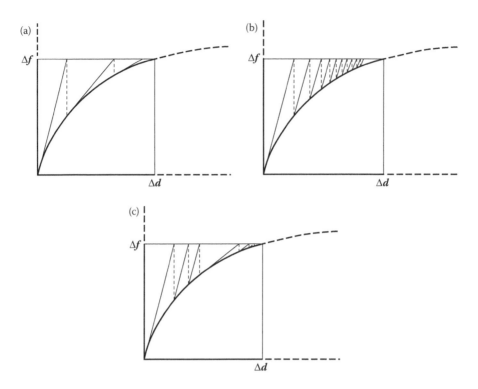

Figure 4.6 Incremental Newton–Raphson methods (INRMs): (a) pure INRM; (b) modified INRM; (c) mixed INRM.

4.3 NON-LINEAR FINITE ELEMENT MODEL FOR STRUCTURAL CONCRETE

4.3.1 Background and scope

On the basis of a readily adjustable modular FE scheme including the quadratic-type FEs discussed in Section 4.1.3, the non-linear numerical modelling of concrete structures may now be outlined. Such modelling consists largely of careful implementation of non-linearities through the gradual updating of element stiffness matrices which, in turn, are essentially described by the matrices $[B]$ and $[D]$. Since the dimensions of ordinary concrete members are such that large displacements are usually prevented throughout most – if not the whole of – the loading regime (so that equilibrium may be formulated in the original configuration), and, in addition, strains being small enough to allow linearised compatibility relations to be adopted, geometric non-linearities are ignored in the present formulation. This simplifies the modelling quite considerably, the matrix $[B]$ remaining constant throughout the analysis. Therefore, consideration need be given only to material behaviour, as the predominant source of non-linearity, the latter being introduced through reformulations of the constitutive matrix $[D]$.

   In describing the way in which the non-linear characteristics of concrete are dealt with in the program, the distinction made in Chapter 2 between micro-cracking and macro-cracking phenomena becomes particularly useful. The gradual and mild nature of the former behaviour, described through the constitutive relations of concrete, makes it relatively easy to implement, in contrast to the sudden occurrence and strong type of non-linearity associated with the latter regime, in which 'structural' cracking takes place once the failure criterion for the material is exceeded. Thus, macro-cracking is generally more difficult to model and its explosive nature may lead to convergence problems. This is why the strategy adopted to describe such cracking will receive separate treatment, and will follow the general outline of the incremental formulation of non-linear concrete behaviour prior to the reaching of the OUFP level which marks the onset of the macro-cracking regime.

The overall strategy to be discussed in the next two sections (one dealing with the modelling of concrete, the second devoted to a brief description of the reinforcing steel) will be illustrated by reference to the 3-D problem. However, it should be pointed out that the strategy developed was initially aimed exclusively at 2-D (plane-stress or axisymmetric) structures (Bedard 1983). Later on, when the basic concepts were extended to 3-D problems, some of the strategies were changed, mainly with a view to improving efficiency, convergence and/or numerical stability (Gonzalez Vidosa 1989), such changes either being necessary in the 3-D context or simply stemming from a natural maturing process built on the experience acquired through the development and use of the 2-D package. The overall strategy was further improved with the inclusion of criteria for 'crack closure' and 'loading/unloading' in localised regions (Kotsovos and Spiliopoulos 1988a, Kotsovos and Spiliopoulos 1988b). Both 'crack closure' and 'loading/unloading' occur due to redistributions of the state of the internal stresses/strains that may result from the cracking process and/or reversals of the external loading. In what follows, the general outline of the developed strategy (see Section 4.3.5) is preceded by a detailed presentation of the strategy's main constituents involving the incremental formulations developed for the description of the behaviour of concrete (up to [see Section 4.3.2], and for the modelling of [see Section 4.3.3], macro-cracking) and steel (see Section 4.3.4), as well as the interaction between concrete and steel.

4.3.2 Incremental formulation up to macro-cracking

4.3.2.1 Incremental Newton–Raphson method

The essence of the non-linear FE procedure is the incremental Newton–Raphson method (INRM) described previously. Irrespective of the updating strategy adopted, its basic formulation may be summarised in flowchart form as depicted in Figure 4.7.

The external-load vector is applied in load steps Δf_e (typically, $\Delta f_e = 5\%$–10% of the estimated failure load), to which the unbalanced nodal forces (i.e., the vector of residual forces, Δf_r) of the previous iteration must be added. Then, a decision on whether or not to update the various D-matrices – and, hence, the incremental stiffness matrices $[k]$ – is made. If the current iteration is an updating iteration, the result is an update of the incremental stiffness matrix of the structure (usually known as the tangent k-matrix – see Section 4.2.2). It should be noted that, although the incremental k-matrix may be the tangent k-matrix, it is not necessary to use fully tangent properties; in fact, any k-matrix derived from initial secant or tangent properties can be used. In the present work, whenever updating of the k-matrix is required, the matrix is only quasi-tangent because it is more convenient to use simply the material tangent moduli which, it will be recalled (see Section 3.1.2.3), expressions (3.41) through (3.43), neglect σ_{id}, so that the true constitutive law is not actually implemented. If the system of equations can be solved, the increments of the nodal-displacement vector Δd are obtained, from which the new increments in strains ($\Delta \varepsilon$) and stresses ($\Delta \sigma$) at all Gauss points are calculated through the matrices $[B]$ and $[D]$, respectively; thus, the total (cumulative) strains (ε) and stresses (σ) may be ascertained. The new total stresses are now balanced, that is, they satisfy, at this stage, equilibrium (namely, $\int [B]^T \sigma' dV = f_i \neq f_e$, where f_i is the set of (nodal) internal forces), but, in general, they are not compatible with the actual material

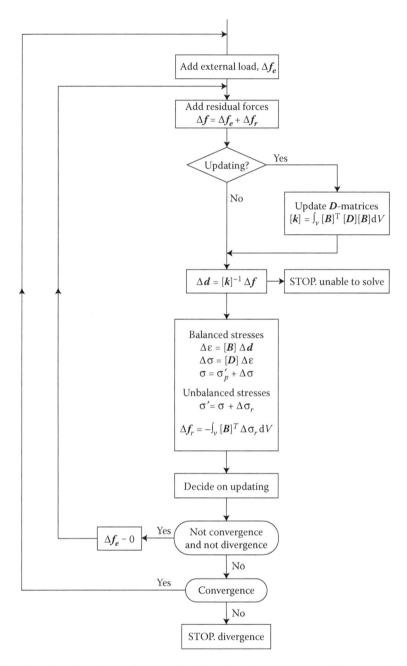

Figure 4.7 Basic flowchart for the non-linear analysis based on the incremental NR method.

stress–strain relationships (unless, of course, convergence has been obtained). Therefore, these equilibrated stresses are corrected so as to satisfy the constitutive equations (at this point, obviously, the coupling effect σ_{id} must formally be taken into account), and this requires the additional stress increments $\Delta\sigma_r$, which lead to the total stresses σ' that are now unbalanced (as equilibrium is no longer satisfied since $\int[B]^T\sigma'dV = f_i \neq f_e$). These corrective stresses $\Delta\sigma_r$ create new residual or unbalanced forces Δf_r which are applied to the structure in the next iteration in order to re-establish equilibrium conditions. If unbalanced forces do not satisfy convergence criteria, the external load is kept constant and further iterations are carried out; otherwise a new external-load increment is applied and the whole procedure repeated. There are only two possible reasons for stopping the analysis: unrealistic solutions to the set of equations (e.g., owing to ill conditioning), or divergence of residual forces. It should be stressed at this point that the preceding outline, as well as much of the subsequent material in the present section devoted to the micro-cracking regime is, clearly, also relevant in the presence of macro-cracking, although, as stated earlier, the specific features of the latter regime will be covered in detail in following section.

An essential requirement of any non-linear package is the ability to store intermediate results which are then used to obtain more accurate values as the iterative solution proceeds and convergence criteria are eventually met. The need for such a storage facility is even more apparent when an incremental procedure is being implemented. Such a procedure imposes on the structure under consideration additional external loading when the iterative solution has converged at a given load level, thus enabling the program to follow automatically a monotonic loading path up to overall failure. It is clear that further storage is necessary here in order to follow the structural response at each external-load level, which implies storing strains, stresses, displacements and cracking/yielding information. The storage of such vast quantities of data might, at first, appear somewhat wasteful, in the sense that, once a given external-load increment has achieved convergence, information on previous load steps could be dispensed with; however, the retention of the data pertaining to the whole of the analysis is useful for post-processing purposes such as, for example, plotting and, more generally, the capacity for studying in detail and at leisure the various parameters at given solution stages. While the necessary storage additions and accompanying programming strategies are beyond the scope of the present discussion, the relevant background may be found in Bedard 1983; Gonzalez Vidosa 1989.

So far, no specific choice of updating strategy has been mentioned, all INRMs sharing the common layout of Figure 4.7, as mentioned previously. However, this question of updating technique must be addressed when certain aspects associated with the convergence and efficiency of the adopted algorithm are being considered. Although the rates of convergence of the modified and mixed INRMs are slower than that corresponding to the pure INRMs (see Figure 4.6), the former methods usually economise on computer time since they cut down on the high cost of the numerical integration of stiffness matrices. Furthermore, if instead of the set of equations being solved by iterative methods, a single reduction or decomposition of the stiffness matrix is carried out, then the former methods (i.e., modified and mixed INRMs) also save in the number of factorisations of the k-matrix of the structure. Now, it must be stressed at this point that the analysis of concrete structures has two sources of high localised non-linearities which make the above reasoning about efficiency of secondary importance: cracking of concrete and yielding of steel at given Gauss points. These, however, will be discussed later (for concrete, see Section 4.3.3) and, before their occurrence, the solution searches corresponding to the micro-cracking regime associated with a much milder form of non-linearity may follow safely the mixed INRM in which the stiffness matrix is reformulated or updated only periodically. This combination of the pure and modified INRMs represents a sensible compromise between the high convergence of the former and the low cost of the latter (Philips and Zienkiewicz 1976; Bedard 1983).

To summarise, therefore, the incorporation, in the solution strategy, of the constitutive relationships throughout the micro-cracking regime consists, very broadly, of the following three specific steps, which are carried out at each integration point:

- At each iteration, a check is made that the state of stress lies within the failure envelope.
- At each iteration, the state of stress corresponding to the state of strain generated by the FE solution is corrected so that the constitutive laws are satisfied.

- When the updating of the stiffness matrix is required, the concrete material properties K_t, G_t and hence E_t, v_t which correspond to the actual state of stress, are obtained.

In order to outline more fully the overall FE procedure described in Figure 4.7, the formulation of the **B**-, **D**- and **k**-matrices needs to be discussed in some detail, as well as the residual-force implementation and the criteria adopted for convergence and divergence. Such aspects are dealt with in the subsequent five sections and, whenever relevant, this will be done, in generic form, by reference to the 3-D brick element.

4.3.2.2 Incremental strain–displacement relationships

The incremental version of Equation 4.1 that link the 3-D displacement field and the nodal displacements is given by

$$\Delta u = [N]\Delta d \tag{4.70}$$

where $\Delta u = (\Delta u_x, \Delta u_y, \Delta u_z)$ are increments of displacements, $\Delta d = (\ldots, \Delta d_{xI}, \Delta d_{yI}, \Delta d_{zI}, \ldots)$ are increments of nodal displacements and $[N]$ is the matrix of shape functions. By reference to brick elements, (4.70) may be written as

$$\begin{bmatrix} \Delta u_x \\ \Delta u_y \\ \Delta u_z \end{bmatrix} = \begin{bmatrix} & N_I & 0 & 0 & \\ \cdots & 0 & N_I & 0 & \cdots \\ & 0 & 0 & N_I & \end{bmatrix} = \begin{bmatrix} \vdots \\ \Delta d_{xI} \\ \Delta d_{yI} \\ \Delta d_{zI} \\ \vdots \end{bmatrix} \tag{4.71}$$

in which N_I is the shape function of the Ith node.

The vector of strains ε (ε_x, ε_y, ε_z, γ_{xy}, γ_{xz}, γ_{yz}) is defined on the assumptions of the linear theory of elasticity (Timoshenko and Goodier 1970). Accordingly, the relevant expressions are

$$\varepsilon_x = \partial u_x / \partial x \tag{4.72}$$

$$\varepsilon_y = \partial u_y / \partial y \tag{4.73}$$

$$\varepsilon_z = \partial u_z / \partial z \tag{4.74}$$

$$\gamma_{xy} = \partial u_x / \partial y + \partial u_y / \partial x \tag{4.75}$$

$$\gamma_{xz} = \partial u_x / \partial z + \partial u_z / \partial x \tag{4.76}$$

$$\gamma_{yz} = \partial u_y / \partial z + \partial u_z / \partial y \tag{4.77}$$

where the (engineering) strain definitions adopted differ from those corresponding to the components of the strain tensor $\varepsilon_{ij} (= (1/2)(\partial u_i / \partial x_j + \partial u_j / \partial x_i))$. By reference to Equation 4.8, the incremental strain-nodal displacement relations are

$$\Delta \varepsilon = [B]\Delta d \tag{4.78}$$

which, in 3-D problems, is obtained by combining Equations 4.71 through 4.77, the result being

$$
\begin{bmatrix} \Delta\varepsilon_x \\ \Delta\varepsilon_y \\ \Delta\varepsilon_z \\ \Delta\gamma_{xy} \\ \Delta\gamma_{xz} \\ \Delta\gamma_{yz} \end{bmatrix} = \begin{bmatrix} \cdots \end{bmatrix} \begin{bmatrix} b_{xI} & 0 & 0 \\ 0 & b_{yI} & 0 \\ 0 & 0 & b_{zI} \\ b_{yI} & b_{xI} & 0 \\ b_{zI} & 0 & b_{xI} \\ 0 & b_{zI} & b_{yI} \end{bmatrix} \begin{bmatrix} \cdots \end{bmatrix} = \begin{bmatrix} \vdots \\ \Delta d_{xI} \\ \Delta d_{yI} \\ \Delta d_{zI} \\ \vdots \end{bmatrix} \tag{4.79}
$$

where $b_{xI}, b_{yI}, b_{zI} (= \vartheta N_I/\vartheta x, \vartheta N_I/\vartheta y, \vartheta N_I/\vartheta z)$ are functions of x, y, z. These derivatives completely define the 6×3 block of the B-matrix corresponding to the contribution of the displacements of node I to the strain increments at a given point within the element. Since, as pointed out earlier, the analysis does not include geometrical non-linearities, such derivatives remain constant throughout the analysis. Therefore, they are calculated for all Gauss points only once (at the first iteration of the analysis). Their calculation, in terms of the local coordinates and the ensuing Jacobian matrix, has already been explained in Section 4.1.3.

4.3.2.3 Incremental stress–strain relationships for uncracked concrete

The increments of stresses and strains are related by the D-matrix adopted. By reference to Equation 4.12, and neglecting initial strains and/or stresses, its incremental counterpart is simply

$$
\Delta\sigma = [D]\Delta\varepsilon \tag{4.80}
$$

For uncracked-concrete Gauss points, the D-matrix may be calculated by reference to a linearly-elastic isotropic material which is usually described in the following concise form

$$
\sigma_{ij} = 2G\varepsilon_{ij} + 3\mu\varepsilon_o\delta_{ij} \tag{4.81}
$$

where G and μ are the shear and Lame's moduli, the former being given by (3.2) while the latter is also related to E and ν by the expression

$$
\mu = \nu E/[(1 + \nu)(1 - 2\nu)] \tag{4.82}
$$

On the basis of (4.78), therefore, the incremental constitutive relations (4.72) through (4.77) for uncracked concrete may be written as

$$
\begin{bmatrix} \Delta\sigma_x \\ \Delta\sigma_y \\ \Delta\sigma_z \\ \Delta\tau_{xy} \\ \Delta\tau_{xz} \\ \Delta\tau_{yz} \end{bmatrix} = \begin{bmatrix} 2G+\mu & \mu & \mu & 0 & 0 & 0 \\ \mu & 2G+\mu & \mu & 0 & 0 & 0 \\ \mu & \mu & 2G+\mu & 0 & 0 & 0 \\ 0 & 0 & 0 & G & 0 & 0 \\ 0 & 0 & 0 & 0 & G & 0 \\ 0 & 0 & 0 & 0 & 0 & G \end{bmatrix} \begin{bmatrix} \Delta\varepsilon_x \\ \Delta\varepsilon_y \\ \Delta\varepsilon_z \\ \Delta\gamma_{xy} \\ \Delta\gamma_{xz} \\ \Delta\gamma_{yz} \end{bmatrix} \tag{4.83}
$$

where $(\Delta\sigma_x, \Delta\sigma_y, \Delta\sigma_z, \Delta\tau_{xy}, \Delta\tau_{xz}, \Delta\tau_{yz})$ are the increments of direct and shear stresses in global coordinates, while G and μ are derived from the tangent shear and bulk moduli described in section 3.1.2.3 (i.e., expressions [3.41] through [3.43]). Clearly, the coefficients of the D-matrix are functions of the state of stress (i.e., $G(\tau_o)$, $\mu(\sigma_o,\tau_o)$), but, at the same time, it is worth noting that the constitutive matrix is isotropic throughout the micro-cracking regime and, hence, invariant with respect to any set of orthogonal axes.

Although the D-matrices for cracked Gauss points will be described in detail in Section 4.3.3, it is convenient to note at this stage that they are anisotropic and that they will be defined with respect to cracked directions. Thus, all their coefficients in global coordinates will, in general, be non-zero, since cracked non-isotropic D-matrices require a transformation from local to global directions (see Appendix B for such a transformation). However, it is evident that axes transformations do not affect the D-matrix in the present case of isotropic behaviour before macro-cracking.

4.3.2.4 Incremental force–displacement relationships

The incremental stiffness matrix of an element connects the increments of nodal forces and nodal displacements. The relevant expression may readily be written down by reference to either Equation 4.21 or 4.22 which relates total displacements and forces up to a given stage of loading, the result being

$$\Delta f = [k]\Delta d \qquad (4.84)$$

with the expression for $[k]$ given by Equation 4.20. As explained previously, this k-matrix is calculated in global directions by numerical integration and, hence, may be written as

$$[k] = \sum_{i=1}^{n}([B]_i^T[D]_i[B]_i J_i)w_i \qquad (4.85)$$

where J_i and w_i are the Jacobian and the weight of the ith Gauss point, respectively, and n is the total number of Gauss points in the element. As was the case with the coefficients of the B-matrix, Jacobians do not change throughout the analysis and, hence, they are calculated only at the first iteration. Furthermore, as $[k]$ is symmetric, it is necessary to calculate only the coefficients of its upper (or lower) triangle.

While $[k]$ can be calculated directly from expression (4.85) without reference to the actual B- and D-matrices, it is worth noting that the expressions for its coefficients can be computed more efficiently by taking into account any special features (e.g., sparsity) of the relevant B- and D-matrices. This will be discussed briefly below, as it leads to efficient procedures for the numerical calculation of $[k]$, and also prepares the ground for the discussion (in Section 4.3.3) of the effect of the smeared representation of cracking on the conditioning of stiffness matrices.

Expression (4.85) can split into blocks of 3×3 coefficients relating to the DOF of pairs of nodes. Let $[k]^{IJ}$ be one such 3×3 block relating to the DOF of nodes I and J

$$[k] = \begin{bmatrix} \cdot & \cdot & \cdot & \cdot & \cdot \\ \cdot & [k]^{II} & \cdot & [k]^{IJ} & \cdot \\ \cdot & \cdot & \cdot & \cdot & \cdot \\ \cdot & [k]^{JI} & \cdot & [k]^{JJ} & \cdot \\ \cdot & \cdot & \cdot & \cdot & \cdot \end{bmatrix} \begin{matrix} \text{node} \quad I \\ \\ \text{node} \quad J \end{matrix} \qquad (4.86)$$

This $[k]^{IJ}$ block is equal to $\sum_{i=1}^{n}[k]_i^{IJ}$, where $[k]_i^{IJ}$ is the contribution of the ith Gauss point to such a block. For the brick elements adopted herein, and for the general case of cracked Gauss points, this contribution is given by the following expression:

$$[k]_i^{IJ} = w_i J_i \begin{bmatrix} b_{xI} & 0 & 0 & b_{yI} & b_{zI} & 0 \\ 0 & b_{yI} & 0 & b_{xI} & 0 & b_{zI} \\ 0 & 0 & b_{zI} & 0 & b_{xI} & b_{yI} \end{bmatrix}_i$$

$$\begin{bmatrix} d_{11} & d_{12} & d_{13} & d_{14} & d_{15} & d_{16} \\ & d_{22} & d_{23} & d_{24} & d_{25} & d_{26} \\ & & d_{33} & d_{34} & d_{35} & d_{36} \\ & & & d_{44} & d_{45} & d_{46} \\ & symmetric & & & d_{55} & d_{56} \\ & & & & & d_{66} \end{bmatrix}_i \begin{bmatrix} b_{xJ} & 0 & 0 \\ 0 & b_{yJ} & 0 \\ 0 & 0 & b_{zJ} \\ b_{yJ} & b_{xJ} & 0 \\ b_{zJ} & 0 & b_{xJ} \\ 0 & b_{zJ} & b_{yJ} \end{bmatrix}_i \quad (4.87)$$

The sparsity of the B-matrix is evident: half of its coefficients are zero and placed at known positions. Thus, the number of computations may be reduced quite significantly once all the multiplications involving zero terms are identified and left out of subsequent numerical operations. The D-matrix in Equation 4.87 is not sparse, as it refers to the general case of a cracked Gauss point expressed in terms of global coordinates. On the other hand, uncracked Gauss points are described by Equation 4.83 and, for such isotropic conditions, further reduction of computing effort is clearly possible. Various ways of achieving such computational savings for both isotropic and anisotropic material descriptions have been explored in Gonzalez Vidosa (1989), with subsequent implementation in the computer program as appropriate.

4.3.2.5 Residual forces

In accordance with the present FE model, the non-linear force–displacement relationships at the structural level arise exclusively as a result of the non-linearities in the stress–strain expressions. The iterative procedure required to follow these non-linear $\sigma - \varepsilon$ laws relies on the residual-forces method, by means of which stress corrections (at the material level) of balanced stresses cause the appearance of equivalent unbalanced nodal forces that must be applied in the next iteration in order to re-establish the equilibrium conditions for the overall structure. This may be summarised through the following expression, in which the equivalence between external and internal forces is implicit:

$$f_e = \int [B]^T \sigma' dV + \int [B]^T (\sigma - \sigma') dV \quad (4.88)$$

where σ and σ' are balanced and unbalanced stresses respectively; and the first term denotes unbalanced (internal) forces (but satisfying the constitutive relations) while the second term re-establishes overall equilibrium conditions (but causing, in turn, lack of compatibility between stresses and strains). Therefore, the residual forces are given by

$$\Delta f_r = -\int [B]^T (\sigma' - \sigma) dV = f_e - \int [B]^T \sigma' dV \quad (4.89)$$

where, it should be recalled, the last term represents nodal internal forces.

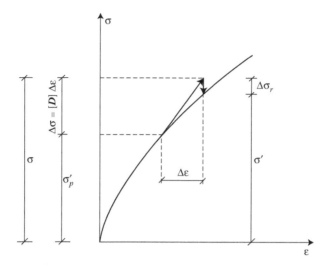

Figure 4.8 Stress correction, by the standard initial-stress technique that achieves satisfaction of the constitutive law but disturbs the equilibrium of the previously balanced stresses σ.

When the constitutive laws are expressed in the form $\sigma = \sigma(\varepsilon)$, as is the case for cracked Gauss points (to be discussed in Section 4.3.3) and, also, for steel Gauss points both before and after yielding, unbalanced stresses are worked out in accordance with such laws in the standard manner shown in Figure 4.8. The residual stresses are given by the components of $\Delta\sigma_r = \sigma' - \sigma$ and must be checked (together with residual forces and, possibly, other criteria) for convergence.

The stress correction for uncracked Gauss points, on the other hand, follows an initial-strain technique, since the constitutive law (for uncracked concrete) is given in the form $\varepsilon = \varepsilon(\sigma)$ (see expressions (3.44) in Section 3.1.2.3). Figure 4.9 summarises schematically the implemented initial-strain technique for uncracked concrete points. First, increments of strains (and, then, the total strains) are computed from increments of nodal displacements (subscripts other than r indicate the iteration number)

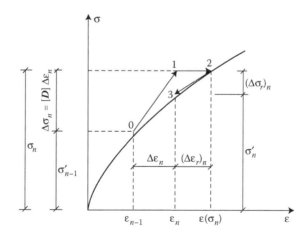

Figure 4.9 Stress correction, by the initial-strain technique, used for uncracked concrete points (for cracked concrete points, and for all steel Gauss points, refer to Figure 4.8).

$$\Delta\varepsilon_n = [B]\Delta d_n \tag{4.90}$$

$$\varepsilon_n = \varepsilon_{n-1} + \Delta\varepsilon_n \tag{4.91}$$

Balanced stresses are next computed using the D-matrix incorporated into the set of equations

$$\sigma_n = \sigma_{n-1} + [D]\Delta\varepsilon_n \tag{4.92}$$

These balanced stresses are corrected by the following expression (see Figure 4.9)

$$\sigma'_n = \sigma_n - [D(\sigma_n)]\{\varepsilon_n(\sigma_n) - \varepsilon_n\} \tag{4.93}$$

where $[D(\sigma_n)]$ and $\varepsilon_n(\sigma_n)$ are, respectively, the tangent D-matrix given by Equation 4.83 and strains, both of these being in accordance with the constitutive law corresponding to the balanced stress level. The components of $(\Delta\varepsilon_r)_n = \varepsilon_n(\sigma_n) - \varepsilon_n$ are residual strains that have to be checked for convergence. It is worth noting that this initial-strain technique converges very quickly in practice owing to the mild nature of the uncracked material non-linearities. (Clearly, while the standard initial-stress method of Figure 4.8 involves the satisfaction of the material law at the end of every iteration, the initial-strain technique satisfies neither equilibrium nor the material law at the end of an iteration unless, of course, convergence has taken place.)

Once stresses are corrected at a given Gauss point, the numerical integration of Equation 4.89 leads to the following contribution to the residual-forces vector

$$\begin{bmatrix} (\Delta f_x)_r \\ (\Delta f_y)_r \\ (\Delta f_z)_r \end{bmatrix}_I = -w_i J_i \begin{bmatrix} b_{xI} & 0 & 0 & b_{yI} & b_{zI} & 0 \\ 0 & b_{yI} & 0 & b_{xI} & 0 & b_{zI} \\ 0 & 0 & b_{zI} & 0 & b_{xI} & b_{yI} \end{bmatrix}_i \begin{bmatrix} (\Delta\sigma_x)_r \\ (\Delta\sigma_y)_r \\ (\Delta\sigma_z)_r \\ (\Delta\tau_{xy})_r \\ (\Delta\tau_{xz})_r \\ (\Delta\tau_{yz})_r \end{bmatrix}_i \tag{4.94}$$

In the above expression, $[(\Delta f_x)_r, (\Delta f_y)_r, (\Delta f_z)_r]_I$ are the residual forces at node I caused by a stress correction $[(\Delta\sigma_x)_r, (\Delta\sigma_y)_r, (\Delta\sigma_z)_r, (\Delta\tau_{xy})_r, (\Delta\tau_{xz})_r, (\Delta\tau_{yz})_r]$ at the ith Gauss point.

4.3.2.6 Convergence and divergence criteria

The convergence of solutions obtained by iterative procedures can be checked in terms of one or more vector increments of various parameters, the norms or elements of which must all be smaller than certain prescribed values. The parameters in question include quantities such as displacements, residual forces, residual stresses and residual strains (Bergan and Clough 1972). In practice, only one of these vector increments is checked, since all of them are interrelated. Displacement and force criteria are usually preferred to stress and strain criteria. With regard to non-linear analyses of structural concrete, it would appear that force criteria have mostly been used (Suidan and Schnobrich 1973; Lin and Scordelis 1975, Philips and Zienkiewicz 1976, Cedolin and Dei Poli 1977, Cristfield 1982).

In the earlier part of the work (Bedard 1983) the two residual-stress criteria

$$\max_i |\Delta\sigma_r(i)| < 0.1 \text{ MPa} \tag{4.95}$$

$$\max_i |\Delta\sigma_r(i)| < 0.01 |\sigma_r(i)| \tag{4.96}$$

were adopted in recognition of the fact that a single criterion might prove unrealistic for the whole of the loading path. Thus, for example, while Equation 4.95 can ensure a reasonable level of accuracy in the early load steps, at more advanced stages of the loading such a residual-stress value might become impractical to achieve, especially as failure is approached, when the numerical solution tends to become unstable. This is why the second convergence criterion – Equation 4.96 – based on residual stresses given as percentages of total-stress values, is more attractive as one nears ultimate-load conditions, and is in keeping with the notion of accepting a larger force imbalance as the total load increases (Cedolin and Nilson 1978). Clearly, the satisfaction, at each Gauss point, of either of the criteria defined by Equations 4.95 and 4.96 is sufficient for convergence.

In subsequent work (Gonzalez Vidosa 1989), several additional convergence criteria were studied, of which the following were implemented in the non-linear procedure. First, a maximum residual strain of 2.5 mm/m, or less than 0.5 of the total strain, was adopted for uncracked concrete, which was found to be slightly more restrictive than the stress criteria adopted earlier (i.e., Equations 4.95 and 4.96). In addition, once cracking was implemented, it was prescribed that no concrete Gauss points should exceed the failure envelope, with a maximum residual stress of 0.1 MPa set as the limit at cracked Gauss points. The latter stress criterion was also adopted for steel Gauss points, and for these any change from one linear branch to another was treated as a lack of convergence requiring further iterations. Now, it was found that these various convergence criteria are usually met on specification of the following (additional) residual-force criterion:

$$\max_i |\Delta f_r(i)| < 0.001 |\Delta F| \tag{4.97}$$

where $|\Delta F|$ is the applied load step (between 5% and 10% of the ultimate experimental load). This last condition is a very restrictive requirement indeed when compared with other reported criteria (see e.g., Suidan and Schnobrich 1973, Lin and Scordelis 1975, Philips and Zienkiewicz 1976, Cedolin and Dei Poli 1977, Cristfield 1982 mentioned previously). However, all the above criteria were kept (including Equation 4.97) in order to avoid convergence in certain situations, such as concrete Gauss points being outside the failure envelope ('cracking' criterion) or new yielding of the steel ('yield' criterion), as may occur with less restrictive convergence criteria. Nevertheless, it should be said that residual forces become negligible as soon as material properties are updated, and no new cracking and/or yield occur.

The incremental process stops either because of divergence of residual forces, which is taken to have occurred when

$$\max_i |\Delta f_r(i)| > 100 |F| \tag{4.98}$$

where $|F|$ is the total external load, or because the system of equations cannot be solved as, with extensive cracking, the degradation of the stiffness matrix of the structure leads to a non-positive or ill-conditioned $[K]$. The former divergence criterion – Equation 4.98 – is very tolerant and thus rarely fulfilled. Hence, most runs stop when [K] becomes sufficiently degraded. As a consequence, all ultimate-load predictions refer to the maximum sustained load (MSL) in the analysis, which corresponds to the last *converged* load step.

4.3.3 Modelling of macro-cracking

The presence of macro-cracking or 'structural' cracks is unavoidable in most RC structures, and hence codes of practice generally take this into account. For example, BS 8110 (British Standards Institution 1985a,b) suggests a maximum crack width of 0.3 mm for RC (Part 1 – thus removing the earlier distinction between normal exposure [0.3 mm] and exposure to particularly aggressive environments [>0.004 times the nominal cover to the main reinforcement] present in CP 110, British Standards Institution 1972), whereas for pre-stressed concrete members (Part 2) this limit is set at 0.1 mm for components exposed to particularly aggressive environments and 0.2 mm for all other units. As an instance, typical load–deflection curves from tests on flexural beams exhibit a distinct non-linear behaviour attributed to the onset of (macro) cracking at load levels as low as 10%–20% of the structural failure level (Burns and Siess 1966). Accordingly, macroscopic cracks are not only present in most RC structures, but they are usually also present during most of the loading history. It is not an exaggeration, therefore, to say that macro-cracking represents the key feature in the non-linear behaviour of structural concrete, governing its failure mechanism. As a result, the successful numerical treatment of cracking – the predominant non-linear effect – is an obvious prerequisite for a reliable FE model. In what follows, a description of crack representation will be outlined, while the gradual (or otherwise) implementation of crack propagation – which may give rise to very serious problems of numerical stability – is fully described elsewhere (Kotsovos and Pavlovic 1995).

4.3.3.1 Nature of structural cracking

As explained in Chapter 3, the behaviour of concrete is dictated by the fracture processes that the material undergoes under increasing load. These fracture processes first take the form of microscopic cracks (some of which exist within the material even before the application of the load) that extend in the plane of the maximum and intermediate compressive stresses, and propagate in the direction parallel to the maximum principal compressive stress (or in the direction orthogonal to the maximum principal tensile stress). Such micro-cracks develop in a stable manner up to the OUFP stress level (their effect on deformation, up to this point, being accounted for by the adopted constitutive relations), which, for practical purposes, can be taken as the failure stress envelope. Once this level is reached, the microscopic cracks link to form a limited number of macroscopic cracks: cracking now becomes an unstable process and concrete suffers a noticeable loss of material continuity (macro-cracking) in the direction orthogonal to the maximum principal tensile (or minimum principal compressive) stress, which leads to a sudden collapse of concrete specimens unless the fracture process is somehow restrained. In this respect, it is worth recalling the fully brittle nature of concrete once the OUFP level has been reached. As demonstrated experimentally (see Section 2.2.1.2), the strain-softening branches observed in compressive tests arise as a result of friction between loading platens and concrete specimens, so that, if friction is eliminated, a complete loss of load-carrying capacity becomes the most realistic description for concrete behaviour in compression. Similarly, with regard to the post-ultimate behaviour of concrete in tension, it has been shown that strain-softening branches can be observed only in tests where the stiffness of the testing machine is steeper than the steepest portion of the falling branch – otherwise, a sudden failure is deemed to occur. Such testing machines impose large restraints to the fracture processes of concrete that are unlikely to hold for concrete within a structure.

The above observations are summarised schematically with reference to a typical uni-axial $\sigma - \varepsilon$ curve (but equally valid for more general – biaxial or triaxial – conditions). Figure 4.10a represents such a characteristic in compression, the rising portion covering the various levels of micro-cracking, while A marks the peak load level (defined by the ultimate surface envelope) at which (macro) cracking occurs. Beyond A, the loss in stiffness proper-ties of the material is immediate and complete. A similar model is applicable to the material characteristic in tension (Figure 4.10b). (The linearity of the stress–strain relationships in tension may be explained by the fact that controlled crack propagation [micro-cracking] up to the OUFP level [which practically coincides with macro-cracking] can occur only under compressive stress states. In the case of tensile stresses, *any* cracking implies failure and, hence, since non-linearity reflects micro-cracking fracture processes, the $\sigma - \varepsilon$ relationships for tension should be essentially linear up to failure [i.e., the start of micro-cracking and macro-cracking coincides]. This may be argued further on the basis that any departures from linearity as a result of other causes [e.g., time effects] in the case of tension are much less pronounced than in compression. Finally, any attempt at refining the linear model should be tempered by the large scatter inherent in tensile-test results.) Despite the realistic description of both compressive and tensile behaviour at the material level, however, it is only the latter characteristic that is of relevance in a concrete *structure* beyond peak-stress levels. As explained in Section 2.4, the peak level in compression is never attained within a structure as tensile failure precedes it (with consequent redistribution of stresses) in adja-cent regions. Therefore, although the true zero-stiffness model beyond A in Figure 4.10a will be adopted in the present work, even the inclusion of (the non-existing) strain soften-ing in compression would not influence the analytical prediction of structural members, as will be seen by reference to the examples of Chapters 5, 7 and 8. Conversely, results will generally be affected by the chosen post-ultimate characteristic in tension, and it must be stressed here that the sudden unloading in Figure 4.10b, which is to be used henceforth, is in marked contrast with most current modelling approaches which tend to assume a gradual unloading (usually linear), that is strain softening in tension, in marked disregard of proven experimental evidence on the one hand, and the difficulties inherent in tensile-test data interpretations on the other.

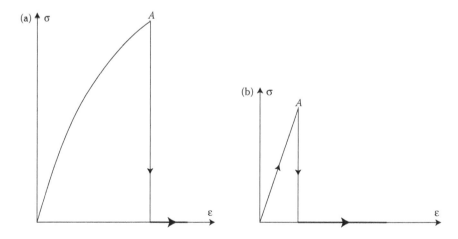

Figure 4.10 Schematic representation of a typical stress–strain curve for concrete: (a) in compression; (b) in tension (i.e., in the direction orthogonal to the first crack).

Figure 4.10b refers to the stiffness characteristics in the direction orthogonal to a crack. With regard to the stiffness of cracked concrete in the direction parallel to macro-cracks, it is a common hypothesis in analyses to assume that concrete keeps its uncracked stiffness in those directions (such as, for example, that defined by the ascending regions in Figure 4.10). Although a certain decay may be expected, it can be shown that the remains of specimens subjected to triaxial tests exhibit uniaxial $\sigma - \varepsilon$ relations similar to those of uncracked specimens. Therefore, it appears realistic to assume that concrete between cracks *in a structure* retains its uncracked stiffness in the direction parallel to the cracks.

Some comments with respect to the stress combinations causing failure at a given point/region within a structure may be made by reference to Figure 4.11. Since cracking causes unloading of the tensile stresses that existed before its occurrence, it leads to states of stress that can be sustained by concrete as they are inside the failure envelope (see zones T–C–C, C–T–T and T–T–T in Figure 4.11, where C stands for compression and T for tension). Therefore, the process can be stabilised at both the material and structural level, provided that those tensile stresses can be redistributed to adjacent steel or concrete elements. For this type of failure, where at least one stress component is tensile, the fracture mechanism involves the presence of a relatively small number of fissures with the same orientation

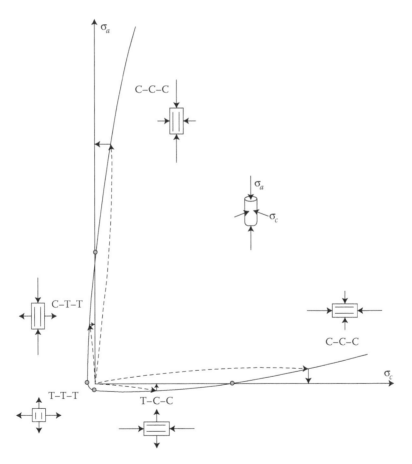

Figure 4.11 Effect of stress combination or 'zone' on the failure of concrete (Note: dashed [arrowed] lines indicate loading paths that eventually hit the failure envelope, at which point sudden unloading [towards the relevant axis σ_a, σ_c as shown by the arrows] leads to stress states that lie inside the failure envelope for cases T–C–V or C–T–T and outside it for case(s) C–C–C).

(orthogonal to the maximum principal tensile stress), which suggests that the material is insignificantly damaged in a direction parallel to these fissures. It may be assumed, therefore, that any change of the stress state will be gradual (from triaxial to biaxial, or from biaxial to uniaxial or from uniaxial to a complete loss of load-carrying capacity). On the other hand, for fully compressive states of stress (zones C–C–C in Figure 4.11), cracking would cause an immediate loss of load-carrying capacity as indicated schematically in Figure 4.11a. Therefore, local failures in compression are best presented as localised failures of the material in the three directions (i.e., loss of stiffness in all directions). After all, at least one stress component has a high compressive value (certainly in excess of the uniaxial strength) and this implies that a relatively large number of fissures are present, suggesting that extensive damage is taking place in the material, which then loses all its load-carrying capacity. This is why a distinction will henceforth be made between 'tensile cracking', termed simply 'cracking', and 'compressive cracking' which is equivalent to, and is thus denoted as, 'local failure of the material'. However, it is important to recall the argument put forward in Section 2.4 (and also mentioned above) that 'concrete in a structure never fails in compression', and, hence, the above discussion on the post-ultimate compressive-stress states (local failure) may be regarded as irrelevant. Nevertheless, as stressed in Section 2.5, the present FE model does not prevent fully compressive stress states from developing at failure (this lack of C–C–C failure modes will be shown in Chapters 5, 7 and 8 to be a *result* of analysis and not an *a priori* imposed condition): thus the need formally to include this possibility as one of the failure criteria.

In view of the preceding discussion, the main features of the fracture mechanism in, and subsequent properties of, concrete relevant to the FE model, may be summarised as follows.

- Under increasing stress, concrete behaviour is that of an isotropic continuum until the stress level reaches a peak value. Beyond this level, either 'local failure' or 'cracking' occurs. Should the former be operative, with the strength envelope reached in compression (zones C–C–C), concrete would suffer a complete loss of stiffness in all directions and hence would no longer be able to sustain any stresses. It is, however, the 'cracking' mechanism which one invariably finds in the course of structural analyses and hence its importance in subsequent work.
- A crack plane forms when concrete reaches the strength envelope in tension (zones C–C–T, C–T–T and T–T–T). Such a plane is orthogonal to the maximum principal tensile stress existing just prior to cracking. The presence of an open crack plane causes a complete loss of material stiffness in its orthogonal direction, and, therefore, direct tensile and, in principle, shear stresses can no longer be transferred across the crack plane (but see subsequent sections for the presence and role of the so-called shear-retention factor). When cracking occurs, a discontinuity is introduced into the hitherto continuous material, and this must be allowed for by suitable changes in the characteristics of the stiffness matrix (by means of either the smeared- or discrete-crack approach – see Section 4.3.3.2).
- Cracking takes place predominantly in zones of high tensile stress and strain concentrations, thus relieving these (see Figure 4.12). As a result, stress redistribution is also associated with cracking; in this way, the tensile stresses that existed before the formation of the crack have to be transformed into equivalent residual forces and redistributed to adjacent concrete and/or steel zones.
- While the cracking process tends to relieve the tensile stress and strain concentrations orthogonal to the crack, as a result of the ensuing stress redistribution, new zones of high tensile stress and strain concentrations are created near the crack tips, orthogonal to the crack-extension path. Depending on the amount of stress 'redistribution' within

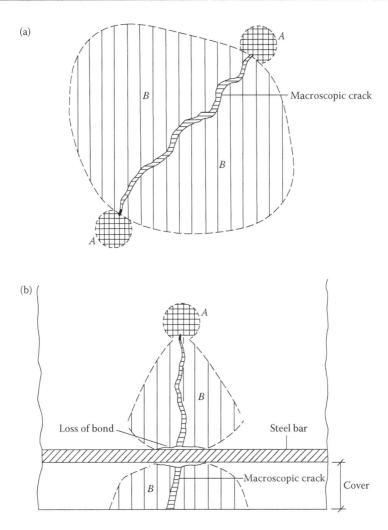

Figure 4.12 Schematic representation of cracking in concrete, where A marks the zones of high tensile stress concentrations near the crack tips, while B represents zones where the tensile stresses orthogonal to the crack have been relieved: (a) plain concrete; (b) reinforced concrete.

the structure, a given crack can be either stable or unstable. At a given load level, a stable crack extends only to a certain length whereas an unstable crack propagates until it induces structural failure.

- When the crack path intersects a reinforcing bar, loss of bond between steel and the surrounding concrete results over a certain length of bar on both sides of the crack, as shown schematically in Figure 4.12b. This question of bond between steel and concrete will be taken up again in Section 4.3.4.

- Finally, it is important to note that the fracture mechanism outlined above applies merely to the initiation of a single crack. Cracking in RC structures under increasing load involves the interaction of many such cracks, each of which initiates and propagates. Such interaction may be quite complex, and many propagation mechanisms are possible, as, for instance: all cracks extend at the same time; a few cracks extend first, with other cracks developing later; single cracks propagate one by one. Any of these processes of crack propagation may be considered to occur in the course of a given applied (finite) load step, and the crack pattern at the end of the convergence run should reflect as closely as possible the actual pattern of cracks that would have resulted in the real structure, bearing in mind that a finite load step in an FE analysis usually represents a quicker rate of loading than that in an actual test. In addition, the three types of crack propagation listed above are also relevant in the course of the various iterations themselves within a given load step, as will be seen in Section 4.4.

4.3.3.2 Smeared-crack approach versus the discrete-crack approach

A characteristic feature of cracking is the inherent discontinuity that arises along its path, referred to in the preceding section. On the other hand, the FE model is essentially a continuum-mechanics technique. Such an apparent incompatibility between problem and modelling tool requires the development of special schemes which, while in keeping with the general notion of 'continuity' in the FE solution, exhibit drastic and sudden changes in the material at the location of cracks. Two basic schemes, that constitute the fundamental alternatives to crack modelling, are the discrete-crack approach and the smeared-crack approach.

The discrete-crack approach introduces an actual gap in the FE mesh at the location of a crack. It achieves this by doubling and separating the nodal coordinates lying along individual crack paths. This implies important changes in the numbering of nodes and element connectivities which, in turn, affect the global stiffness matrix. The first discrete-crack model appears to have been used by Ngo and Scordelis (1967). Early versions of the discrete approach to crack modelling were relatively crude since the crack-propagation path had to follow the boundaries of existing elements. Later on, a technique which automatically redefines the mesh as cracking propagates, apparently without bias in respect of the orientation of cracks, was proposed by Saouma and Ingraffea (1981). This appears to require intricate programming techniques and still to be significantly restricted in use since complicated mesh refinements become necessary to accommodate the propagation of only a few discrete cracks. Figure 4.13 shows a representative FE mesh necessitated by the presence of only six discrete cracks. This mesh generated near failure should be compared with the relative simplicity of the original mesh. Therefore, it is evident that, owing to practical limitations on the number of FEs, discrete-crack predictions are constrained by FE mesh sizes, in addition to the large effort required for the constant redefinition of the mesh topology, even for plane-stress cases (see Figure 4.13); in this respect, it is important to note that the discrete-crack technique does not seem to have been applied to 3-D cases (or even to 2-D axisymmetric problems), for which practical difficulties can easily be appreciated. Two further possible complications may be envisaged as cracking progresses in the course of discrete modelling: one is the formation of 'elongated' FEs, that is those possessing large ratios between sides (as in the case of many of the elements making up the mesh in Figure 4.13b), with consequently poor(er) numerical performance; the other possible difficulty concerns the subdivision of the mesh (in parts at least) to such a degree that the material data used for the analysis (and based on the average properties of a test specimen of finite size) become of doubtful validity.

It might be argued that the discrete-crack approach allows the natural description of individual cracks and, also, it would seem, of the phenomenological processes that are deemed to occur between these cracks. Therefore, the approach appears to be quite suitable, at least, to localised phenomena, such as aggregate interlock, concrete–steel bond and dowel action (see e.g., Tassios and Scarpas 1987). However, meaningful constitutive relations for these local phenomena are very sparse and, in addition, it appears that such effects play a secondary role on the mechanisms of resistance to applied load in concrete structures (Kotsovos and Pavlovic 1986, Kotsovos et al. 1987). What is important in a discrete-crack approach

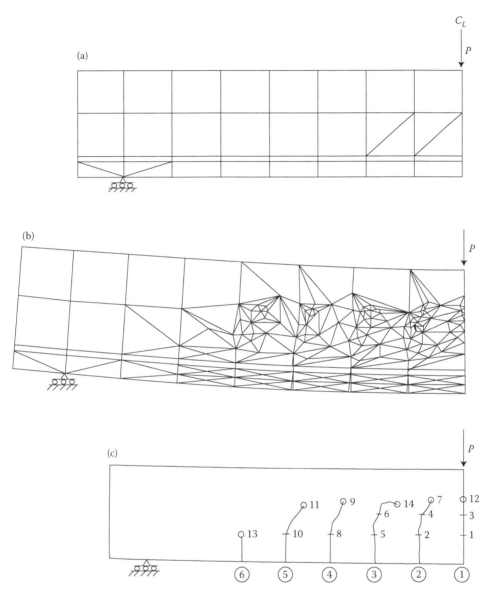

Figure 4.13 Examples of application of the discrete-crack approach: (a) initial FE mesh; (b) FE mesh configuration near failure; (c) crack pattern near failure.

is the ability to model the high tensile stresses taking place at the crack tips, and it is in this respect that a major disadvantage of the scheme (or of the alternative approach, where a single crack within an element is dealt with by fracture mechanics – see Shenglin et al. 1987) manifests itself. For, there is simply no reliable experimental data on the material behaviour in the region of the crack tips, where high stress concentrations govern the initiation and subsequent spread of cracking; and this problem is further compounded by the fact that, at such microscopic level, the true heterogeneous nature of concrete can no longer be ignored, thus posing formidable problems even to the powerful analytical tools afforded by present-day fracture-mechanics theory.

The above outline of the discrete-crack approach was undertaken in order to show why its alternative, the smeared-crack method, adopted in the present FE model, represents, in the current state of knowledge (as regards, primarily, material data, but also numerical capabilities), the more realistic – although, admittedly, less ambitious – technique for crack modelling. The method, whose first practical application seems to have been the predicted response of a nuclear pre-stressed vessel under inner pressure (Rashid 1968), and which has subsequently formed the basis for most FE analyses of structural concrete, makes use of the drastic material property changes caused by, and at the location of, cracking as a means of simulating discontinuity. Such changes are achieved by removing or reducing the stiffness properties in the direction orthogonal to the crack, without introducing any gap in the initial mesh, and leaving the latter unchanged throughout the analysis. Since material properties are evaluated only at specific points in an element, such as the integration points or the nodes, the alteration of material properties due to cracking consequently affects the contributing region from which these properties are evaluated, hence *smearing* the effect of cracking over the whole of that region. In fact, a 'single' crack represents an infinite number of parallel fissures throughout that part of the element related to an integration point or a node. As the present work is based entirely on the isoparametric formulation, the points around which the smearing of cracks takes place are always the integration points. This 'smearing' of material properties over a finite region of the element models the relief of tensile stresses orthogonal to a crack (see Figure 4.12).

The essence of smeared modelling, therefore, is the setting up of cracked areas by modification of the stiffness properties and stresses (equilibrium conditions) at the relevant Gauss points, that is, those points used for the numerical integration of stiffness matrices and for the calculation of residual forces. Thus, while discrete models account for stiffness losses by doubling the nodes lying on individual cracks, smeared models simply replace uncracked *D*-matrices by cracked ones. (Equilibrium conditions are lost in both cases either as a result of mesh modifications [discrete models] or directly through stress corrections [smeared models]; subsequently, such equilibrium is re-established by means of residual forces in both cases.) The fact, that no mesh modifications are required throughout an analysis, based on the smeared-crack approach not only constitutes a signal computational advantage in this technique, as mentioned already, but also allows the element size to be kept similar to that of the specimens from which the constitutive relations and fracture criteria were derived (the latter, an important practical consideration, will be touched upon later, when the effect of FE mesh size on numerical predictions is considered).

4.3.3.3 Incremental stress–strain relationships for cracked concrete

This section deals with the *D*-matrices for cracked concrete for 3-D analysis, while the cases of 2-D (both plane and axisymmetric) problems have been fully discussed in Kotsovos and Pavlovic (1985). While, before cracking, the *D*-matrix is isotropic, obeying the constitutive laws applicable throughout the ascending branch (at the material level – see Figure 4.10), after cracking, the material becomes anisotropic and the previous constitutive relationships can no longer be used, as they were derived for concrete in the uncracked state. Since there are few reliable data on the deformational characteristics of cracked concrete, it will henceforth be assumed that the properties of a *cracked D*-matrix remain constant during further loading. (These properties are updated only if and when a second crack develops at the same Gauss point, and then, again, the new stiffness is maintained constant; should a third, and final, crack occurs – as, possibly, in 3-D or axisymmetric problems – it would lead to the last updating of the *D*-matrix.) Such an assumption is both necessary (in the absence of relevant experimental information) and reasonable (since the main governing factor once cracking occurs is further cracking rather than any detailed/accurate description of deformation in between two consecutive cracks at a given location where, in any case, load is likely to be shed towards uncracked regions).

The 3-D isotropic **D**-matrix relating increments of stresses and strains before cracking, is given by Equation 4.83. When the state of stress at a Gauss point reaches the triaxial envelope involving at least one principal tensile component for the first time, a crack plane is assumed to form in the direction orthogonal to the maximum principal tensile stress. As explained previously, such a tensile stress is set to zero and transformed into equivalent unbalanced forces (to be distributed throughout the surrounding zones), and the adopted incremental constitutive relationships in local axes (defined with respect to the cracked plane – see the system (x', y', z') in Figure 4.14) are subsequently given the following matrix expression:

$$
\begin{bmatrix}
\Delta\sigma'_x \\
\Delta\sigma'_y \\
\Delta\sigma'_z \\
\Delta\tau'_{xy} \\
\Delta\tau'_{xz} \\
\Delta\tau'_{yz}
\end{bmatrix}
=
\begin{bmatrix}
2G + \mu & \mu & 0 & 0 & 0 & 0 \\
\mu & 2G + \mu & 0 & 0 & 0 & 0 \\
0 & 0 & 0 & 0 & 0 & 0 \\
0 & 0 & 0 & G & 0 & 0 \\
0 & 0 & 0 & 0 & \beta G & 0 \\
0 & 0 & 0 & 0 & 0 & \beta G
\end{bmatrix}
\begin{bmatrix}
\Delta\varepsilon'_x \\
\Delta\varepsilon'_y \\
\Delta\varepsilon'_z \\
\Delta\gamma'_{xy} \\
\Delta\gamma'_{xz} \\
\Delta\gamma'_{yz}
\end{bmatrix}
\qquad (4.99)
$$

where the third local axis (z') is orthogonal to the crack plane and the first and second Cartesian axes (x', y') lie on it (see Figure 4.14). (It should be noted that, as in previous instances, the above stress and strain increments are due to *subsequent* load increments and,

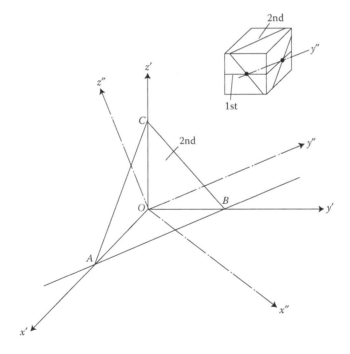

Figure 4.14 Local axes for a Gauss point with one and/or two cracks (first crack parallel to plane OAB, second crack parallel to plane ABC; it should be noted that, following the second crack, y'' defines the direction of the only remaining direct stiffness).

hence, even though they are computed by reference to the principal directions existing at cracking, the total stresses and strains in such directions are no longer principal.) The constitutive coefficients at the relevant Gauss point are set equal to the corresponding coefficients of the last uncracked D-matrix (under triaxial conditions) at that point, even though it could be envisaged, for zones of compression with one crack, that the use of biaxial stress–strain laws on planes parallel to the crack might be more rational (in a *smeared-crack* model) until a second crack forms; however, this represents too refined an approach, especially in view of the lack of knowledge on the constitutive relations for cracked concrete, as pointed out earlier. Besides, the presence of the βG terms (representing residual shear stiffness across the plane of the crack) implies that the conditions along the crack's plane are not quite plane-stress ones, but that the region still represents some sort of 'continuum'. Therefore, it is for these reasons that the coefficients $2G + \mu$, and μ, may remain unchanged as Equation 4.83 becomes Equation 4.99.

If the state of stress at a given Gauss point reaches the triaxial envelope in tension a second time, then a second crack plane is assumed to form (unless the two planes of fracture are near-coincident). This new crack plane is now orthogonal to the current maximum tensile principal stress and, consequently, is not necessarily orthogonal to the first crack plane. The combination of the two fracture planes only leaves stiffness in the direction of the intersection of both planes (direction y'' or AB in Figure 4.14). As for the one-crack case, some residual shear stiffness has to be kept in order to improve the conditioning of the cracked stiffness matrices. Explicitly, the incremental constitutive relationships adopted for this case are as follows:

$$
\begin{bmatrix} \Delta\sigma''_x \\ \Delta\sigma''_y \\ \Delta\sigma''_z \\ \Delta\tau''_{xy} \\ \Delta\tau''_{xz} \\ \Delta\tau''_{yz} \end{bmatrix} = \begin{bmatrix} 0 & 0 & 0 & 0 & 0 & 0 \\ 0 & 2G+\mu & 0 & 0 & 0 & 0 \\ 0 & 0 & 0 & 0 & 0 & 0 \\ 0 & 0 & 0 & \beta G & 0 & 0 \\ 0 & 0 & 0 & 0 & \beta G & 0 \\ 0 & 0 & 0 & 0 & 0 & \beta G \end{bmatrix} \begin{bmatrix} \Delta\varepsilon''_x \\ \Delta\varepsilon''_y \\ \Delta\varepsilon''_z \\ \Delta\gamma''_{xy} \\ \Delta\gamma''_{xz} \\ \Delta\gamma''_{yz} \end{bmatrix} \tag{4.100}
$$

where the new axes are now defined by the double primes in Figure 4.14. In line with previous discussion, $2G + \mu$ is retained at present for the second diagonal term instead of simply E (for $v = 0.2$ e.g., these two terms differ only by a factor of 1.1).

If the state of stress reaches the failure envelope either in compression (i.e., all three principal stresses – present in the global axes because of the 'continuum' implied by the βG factors – being compressive) or for a third time in tension, two options may be considered. The first option is to impose a zero constitutive matrix, that is, a complete loss of stiffness in all directions (Z3 matrix), and the second is to use a matrix with only residual values of βG for the three shear-stress components (G3 matrix). The latter option arises again for the same reason as for one-crack and two-crack cases, its chief aim being the improvement of numerical stability.

The matrices in Equations 4.99 and 4.100 are valid only in the local set of axes shown in Figure 4.14 and, hence, they require a transformation to global coordinates (see Appendix B). Therefore, all the coefficients of the constitutive D-matrix for the stress–strain relationships for one-crack and two-crack Gauss points with respect to 3-D global axes (x, y, z) will, in general, be non-zero. On the other hand, both Z3 and G3 matrices corresponding to a third cracking (or, theoretically, full compressive failures at any stage) are clearly isotropic.

The above outline completes the description of cracking at a single Gauss point. The updating of the stiffness properties in a structure when several cracks occur simultaneously at various locations upon incrementing the load will be considered in Section 4.4 as the need for adequate crack-propagation procedures and stiffness-updating strategies becomes evident.

4.3.3.4 Shear-retention factor: Its role on the conditioning of stiffness matrices and actual physical evidence

In Section 4.3.2.4, it was shown how the use of numerical integration leads to expression (4.85) for the element stiffness matrix. It was also pointed out there that the 3×3 block of coefficients of the k-matrix relating to the DOF of any pair of nodes (block $[k]^{IJ}$ in Equation 4.86) is the addition of the contributions of the n Gauss points of the finite element ($[k]^{IJ} = \sum_{i=1}^{n} [k]_i^{IJ}$). Although these blocks are not calculated in practice, such blocks could be obtained by means of expression (4.87).

Consider now the contribution of a Gauss point at which one crack has appeared. If, for convenience, it is assumed that the local set of axes (fixed to the crack plane as shown in Figure 4.14) coincide with the global set of axes, substitution of the adopted D-matrix (as given by Equation 4.99) into expression (4.87) leads to the following expression for block $[k]^{IJ}$:

$$[k]_i^{IJ} = w_i J_i \begin{bmatrix} a_1 n_1 m_1 + G(n_2 m_2 + \beta n_3 m_3) & a_2 n_1 m_2 + G n_2 m_1 & \beta G n_3 m_1 \\ a_2 n_2 m_1 + G n_1 m_2 & a_1 n_2 m_2 + G(n_1 m_1 + \beta n_3 m_3) & \beta G n_3 m_2 \\ \beta G n_1 m_3 & \beta G n_2 m_3 & \beta G(n_1 m_1 + n_2 m_2) \end{bmatrix}_i$$

(4.101)

where, for clarity, $n_1, n_2, n_3 \equiv b_{xI}, b_{yI}, b_{zI}$; $m_1, m_2, m_3 \equiv b_{xJ}, b_{yJ}, b_{zJ}$; $a_1 \equiv 2G + \mu$; $a_2 \equiv \mu$.

If, next, a zero value for β is assumed, the contribution of the Gauss point to the $[k]^{IJ}$ block of the stiffness matrix of the element reduces to the following:

$$[k]_i^{IJ} = w_i J_i \begin{bmatrix} a_1 n_1 m_1 + G n_2 m_2 & a_2 n_1 m_2 + G n_2 m_1 & 0 \\ a_2 n_2 m_1 + G n_1 m_2 & a_1 n_2 m_2 + G n_1 m_1 & 0 \\ 0 & 0 & 0 \end{bmatrix}_i$$

(4.102)

This expression shows formally that, if $\beta = 0$ and the local axes coincide with global axes, a one-crack Gauss point makes no contribution to the coefficients that relate forces and displacements in the third global direction; this is a consistent consequence of the complete loss of stiffness in the direction orthogonal to the crack plane that the D-matrix in Equation 4.99 implies. It follows, therefore, from Equations 4.101 and 4.102 that, in the absence of a shear-retention factor (SRF), and for an element with all the integration points cracked in parallel planes, the D-matrix in Equation 4.99 leads to either zero values in the diagonal of the element stiffness matrix (if the cracks are parallel to one of the global planes) or to ill-conditioning (otherwise). In practical situations, it is common to have whole concrete FEs in tensile zones so that all their integration points may crack. Admittedly, such cracks will seldom be exactly parallel. However, small pivots must be expected. This explains why the use of a non-zero β is *numerically* essential for brittle modelling regardless of its physical meaning (Gonzalez Vidosa et al. 1988). The present argument is reinforced by some eigenvalue tests of the k-matrix for a plain-concrete element, modelled by brick elements of the serendipity (20-node element, used with both under-integration ($2 \times 2 \times 2$) and full integration ($3 \times 3 \times 3$)) and Lagrangian (27-node element) types. The results are summarised in Table 4.1. (The results in Table 4.1 are, of course, independent of the adopted set of axes, as the eigenvalues of a k-matrix are invariants, since any k-matrix is a second-order tensor.) It is evident that the number of mechanisms increases as one goes from the uncracked to the cracked ($\beta = 0.1$) case and, significantly, this increase is even more drastic when β is reduced from 0.1 through 0. (It may also be of interest to note that the ratio of the mechanisms for the cracked [$\beta = 0.1$] and uncracked FEs is the same for all three FEs used and that, therefore, no conclusion about their relative performance in the present context of numerical stability can be reached.)

As mentioned in Section 4.1.3, the mechanisms of the stiffness matrix are those modes of deformation of the DOF of the FE that do not require any strain energy to occur (Cook 1981). The elastic strain energy U associated with a given set of nodal displacements d is given by

$$U = (1/2)f^T d = (1/2)d^T[k]d \tag{4.103}$$

where f is the corresponding set of nodal forces (i.e., $f = [k]d$, as in Equation 4.22). Therefore, if d is an eigenvector of the element stiffness matrix (i.e., $[k]d = ad$) of zero eigenvalue, it follows from Equation 4.103 that $U = 0$. Hence, the number of zero eigenvalues is the number of mechanisms, and their corresponding eigenvectors are the mechanisms of the element stiffness matrix. Since a solid admits three translations and three rotations without straining (rigid-body motions), it is usually considered that the k-matrix of a brick FE should have six mechanisms (six zero eigenvalues); and any mechanisms in excess of six are referred to as spurious mechanisms. As may be seen in Table 4.1, even uncracked under-integrated elements present some spurious mechanisms; nevertheless, they are often regarded as better elements than their fully integrated counterparts (Zienkiewicz et al. 1971). However, as cracking sets in, the likelihood of (additional) spurious mechanisms increases, especially if the SRF is set to zero; and it is, therefore, to minimise this increase that the brittle nature of the material modelling requires the assignment of a non-zero value to β. (Thus e.g., Hand et al. 1973 [who referred to spurious mechanisms simply as 'unstable cracked configurations'] pointed out empirically that a zero SRF could not be used. In this respect, expressions (4.101) and (4.102) give a formal rational explanation for this, irrespective of the order of integration.)

With regard to the conditioning of the k-matrix of the structure, the decay of its coefficients tends to be less important than that of the k-matrix of the cracked elements, since

Table 4.1 Number of zero eigenvalues of brick-element stiffness matrix (plain concrete, aspect ratio $1 \times 1 \times 1$)

		Quadratic 3-D element		
Number of parallel cracks	SRF	Serendipity (20 nodes) $2 \times 2 \times 2$	$3 \times 3 \times 3$	Lagrangian (27 nodes) $3 \times 3 \times 3$
None	–	12	6	6
All Gauss points	0.1	20	10	10
All Gauss points	0	36	29	36

Source: Gonzalez Vidosa F., Kotsovos M.D. and Pavlovic M.N., 1988, *Communications in Applied Numerical Methods*, 4, 799–806.

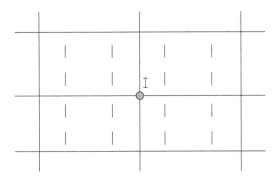

Figure 4.15 Disconnection of the horizontal DOF of node I due to (near) parallel cracking of surrounding FEs.

the presence of adjacent uncracked elements and/or steel elements improves the numerical conditioning. Nevertheless, Figure 4.15 shows an example in which the horizontal displacement of a node is rather loosely connected to the structure ($\beta \neq 0$) or fully unconnected ($\beta = 0$).

What is the value actually assigned to β? Before answering this question, it should be pointed out that the SRF is a parameter clearly associated with the notion of 'aggregate interlock'. Now, contrary to the widely held view that 'aggregate interlock' plays an important role in transferring shear across cracked areas, experimental evidence points to the opposite conclusion (Kotsovos et al. 1987). The latter has been reinforced by FE modelling itself coupled with earlier – but, originally, misinterpreted – experimental data (Kotsovos and Pavlovic 1986), and this has been presented in Kotsovos and Pavlovic (1995). Therefore, a large value of β (say between 0.5 and 1) is physically unrealistic; moreover, its effect would be to (spuriously) attract load on account of the large (fictitious) stiffness associated with it (the numerical implications of which are fully discussed in Kotsovos and Pavlovic 1995). On the other hand, it has just been shown that zero or small values (say $\beta \approx 0.01$) of SRF lead to ill-conditioning of the stiffness matrices of the cracked elements. Results of numerical experimentation [Bedard 1983, Bedard and Kotsovos 1986] suggest that, in general, analytical predictions are fairly insensitive to values of the SRF in the range $0.1 \leq \beta \leq 0.5$. As a consequence, therefore, it appears that a suitable value for β is around 0.1, which is in keeping with the negligible effect of 'aggregate interlock', on the one hand, while meeting the constraints imposed by the requirements for numerical stability, on the other.

4.3.3.5 Macro-crack closure

Macro-crack closure is considered to occur when the strain normal to the plane of a crack becomes compressive and larger than the value calculated at crack formation (as already discussed, compression is assumed to be positive). As for the case of crack formation, crack closure is numerically described through the modification of the **D**-matrix of cracked Gauss points at which crack closure occurs and the calculation of the residual forces that must be applied in the next iteration in order to re-establish the equilibrium conditions for the overall structure.

For the case of a Gauss point with a single crack (see Equation 4.99), crack closure is effected by replacing the relevant **D**-matrix terms with their values describing the behaviour of concrete at its uncracked state (see Equation 4.83), whereas the calculation of the residual forces is carried out as described in Section 4.3.2.5. Similarly, the closure of a crack at a Gauss point with two (see Equation 4.100) or three (see also Equation 4.100, but, with zero diagonal terms for all normal stress components) cracks is described by restoring material continuity only in the direction normal to the closing crack by following an approach opposite to that adopted for crack formation (i.e., the relevant diagonal terms of the relevant D-matrix are replaced with their values describing the behaviour of concrete at its state before the occurrence of the [now] closing crack).

4.3.4 Description of the reinforcing steel

On the basis of deformational and yield characteristics of the steel reinforcement (Section 3.3), the constitutive properties and failure criterion for the FEs representing the steel bars may readily be imposed. This is done by correcting the stress values (in accordance with the initial-stress technique – Section 4.3.2.5) so as to satisfy the constitutive relationships, providing updated stiffness properties (E_t) corresponding to the current state of stress, and performing failure-criterion checks based on the ultimate strain value. As explained in Chapter 3, perfect plasticity is never imposed on the final branch of the stress–strain plot even in the case of mild steels which approximate such an extreme condition (in terms of average, but not actual, strains, bearing in mind the dynamic nature of steel yielding, Pavlovic and Stevens 1981); instead, a small, but finite, value is given to E right up to the ultimate strain ε_u.

In what follows, a brief outline of the incremental relations for steel is given by reference to the more general 3-D case. Additionally, the interaction between steel and concrete is considered by reference to bond and tension stiffening, and, here, it will be shown how the present simplified treatment – in the context of smeared modelling – avoids the unjustifiably laboured efforts often adopted in respect to these phenomenological parameters.

4.3.4.1 Incremental relations for the three-node uniaxial element

For a curved-line element oriented with respect to the (x, y, z) set of Cartesian global coordinates, let s represent the length measured along its axis, so that $t = (dx/ds, dy/ds, dz/ds)$ is then the tangent vector. On the assumption that the displacements are kept small, the longitudinal displacement, u_s, can be obtained from the scalar product of the tangent vector and the vector of global displacements $u = (u_x, u_y, u_z)$

$$u_s = t\, u \tag{4.104}$$

Since only the axial stiffness of the reinforcement is deemed to be of significance, the single strain component of interest is the direct strain along the element. This is simply the equivalent of Equation 4.72 (but, now, as the total derivative)

$$\varepsilon_s = du_s/ds \tag{4.105}$$

Global displacements are still related to nodal displacements as in expression (4.71) (this being a generic type of relation for any number of nodes with three DOF). Then, by combining expressions (4.104), (4.105) and (4.71), the incremental strain–nodal displacement relationship is obtained, that is, $\Delta\varepsilon_s = [b]\Delta d$ (see Equation 4.78, but with $[B] \equiv [b]$, the latter serving as a reminder that the problem under consideration refers to a (curved) line element). Thus

$$\Delta\varepsilon_s = [\cdots(dx/ds)(dN_I/ds) \quad (dy/ds)(dN_I/ds) \quad (dz/ds)(dN_I/ds)\cdots] \begin{bmatrix} \vdots \\ \Delta d_{xI} \\ \Delta d_{yI} \\ \Delta d_{zI} \\ \vdots \end{bmatrix} \qquad (4.106)$$

In the spirit of previous work relating to concrete, the coefficients of $[b]$ in Equation 4.106 are derived by making use of the isoparametric formulation. Thus, it should be recalled that $x_i = N_I x_{iI}$, where x_i and x_{iI} are the ith global coordinate and the ith global coordinate of the Ith node, respectively. Therefore, $dx_i/ds = x_{iI} N_I/ds$. In addition, it is also necessary to calculate $ds/d\xi$ in order to express derivatives with respect to the length coordinates in terms of derivatives with respect to the non-dimensional coordinate ξ (i.e., $(dN_I/d\xi)(d\xi/ds)$). The derivative $d\xi/ds$ follows from the following expressions:

$$ds/d\xi = \left[(dx/d\xi)^2 + (dy/d\xi)^2 + (dz/d\xi)^2\right]^{(1/2)} \qquad (4.107)$$

and hence

$$ds/d\xi = \left\{[(x_I)(dN_I/d\xi)]^2 + [(y_I)(dN_I/d\xi)]^2 + [(z_I)(dN_I/d\xi)]^2\right\}^{(1/2)} \qquad (4.108)$$

Increments of stresses and strains are related by Young's modulus

$$\Delta\sigma_s = E_t\Delta\varepsilon_s \qquad (4.109)$$

while the incremental stiffness matrix is given by the following integral, where A_s is the area of the cross section of the bar element

$$[k] = \int([b]^T E_t[b])(A_s)(ds/d\xi)d\xi \qquad (4.110)$$

Finally, stress conditions give rise to residual nodal forces which are calculated as follows:

$$\Delta f_r = -\int([b]^T \Delta\sigma_r)(A_s)(ds/d\xi)d\xi \qquad (4.111)$$

Expressions (4.110) and (4.111) are calculated by means of Gaussian integration.

4.3.4.2 Concrete–steel interaction

This comprises two main aspects.

Bond. The question of bond modelling between steel and concrete is often given prominence in the course of development of an FE model for structural concrete. Symptomatic of this attitude are the early attempts to analyse concrete structures by FEs, in which spring linkages were introduced to mimic bond slip through linear (Ngo and Scordelis 1967) or non-linear (Nilson 1968) laws. Although many subsequent investigations were based on the simplifying assumption of perfect bond (e.g., Philips and Zienkiewicz 1976), much effort is still being expended in implementing bond-slip laws. In the past, such 'laws' have ranged

from arbitrarily assigned 'spring' constants to 'theoretical' relationships, the latter sometimes incorporating a certain degree of experimental feedback from the scarce, limited and
often dubious data currently available.

The present model adopts the perfect-bond assumption, implicit in the coincidence, throughout the analysis, of the steel-element nodes with the nodes of the corresponding (adjacent)
concrete-element(s) edges. (Since all elements are of the second order and a unique parabolic
interpolation can be adjusted between the three nodes of a given edge, the coincidence between
steel and concrete-element edges guarantees the same displacements [i.e., perfect bond] for concrete and steel at the same location.) Three arguments may be adduced for such an approach.

Firstly, the notion of perfect bond is compatible with the thinking behind the smeared-
crack model, in the sense that detailed description of localised effects is avoided. In this
respect, it is important to stress that the earlier mention of bond loss between steel and concrete in the vicinity of a crack (see Section 4.3.3.1 and Figure 4.12b) does not contradict the
perfect-bond assumption since the smeared-crack approach spreads the effect of cracking to
such an extent that integration points in bar elements represent, generally, bar lengths which
usually far exceed the localised regions where bond slip occurs, thus precluding a minute
account of the steel–concrete interaction.

Secondly, reliable experimental information of wide applicability on bond-slip characteristics is, as suggested above, not available. In fact, experiments attempting to determine
bond–stress against slip–displacement curves exhibit, typically, a large scatter of data as
regards both individual investigations (see e.g., Edwards and Yannopoulos 1978) and the
overall picture that emerges from a comparison of characteristics proposed by various
authors (see Figure 4.16 taken from Gonzalez Vidosa's [1989] adaptation of the relevant
data in the ASCE Task Committee's state-of-the-art report on the FE analysis of RC structures [ASCE Task Committee on Finite Element Analysis of Reinforced Concrete Structures
1982]). It is not surprising, therefore, that theoretical bond–slip curves used for non-linear
analysis often differ significantly from each other, as may be seen, for instance, by reference
to the characteristics adopted by Nilson (1968) and Labib and Edwards (1978), depicted in
Figure 4.17a and b, respectively. When such variability and the fact that slip transverse to
the reinforcement is governed by different characteristics are considered, it is evident that

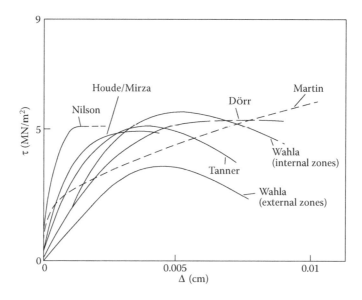

Figure 4.16 Some of the proposed bond–stress versus slip–displacements.

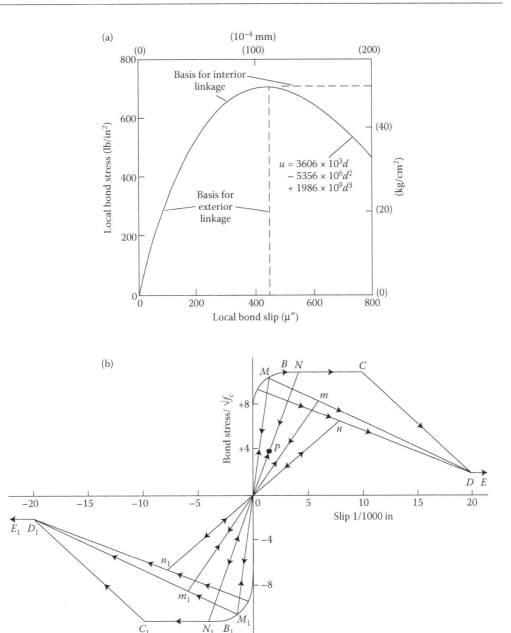

$$u = 3606 \times 10^3 d - 5356 \times 10^6 d^2 + 1986 \times 10^9 d^3$$

Figure 4.17 Some of bond–slip curves used in non-linear FE analysis: (a) Nilson (1968), (b) Labib and Edwards (1978). (From Nilson A. H., 1968, *ACI Journal*, 65, 757–766; Labib F. and Edwards A. D., 1978, *Proceedings of the Institution of Civil Engineers, Part 2*, 65, 53–70.)

any refinement implicit in the departure from the perfect-bond condition cannot be accepted with a sufficient degree of confidence.

Thirdly, and most importantly, there is the actual value of the bond stress itself. On the basis of Figure 4.16 (and, of course, Figure 4.17a), an order of magnitude for the maximum bond stress of around 5 MPa appears to be relevant. For the failure envelope in tension (depicted in Figure 3.27) to reach such a value, the concrete strength f_c must exceed, approximately, the 80 MPa mark. It is evident, therefore, that the concrete in the vicinity of the concrete–steel interface will usually have cracked in tension before the maximum bond stress could be developed. Similar conclusions may be reached by reference to the characteristic of Figure 4.17b, which suggests that no slip occurs below a bond stress of $8(f_c)^{(1/2)}$ (with f_c in lb/in²) or, equivalently, $8(f_c/12)^{(1/2)}$ (with f_c in N/mm²) (Labib 1976). This implies that, for slip to take place, the bond stress must be around twice the tensile strength of concretes with an f_c of around 20–30 MPa; and, once again, parity between maximum bond stress and tensile strength is attained only in the high-strength concrete range of $f_c \approx 80$ MPa and above. Hence, it may be concluded that, for most structural concretes, tension failure occurs in the reinforcement region before the development of maximum bond stresses there, so that the assumption of perfect bond in the analysis is justified on the basis that there is no need to model concrete once macro-cracking has caused it to lose its stiffness. (In fact, since there is no guarantee that the bond-slip models in Figures 4.16 and 4.17 necessarily hold for $f_c > 80$ MPa, the present argument justifying perfect bond might even be extended to encompass the upper range of high-strength concretes.)

Tension stiffening. In the present model, once cracking in the vicinity of, and orthogonal to, the reinforcement takes place, the stiffness of the concrete is set to zero, as for the case of plain concrete. Such a simplification ignores the so-called 'tension stiffening' effect, which may be illustrated by reference to a steel bar encased in concrete and subjected to tension, as depicted schematically in Figure 4.18. The stiffness of such an arrangement exceeds that

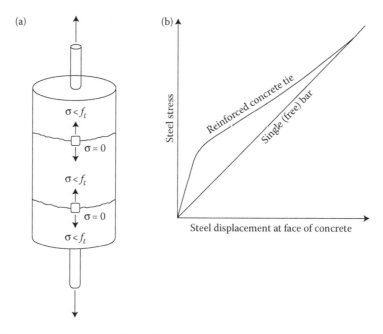

Figure 4.18 RC tie subject to tension: (a) steel bar encased in concrete with the latter continuing to carry load away from macro-cracks; (b) 'tension stiffening' represented by the difference in stiffness of the two systems, namely the composite structure and the bare bar, respectively.

of the steel alone even after macro-cracking, which points to the fact that the concrete in between cracks (where the stresses σ are below f_t, the tensile strength) contributes to the overall stiffness of the system, as the tensile stresses are transferred from steel to concrete by bond action. Clearly, with progressive cracking and bond destruction, the stiffness of the RC tie tends towards that of the bare bar. Specific plots following the generic trend of the one shown in Figure 4.18b may be found in a number of references, some of them based on analytical modelling (e.g., the short eccentric tensile member analysed in Nilson 1968), others derived from actual experiments (e.g., the tension test described in Kollegger and Mehlhorn 1987).

The neglect of tension stiffening is most unlikely to influence ultimate-load predictions (Lin and Scordelis 1975, Greunen 1979, Gonzalez Vidosa 1989), especially for structures exhibiting ductile behaviour, for which the 'tie' effect has nearly been reduced to that of the steel acting alone (see Figure 4.18b). Deflection predictions might be more sensitive to such a simplification, but the complexity that would result if concrete elements adjacent to steel bars were treated differently from 'plain concrete' elements elsewhere (with consequent problems of mesh (re)definition) would not justify the extra refinement. Therefore, while recognising that local stress redistributions could be affected by the lack of allowance for tension stiffening, this simplification appears to be justified on the grounds of both ease of modelling and the requisite accuracy of predictions (as will be seen in subsequent chapters).

Finally, a small but pertinent point on the terminological confusion that is often encountered in the literature seems called for. Models for the tensile characteristics of plain concrete which do not recognise the brittle nature of the material but use descending branches of the type shown in Figure 4.19 (which, if at all present in actual material tests, are likely to be caused by the interaction between testing device and specimen, as pointed out in Section 2.2) sometimes have these descending (i.e., 'strain softening') characteristics referred to as 'tension stiffening'. The reason for this appears to be that, historically, the

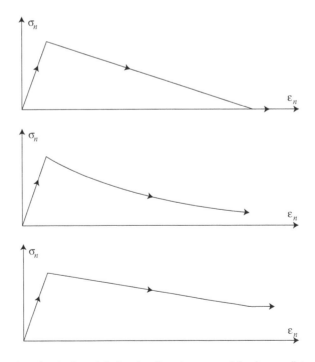

Figure 4.19 Typical 'strain softening' models for the direction normal (orthogonal) to a crack plane.

concepts associated with 'tension stiffening' proper (as illustrated in Figure 4.18) were also gradually – and, it would seem, largely imperceptibly – adopted for plain concrete. The possible confusion – or, at least, terminological interchangeability – of what is implied by 'tension stiffening' should be borne in mind so as to distinguish between two quite distinct phenomenological notions.

4.3.5 Overall non-linear strategy

The formulations described in Sections 4.3.2 through 4.3.4 formed the basis for the development of the non-linear strategy summarised in flowchart form as described in Figure 4.20. The important feature of the strategy is that it comprises the following three distinct stages:

- Initially the k-matrix is updated by using the initial material properties in the D-matrix (see Equation 4.83) and, for each Gauss point, it is established whether concrete and/or steel are in a state of loading or unloading (as defined in Section 3.1.2.2). Once this is established, this state remains unchanged to the end of the iterative procedure of each load increment.
- The second stage makes use of the NR iterative method and allows only crack closure (see Section 4.3.3.5), with the iterative procedure continuing until all cracks due for

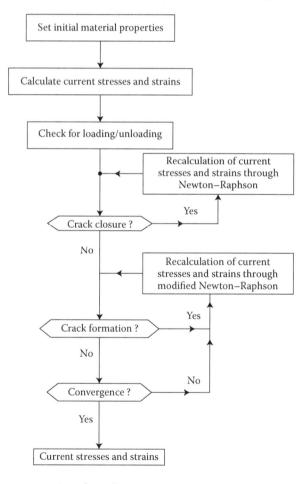

Figure 4.20 Schematic representation of non-linear strategy.

closure do, in fact, close. Within thin stage, only one crack closure is allowed to occur per iteration for the reasons explained in the next section. Following crack closure the relevant *D*-matrix is updated and the residual forces calculated and added to the current load increment.

- Finally, within the third stage, the iterative procedure makes use of a modified version of the NR method, in that the *D*-matrices are updated as soon as 'strong' non-linearities occur, and allows only crack formation, with the iterative procedure continuing until either a convergent solution is obtained or the run stops due to ill-conditioning. Here, it should be pointed out that, as discussed in the next section, cracks are allowed to form in a small (predetermined) number within each iteration.

Dividing the above iterative procedure into three distinct stages has been found to prevent numerical instabilities which are otherwise inevitable due to the built up of large residual forces and the rapid development of spurious mechanisms (Kotsovos and Spiliopoulos 1988a,b).

4.4 MATERIAL AND PROCEDURAL FACTORS INFLUENCING FE PREDICTIONS

Following its appearance in the mid-1950s (Clough 1980), the FE method has become the most widely used numerical technique in engineering analysis. A decade after its inception, this powerful computational tool began to be applied to the non-linear analysis of concrete structures (Ngo and Scordelis 1967, Rashid 1968). However, despite the relatively rapid proliferation of works in this area by the early 1980s (ASCE 1982), it was quickly recognised that the use of non-linear FE packages for structural concrete was seldom adopted in design (ASCE Task Committee on Finite Element Analysis of Reinforced Concrete Structures 1981), in spite of the gradual – but ever widening – acceptance of the limit-state philosophy in codes of practice; and, now, more than three decades later, the attitude of the practising engineer remains largely unchanged. The main reason for this appears to be the general lack of confidence in the overall reliability of the predictions of the various FE models proposed for concrete, as most of these are notoriously deficient with regard to evidence of their applicability, namely the assurance with which they may be used to analyse arbitrary structural forms rather than the specialised structure(s) for which they have been devised and/or against which their predictions were 'tuned' or simply tested. Such an outcome stems from three major factors. Firstly, there is often scant attention given to the true response of concrete at the material level, either because of little regard to actual experimental data or on account of dubious interpretation of experimental results in an attempt to fit preconceived constitutive theories. Secondly, most of the numerical packages proposed to date tend to be lacking in a thorough study of their *objectivity*, namely of the effect on predictions of changes in the various parameters that made up the basic model; while some of these parameters have been discussed already, all of them will be enumerated in this section. Thirdly, many of the existing packages are also notoriously deficient with regard to their *generality*, that is, their applicability to arbitrary structural forms rather than to the specialised structure(s) for which they have been devised and/or against which their predictions were tested: that the present FE model does possess such a feature will become apparent through a study of the wide range of problems described in Chapters 5, 7 and 8.

In order to overcome the first of the above shortcomings, special care was taken to incorporate into the proposed FE model a realistic description of material behaviour (see Chapter 3) based purely on empirical data obtained through the testing of concrete under generalised, triaxial and – above all – *definable* stress conditions (see Chapter 2); and a signal advantage of the material model is the fact that both failure and constitutive laws require only a single parameter (f_c) for their full definition which, in addition to being readily obtainable, presents less scatter than any other material characteristic. In what follows, before attempting – in Chapters 5, 7 and 8 – to demonstrate the generality of the proposed model, attention is focused on ascertaining the key components or parameters that make up the basic model and have a dominant effect on the analysis predictions.

In the evaluation of the objectivity of the model, those components or parameters of the analysis may be distinguished on which there is a broad agreement from others on which there is no obvious general consensus at this stage of development (see e.g., Gonzalez Vidosa 1989; Gonzalez Vidosa 1989 et al. 1991). In the first category, the following can be included: the two well-established techniques of crack modelling, namely smeared and discrete cracking; the use of residual forces so as to follow the non-linear constitutive relations while adhering to the equilibrium requirements of the structure; the strength envelopes for concrete under multiaxial stress conditions; and the non-linear stress–strain relations *up to about failure*. Clearly, these well-established model components are of a basic or 'primary' nature on account of their cardinal significance to successful numerical analysis. By contrast, many of the components making up the second category of parameters, which are often still the object of widely differing views, sometimes tend to be regarded as being of secondary importance although some of them may, in fact, play a key role in certain situations. In any case, a rigorous study of a model's objectivity requires that careful consideration should be given to these 'secondary' parameters, although their large number precludes an exhaustive investigation of each one of them. Such 'secondary' model components may conveniently be subdivided into material parameters (post-peak characteristics of the stress–strain relations, 'aggregate interlock', 'tension stiffening', 'dowel action', bond between steel and concrete), and parameters associated with procedural effects stemming directly from the numerical scheme adopted (iterative method, number of cracks allowed to form or close per iteration and/or cracking sequence followed, FEs chosen, numerical-integration rule used, mesh definition/size, loading path).

The presentation of the subject matter in hand, namely the 'secondary' parameters, will follow the above subdivision between 'material' and 'procedural' factors. Clearly, while some of these factors might sensibly be ascertained on the basis of existing evidence, conclusions regarding others have been reached with reasonable certainty through a set of numerical experiments (as described in Kotsovos and Pavlovic 1995).

4.4.1 Material parameters

Besides the key source of objectivity of the material model (i.e., the choice of f_c to define behaviour up to failure and failure itself), 'strain softening' and 'aggregate interlock' appear to be the two parameters most likely to affect predictions. The complete lack of post-peak characteristics is in keeping with both a proper interpretation of available meaningful experimental evidence (as discussed at length in Chapter 2), and the fact that the peak stress in compression is never actually reached in a structure (as explained in Sections 2.4 and 2.5, and also borne out by all the case studies gathered in this book). Thus, even if any hypothetical 'strain softening' relations in compression were to be included in the analysis, they would make no difference to the results, and, hence, the question of their objectivity need not arise at all. Descending branches in tension, on the other hand, which are also discarded in the present model (as opposed to many other packages), not only do not have a reliable experimental basis either, but may even lead to numerical difficulties (Kotsovos and Pavlovic 1995). The lack of any strain-softening branches is, of course, a consequence of the brittle-material model followed. Another consequence is the notion that cracked concrete makes only a minor contribution to overall structural strength and, in this respect, the low value of SRF adopted accords with the view (espoused in Section 4.3.3.4) that 'aggregate interlock' plays an insignificant role in the load-carrying capacity of a member. That the chosen β value(s) is (are) capable of reliable predictions while, at the same time, providing the necessary degree of numerical stability has been justified in Kotsovos and Pavlovic (1995).

There are three factors that ought to be considered in the modelling of those regions of the structure where reinforcement is located, and of these the first concerns the so-called 'dowel action' of the bars themselves. Here, the sole stiffness component along the axis of the reinforcement is compatible with the view that the effect of any 'dowel action' is insignificant, as described elsewhere (Kotsovos et al. 1987, Jelic et al. 1999). The second factor is the question of the bond between steel bars and surrounding concrete, the justification for perfect (full) bond having been given already (Section 4.3.4.2). Finally, there is the 'tension stiffening' factor, an effect that is not considered in the present model which, in accordance with the smeared-crack philosophy, assumes all the material in the region of the cracks to have lost its stiffness in the direction orthogonal to the cracks. The arguments for such a simplification have also been outlined in Section 4.3.4.2, while recognising the fact that this might have some bearing on the accuracy of computed deformations by underestimating the stiffness of the relevant parts of the structure.

4.4.2 Procedural parameters

As regards the six procedural factors listed previously, past experience has shown that the use of arbitrary non-linear algorithms in the FE analysis of concrete structures is, generally, not feasible, and that specific strategies must be evolved which allow for the special characteristics of the material model. Therefore, the optimum choice of non-linear strategy would depend, for example, on whether brittle or strain-softening features were adopted and this is closely related to the decision on what cracking sequence and/or number of simultaneous cracks is to be adopted. If the material modelling includes post-cracking strain softening, stress release and redistribution take place gradually as cracks open. Thus, numerical procedures associated with such material modelling tend to be more stable than those involving the brittle material model, with which the sudden and complete release of tensile stresses as a result of cracking can by itself lead to divergence of the numerical algorithm. In this respect, account must be taken of the fact that the residual forces caused by a 'single' crack may be of the order of magnitude of the external load step being applied. Hence, if all prospective cracks are allowed to occur (or close) at once (the 'total-crack approach' [TCA]), numerical divergence is very likely to be affected. However, the strategy adopted is one by which cracks are allowed to occur (or close) one by one or in small numbers (the 'single-crack approach' [SCA]), while other prospective new cracks stand by until the numerical process is Stabilised; as a result, it causes a gradual release of residual forces and, therefore, improves the stability of the numerical procedure.

In order to further improve numerical stability, the SCA approach was combined with NR technique, but the tangent stiffness updating was implemented only for those elements that contained new cracks forming (or closing) in the course of the particular iteration of that load step (as well as at steel locations where yielding takes place). This dealt successfully with the high non-linearity associated with crack formation (or closure) and yielding steel, and avoided updating of the stiffness matrix on the basis of the remaining elements (uncracked or cracked previously) as the non-linearities in these are of a mild nature. Once the crack was stabilised, the iterations proceeded with a constant stiffness matrix until either a new crack appeared (or closed) (in which case the procedure updated the stiffness of the relevant element) or the next load step was reached (and then global updating took place).

The use of the smeared-crack modelling of the SCA, which allows the implementation of critical cracks to occur (or close) first, is also compatible with the physical nature of crack propagation (or closure) within a structure, where stress redistributions subsequent to crack initiation affect the stress conditions in neighbouring zones. In principle, if the load steps were to be infinitesimally small, sequential single cracking would generally result unless different locations were to attain the same level of (maximum) stress simultaneously. In view of the above arguments, it would seem that the allowance of all potential cracks to occur at once does not model adequately (or, at least, 'rationally') the actual nature of structural fracture.

The opinion has been expressed that the SCA is dangerous from a *numerical* standpoint in that the results can heavily depend on the iteration procedure adopted (de Borst in Gonzalez Vidosa et al. 1989). However, this may also be true of any other procedure, namely those procedures where several cracks are allowed to form simultaneously in a given iteration. In fact, it may be readily argued that, since the formation of each crack usually gives rise to large residual forces, the SCA is, in general, likely to be more stable (numerically) than other procedures. After all, the SCA is simply a means of reducing the risk of numerical instability. This is quite acceptable, since, in non-linear analysis, it is permissible to iterate in any manner, provided that convergence is achieved with all constraints (equilibrium, compatibility, constitutive relations, failure criteria, etc.) satisfied at the end of the algorithm process adopted. Finally, despite claims to the contrary (Philips in Gonzalez Vidosa et al. 1992), the SCA does not lack objectivity since it always allows a crack to occur at the most critical stress-combination location independently of load-step size (allowance of more than one crack is simply an economic option in the case of a large number of Gauss points). Furthermore, the load level in actual tests is increased until unloading occurs as a result of crack propagation, that is, until failure (cracking) conditions are met at the most highly stressed part of the structure; therefore, the SCA appears to be quite realistic even from a phenomenological point of view.

The general suitability of the various quadratic elements chosen is well established for linear problems. In the case of non-linear analysis of concrete structures, the second-order elements of the serendipity family have also been widely used (in addition to the constant-strain triangle) by investigators in this field (see e.g., Gonzalez Vidosa 1989). However, the performance of the latter isoparametric-element types may be influenced by the integration rule adopted. As pointed out in Section 4.1.3, major economies in computing time result from the use of under-integration, and predictions may even be improved over those of their more formal fully integrated counterparts. It is for these reasons that the analyses of 2 D problems have been carried out by means of under-integrated elements (Bedard 1983). The use of under-integration has been justified elsewhere (Gonzalez Vidosa 1989, Kotsovos and Pavlovic 1995), where the objectivity of predictions have been looked at by reference to the numerical-integration rule employed for the serendipity elements; it has been shown there that the warnings against under-integration in non-linear FE modelling of structural concrete that are sometimes raised by analysts need not cause undue concern. More recently, although the Lagrangian element requires a larger computational effort than the 20-node FE, its use has been adopted not only because it may improve 20-node predictions by reducing round-off errors (Fried 1971), but also because it allows a more flexible distribution of bars due to its mid-face and centre nodes. This is particularly the case for the FE modelling of structural-concrete members designed in accordance with current codes for the design of earthquake-resistant structures. As will be seen in subsequent chapters, such structural members are characterised by very dense reinforcement arrangements.

Mesh configuration and FE size are potentially important factors in both linear and non-linear problems, as they govern the degree of approximation to the continuum. For linear stress analysis, it is usually agreed that, as the mesh is refined, accuracy improves: such mesh refinement is especially recommended in zones where localised effects take place (e.g., load concentrations, sudden changes in geometry, etc.); and although, in principle, accuracy can increase only up to an optimum number of DOF, after which a substantial decay in accuracy follows (Fried 1971), set by the practical limitations connected with unavoidable round-off errors in the course of solution of large equation sets, present-day computing power is usually sufficiently large to ensure that quite fine meshes can be used even for the more complex structural problems.

In the case of non-linear problems in which failure is associated with cracking, the latter can induce deceptive effects in those local regions that are either subject to stress concentrations/singularities or contain previous cracks. For such instances, the finer the mesh, the sooner crack initiation occurs for a given value of applied load so that, in principle, FE analysis involving cracking based on a strength criterion would, in the limit of mesh refinement, converge – incorrectly of course – towards a zero-load prediction (Bazant and Oh 1963; Bedard 1983). Even for finite (but, 'sufficiently' small) mesh sizes, stress redistributions induced by early cracking in some regions can affect numerical stability locally, precipitating premature structural collapse as successive iterations spread these local disturbances throughout the structure. One possible way of averting such difficulties is to stiffen the mesh locally as, in fact, engineers and/or experimentalists actually do when designing suitable construction details at support or anchorage zones. A more general and elegant approach to avoiding mesh-size dependency, however, is simply to adopt a 'coarse' mesh, thus bypassing the possible predominance of localised effects over the overall behaviour of the structure which, after all, is of primary interest. This notion might be difficult to accept from an analyst's point of view but will readily be acceptable to an engineer. Moreover, it is in keeping with the thinking permeating this book, namely a realistic modelling of structural concrete on the basis of existing and reliable material data. Bearing in mind that the experimental conditions under which the constitutive relations (at the engineering level) were derived involved measurements taken from strain gauges that were approximately two to three times larger than the maximum aggregate size used in the concrete mix, it is clear that a lower-bound limit to the size of an FE is provided by such a gauge length (see Section 2.1.1). This also ensures that the assumptions of homogeneity and isotropy of the material might safely be invoked; otherwise, small FE sizes would require different constitutive relations for the aggregate particles and cement paste, involving also separate discretisations of these, as well as adhesion and/or bond characteristics between the two separate material constituents. Were such complex (microscopic) relationships even currently available, it is evident that the adoption of mesh sizes lower than the suggested 'two-to-three times the aggregate size' would quickly limit the range of structures which it would be feasible to model from a computing perspective. The preceding philosophy also has implications for the generally accepted notion of convergence to the correct solution as the mesh is refined: in the present context, this is not considered to be meaningful, as the material characteristics stem from a cylinder or cube test which is sufficiently large to enable the *average* material properties of what is really a heterogeneous mix to be sensibly derived, interpreted, and eventually incorporated into (large-scale) structural analysis.

While the validity of the material data establishes the lower bound of FE dimensions, the upper bound that would be needed for adequate accuracy remains largely a somewhat subjective choice to be based mainly on the analyst's engineering judgement. As Bedard (1983) suggests, the upper bound should be compatible with the maximum size that is considered representative of structural behaviour. He goes on to give two illustrative examples, namely a plain-concrete prism compressed concentrically by strip loading (requiring a minimum of two 'column' elements on either side of the (symmetrical) centreline so as to allow for a reasonable stress-gradient description), and an RC beam under two-point loading (demanding at least two to three 'layers' of elements above the reinforcement in order to detect the stress distribution throughout the compressive zone and depth). Bedard also points out the desirability of ensuring that the 'shape ratio' (i.e., the ratio of longest to shortest sides of an element) does not exceed about 3 (although 'shape ratios' above this value have also been used in some problems, without apparent difficulties), and argues that as regular a grid as possible throughout the structure should generally be aimed at on account of both accuracy and ease of automatic mesh generation (Bedard 1983).

Finally, among the procedural factors that need to be considered, there remains the effect of loading path on predictions. In discussing this parameter, it is useful to distinguish between load increments that precede first cracking and those that are applied subsequently. In principle, a single load increment should suffice up to the formation of the first crack(s) in the structure owing to the fairly 'mild' non-linear nature of the (isotropic) constitutive relationships corresponding to the ascending stress–strain branches, and also because the failure envelope is not stress-path dependent (see Section 2.3.3.2). Therefore, it is clear that the onset of cracking will not be affected by the loading path and that a relatively small number of iterations will suffice for convergence to the correct solution (corresponding to initial cracking) in a single load step. An estimate of the size of load step required could be provided by a preliminary linear computer run that will indicate the location of the 'critical' tensile stress and the magnitude of the applied load at which this would occur.

That recourse to incremental techniques ought to be made after cracking has begun is apparent on account of the 'strong' nature of the non-linearities associated with sudden crack formations. The sudden appearance of many simultaneous cracks could produce results that depend on the size of the load step used and, furthermore, could easily give rise to numerical instability. Ideally, extremely small load steps, which would ensure that each time load is incremented a single crack would form (or, in special cases, a small number of cracks at different locations where the failure envelope happens to have been reached simultaneously) would remove most of the uncertainties associated with possible load-path dependency. Such an option is clearly unrealistic, however, with regard to the computational effort required. Instead, relatively few load steps are applied and, in the course of each, several cracks usually take place; it is for this reason that care must be exercised in following a suitable crack-propagation strategy within each load step and, also, iteration.

In practice, it is found convenient to adopt a constant size of load step throughout the entire loading history. The load-increment size is usually taken to be between 2% and 10% of the estimated ultimate load, which ensures that only 10–50 load steps are needed for the case of monotonic loading, while adequate accuracy is achieved for all practical purposes.

4.4.3 Tentative recommendations

As regards the material parameters, complete lack of 'strain softening', 'aggregate interlock' and 'dowel action' are in keeping with a proper interpretation of available meaningful experimental evidence as discussed at length in Chapters 2 and in preceding sections of the present chapter. Such evidence shows that the post-peak behaviour of concrete is characterised by a complete and immediate loss of load-carrying capacity, whereas the notions of 'aggregate interlock' and 'dowel action' also contrast the mechanism of crack extension which does not involve any 'shearing' movement of the crack's face that may result to 'aggregate interlock' and 'dowel action'; a crack extends in the direction of the maximum principal compressive stress and opens in the orthogonal direction. On the other hand, the condition of 'perfect bond' adopted is compatible with, not only the thinking behind the smeared-crack model, but also with the large variability of experimentally established bond–slip curves, which precludes any refinement implicit in the departure from the perfect-bond condition. Finally, 'tension stiffening' is ignored on the grounds that, although its effect may have an effect on deflections under service loading conditions, the complexity that would result if concrete elements adjacent to steel were treated differently from 'plain-concrete' elements elsewhere would not justify the extra refinement.

The parameters associated with procedural effects stemming directly from the numerical scheme adopted have been ascertained on the basis of existing evidence, a part of which has resulted from numerical experiments. With regard to the procedure for the modelling of crack propagation, although the results obtained from plain-concrete prisms indicate that the predictions of the 3-D model is rather insensitive to the adopted procedure, the SCA is recommended as a general strategy. With respect to the type of the FE, the use of both under-integrated 20-node and Lagrangian elements appears to lead to practical approximations to ultimate load, crack pattern and load–deflection curves. With regard to the iterative method, the comparison of computer runs for RC beams failing in shear shows that the results may be sensitive to adopted iterative technique, on account of lack of convergence (modified NR method) or early divergence due to the propagation of spurious mechanisms (mixed NR method); the immediate update of strong but – usually very localised – non-linearities ('selective' NR method) is definitely recommended. The retention of the residual βG (shear) stiffness, even when the third cracking takes place, is desirable from a numerical viewpoint. With regard to mesh-size dependency, it is recommended to adopt a 'coarse' mesh; the smaller size of the Gauss point region should be taken equal to approximately two to three times larger than the maximum aggregate size used in the concrete mix. Finally, it is found convenient to adopt a constant size of load step throughout the entire loading history. The load-increment size is usually taken to be between 2% and 10% of the estimated ultimate load, which ensures that only 10–50 load steps are needed for the case of monotonic loading, while adequate accuracy is achieved for all practical purposes.

4.5 A BRIEF OUTLINE OF THE SMEARED-MODEL PACKAGE

This section contains a brief description of the main features of the non-linear FE program developed for the analysis of structural concrete. The outline is drawn from the much more detailed presentation of the 3-D model by Gonzalez Vidosa (1989). The non-linear features of the analysis have been implemented within a standard linear package (FINEL) (Hitchings 1972, Hitchings 1980). In addition to the smeared analysis under static loading thus incorporated into the modular structure of FINEL, two post-processing programs were developed for plotting, and these constitute an important part of the non-linear package as will become apparent when the deflected shapes and crack patterns of the various problems to be discussed in subsequent chapters are presented. The present section, however, is concerned with the iterative flow of modules of the smeared analysis, describing also the main modifications introduced in the linear stress modules so as to account for material non-linearities, number of cracks allowed per iteration, and NR updating strategies.

The linear stress analysis of FINEL provides the backbone for the smeared analysis, as is evident from Figure 4.21, which shows the flow of modules of the non-linear strategy. A typical input file required to run an actual non-linear example appears in Figure 4.22: the problem in question represents a plain-concrete prism acted upon concentrically by strip loading across half its end faces. It may be seen that options can be input by using the Jn and Rn register facilities (e.g., LET J4 020202 specifies $2 \times 2 \times 2$ Gauss points). By reference to Figure 4.21, it is clear that, once an iteration is performed, the program returns either

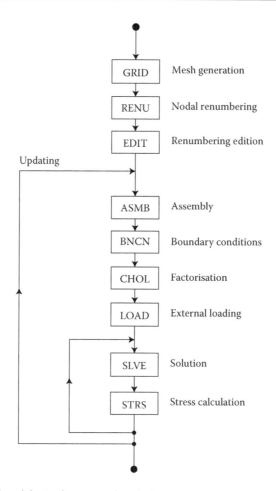

Figure 4.21 Flowchart of modules in the smeared analysis.

to the assembly module (updating iteration) or to the solution module (non-updating iteration) depending on the NR strategy adopted. Updating iterations require reformulation of D-matrices (and hence also of k-matrices) while non-updating iterations follow a constant-stiffness strategy. The iterative flow of modules is achieved by executive commands in the non-linear input file (see Figure 4.22). The input data required for the first iteration goes up to the first END STRS card, while the next three sets of cards are for a non-updating iteration, an updating iteration and a new load-step increment respectively. Before the modifications of the linear modules through the editing of their executive subroutines are discussed, it should be pointed out that some additional storing facilities had to be implemented in order to save intermediate results which are required at different stages of the analysis. In this way, the following could be dealt with data in respect of the analysis (such as the NR strategy, convergence variables, number of cracks allowed per iteration, number of new cracks); data relating to the FE and its Gauss points (i.e., nodal numbering, strains, stresses, cracking information); and storage of some long arrays (e.g., increments of applied nodal forces, increments of nodal displacements, total nodal displacements, total reactions, residual nodal forces, various statistics [especially cracking-information arrays] for each load step, etc.). Also, and as mentioned earlier, two programs were written for plotting purposes (it should be noted that any input file of these two is an output file of a previously executed

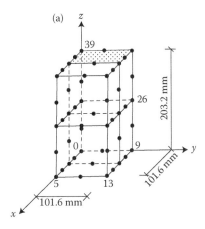

(b) Analysis stress-smeared

```
*
* concrete prism under a strip load
*
LET J4        020202    ; 2 × 2 × 2 Gauss points;
LET J5        000202    ; SRF D-mtx, 2 cracks/iter, "single crack";
LET J6        035001    ; 3 iter/up, 50 load steps, MIXNR;
LET R0           0.1    ; 0.1 MPa: maximum residual stress (convergence);
LET R1          0.01    ; maximum relative residual stress (convergence);
LET R2         100.     ; maximum residual force (divergence);
LET R3         0.005    ; maximum residual force (convergence);
LET R4           0.1    ; Shear retention factor;
*
* mesh generation
*
REGI COOR       0.    0.      0.        101.6   0.      0.
                0.  101.6     0.        101.6  101.6    0.
                0.    0.    203.2       101.6   0.    203.2
                0.  101.6  203.2        101.6  101.6  203.2

ELEM HX20
MATE STIF   20000.    0.2
MATE STRENGTH 26.9
REGI SHEX   1   2   3   4   5   6   7   8     2   1   2
*
* first iteration
*
COOR GENE
FIXE FACE   1   2   1      1    2    6      3
FIXE FACE   1   4   1      1    2   14      2
FIXE FACE   1   3   2      1    6   14      1
FIXE FACE   3   3   1     39   40   44      1
FIXE FACE   3   3   1     39   40   44      2
LOAD UDSL   −1.    3     3    3    1     39   40  44
END STRS
*
* non-updating iteration
*
RE-E SLVE
END STRS
*
* updating iteration
*
RE-E ASMB
END STRS
*
* new load-step
*
RE-E ASMB
LOAD UDSL   −1. 3     3     3   1   39   40  44
END STRS
*
END JOB
```

Figure 4.22 Example of application of the non-linear program to static ultimate-load analysis: FE discretisation; (b) smeared-analysis input file.

program). These programs provide plots of deflected shape and crack pattern of a reference section, one plot per converged load step plus one plot of the last performed iteration. The first program reads and processes deflection and cracking information of converged load steps and, also, of the last performed iteration. It then drops some redundant information and provides an output file with the strictly necessary data for the plots. The second program queues this file to a plotter; as the latter program is the one that includes calls to plotting subroutines, it is plotter dependent.

It is evident from Figure 4.21 that the first three modules (the one for the mesh generation and the two renumbering modules) are equally suited to both linear and non-linear analyses. Therefore, the actual modifications necessary to implement the smeared-analysis strategy are carried out only for the remaining six linear modules. A brief summary of the main features of such modifications is given below. It should be noted that, since the smeared modelling of cracking requires the *D*-matrices of the Gauss points, most of the changes relate to the assembly and the stress-computation modules (ASMB and STRS, respectively). The assembly module ASMB performs the numerical integration of the tangent stiffness matrices of the elements and then it assembles the tangent stiffness matrix of the structure; it also includes several initialisation subroutines. In fact, the first time the subroutine is called, that is, at the first iteration of the analysis, it performs some initialisation tasks, including printing of mesh connectivities and performing a dummy assembly to setup some assembly variables, such as the block size by which the program splits the stiffness matrix of the structure. Other tasks include the reading of any relevant data specified by the In and Rn registers in the input file (see Figure 4.21), including options such as number of Gauss points, NR updating strategy and number of cracks allowed per iteration. There is also initialisation of the *D*-matrices of all concrete Gauss points to values in accordance with the material modelling described in Chapter 3 (therefore, any Young's moduli and Poisson's ratios specified in the input file for concrete are ignored). The initialisation stage is completed by the calculation of *B*-matrices and Jacobians for all concrete and steel Gauss points. It should be recalled that, since the analysis assumes small displacements and strains, both *B*-matrices and Jacobians keep their initial values (calculated in global coordinates by the relevant element subroutines) throughout the analysis.

The numerical integration of the tangent stiffness matrices of the elements is performed by either of two subroutines. One of these (which includes two optional integration schemes (Gonzalez Vidosa 1989)) recalculates stiffness matrices for all elements (concrete and steel), while the other recalculates stiffness matrices of only those concrete elements that contain newly cracked Gauss points plus the stiffness matrices of all steel elements (since the latter elements require relatively little computational effort, their stiffness matrices are always recalculated). The choice between the two subroutine options relates to the chosen NR updating strategy as discussed in Section 4.3.5. Finally, it should be noted that, although it would seem natural that the updating of *D*-matrices should be carried out in this module (ASMB), it is actually done in module STRS. That fact is due to the possibility of selective updating of *D*-matrices at the end of module STRS, since the selective updating decision is taken just after the end of the current iteration (module STRS) so that any related information is readily accessed in that module.

The BNCN module sets up two vectors that relate to boundary conditions: the first vector contains the numbers of DOF being restrained, while the second contains the values of these restraints (non-zero if there are applied displacements). This information is specified by command FIXE in the input file (e.g., FIXE FACE in Figure 4.22). The executive subroutine of the non-linear module differs from the one of the standard FINEL only in that the former is constantly bypassed following the first iteration of the analysis.

The module CHOL factorises the tangent stiffness matrix of the structure by means of Choleski's method ($[K] = [L][U]$). As for the boundary-conditions module, its executive sub-routine is practically identical to the standard subroutine of FINEL. The difference lies in that the latter performs the factorisation provided that the stiffness matrix is positive definite, even though the matrix may be ill-conditioned, whereas this is not the case in the non-linear version. This difference was implemented in order to avoid an excessive number of iterations for the last non-converged load step (which, in spite of the introduced 'short-cut' usually still takes about one-third of the total number of iterations), and to ensure that any converged load step was attained without ill-conditioning of the tangent stiffness matrix.

It should be recalled that the increment of nodal forces is the addition of applied (external) forces and residual forces. The latter appear for equilibrium purposes as a result of stress corrections introduced at the previous iteration in order to follow a non-linear stress–strain relationship. The external-loading module LOAD, however, calculates only those equivalent nodal forces due to applied loading, that is, loading caused by either applied forces or non-zero imposed displacements ($-[K_{ur}]\Delta d$, as will be shown below). Residual forces are calculated in module STRS (previous iteration), where stress corrections take place, and added to applied forces (if any) in module SLVE. Therefore, the executive subroutine of the module LOAD is executed only at the first iteration of each load step.

The module SLVE calculates increments of nodal displacements and increments of reactions, and updates their total values. Subscripts u and r denote unrestrained and restrained displacements (and forces associated with them), respectively. Upon rearrangement of coefficients, the set of equations of a given iteration can be partitioned as follows:

$$\begin{bmatrix} [K_{uu}] & [K_{ur}] \\ [K_{ru}] & [k_{rr}] \end{bmatrix}\begin{bmatrix} \Delta d_u \\ \Delta d_r \end{bmatrix} = \begin{bmatrix} \Delta f_u \\ \Delta f_r \end{bmatrix} \tag{4.112}$$

where $[K_{ij}]$ are partitions of the tangent stiffness matrix of the structure ($[K_{uu}]$ is the deflated tangent stiffness matrix, that is, the matrix relevant to the case when $\Delta d_r = 0$ and Δf_r is not required); Δd_u, Δd_r, are increments of displacements; and Δf_u, Δf_r are increments of nodal forces (f_r being the reactions). From the above equation, Δd_u, Δf_r are found through

$$\Delta d_u = [K_{uu}]^{-1}(\Delta f_u - [K_{ur}]\Delta d_r) \tag{4.113}$$

$$\Delta f_r = [K_{ru}]\Delta d_u + [K_{rr}]\Delta d_r \tag{4.114}$$

While the standard executive subroutine SLVE includes the calculation of Δd_u by Choleski's method ($[K_{uu}]$ is factorised and stored in the ZFILE in module CHOL), the non-linear sub-routine version also includes the computation of reactions f_r.

It is the stress module STRS which contains most of the smeared-analysis tasks and hence it differs fundamentally from its linear counterpart. Broadly, two different parts make up its executive subroutine. The first part is a block of stress computations: it calculates and corrects total stresses, sets up Gauss points where new cracks form or close, and calculates the residual forces due to stress corrections. The second part checks whether or not convergence is achieved for the current load step and then a decision is made on the type of updating for the next iteration.

The role of the first part of the STRS module is quite substantial. Increments of strains and stresses, stress corrections and residual forces are calculated from increments of nodal displacements in two subroutines, one for concrete and the other for steel Gauss points. While the subroutine for concrete Gauss points will be described in detail later on, it should be noted at this stage that it checks whether or not the states of stress at these Gauss points are outside the failure envelope described in Section 3.2. Those concrete Gauss points that exceed the envelope form the set of prospective crack locations; however, the present subroutine neither sets up any new crack(s) nor corrects any state(s) of stress outside the failure envelope. With regard to steel Gauss points, its subroutine also checks whether or not any of them change from one branch to another of the trilinear stress–strain relationship (Section 3.3). The next step depends on the crack-propagation criteria adopted. If the 'total crack' criterion is specified in the input file, then all prospective cracks are setup (the actual crack implementation is carried out by a subroutine especially earmarked for this task), that is, the number of new cracks is equal to the number of prospective cracks. On the other hand, if a 'single crack' criterion is specified, then only one (or a small number of) new crack(s) is allowed in the current iteration. In the latter instance, prospective cracks are sorted out according to the percentage by which their state of stress exceeds the failure envelope (i.e., τ_o/τ_{ou} – see Section 3.2), so that only those which exceed failure conditions by the largest margin are actually setup in the current iteration. Clearly, while there are prospective cracks, convergence cannot be fulfilled and the external load is kept constant, as will be discussed below. It should be borne in mind that any stress corrections in the preceding subroutine components within module STRS give rise to residual forces that are stored for the next iteration.

The above two subroutines for concrete Gauss points, dealing with failure criteria and crack implementation, respectively, require further description. The first of these subroutines, based on checks for possible macro-cracking or crack closure, begins by calculating increments of global strains from increments of nodal displacements, and then updates total strains. If a Gauss point is uncracked, that is, not cracked in previous iterations, concrete is assumed to be isotropic and increments of stress can be directly calculated in global directions, since its D-matrix is invariant for any set of axes. On the other hand, if the Gauss point is already cracked, then the material is anisotropic and its D-matrix is only valid in local (cracked) directions, as discussed in Section 4.3.3; hence, increments of strains are transformed to local directions in order to compute stress increments and total stresses in these local directions. The next step depends on the type of NR updating strategy for the current run. For example, if a given Gauss point is already cracked but its D-matrix has not been updated to allow for this crack occurrence (i.e., an isotropic D-matrix is kept), then its state of stress has to be corrected (namely by setting to zero the stress orthogonal to the crack plane); otherwise, the stress state of the cracked Gauss point would not be in accordance with the new situation implied by cracking. (That the use of iterative procedures which keep D-matrices either permanently or temporarily uncracked (i.e., modified or mixed INRM, respectively) attracts direct tensile stresses across crack planes in subsequent iterations which have to be set equal to zero is illustrated schematically in Figure 4.23. (It should be noted that, even though the assumption of uncracked properties is implicit in the lack of the updating of the D-matrix, the initial-stress approach is used in recognition of the fact that cracking has been signalled in the iterative procedure.) This gives rise to successive residual forces which, sometimes, may lead to non-convergent solutions (modified INRM) or to early divergent analyses as a result of propagation of spurious mechanisms (mixed INRM). This task of stress correction across cracks is performed by a special subroutine when either constant-stiffness or mixed NR strategies are followed (i.e., Figure 4.6b and c). Once the state of stress in local directions has been corrected, stresses are transformed to global directions. It is then that a check is performed to ascertain whether or not the state of stress is outside the failure envelope: this requires the calculation of principal and octahedral

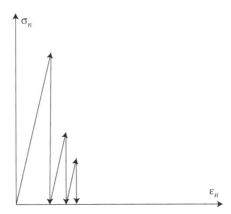

Figure 4.23 Residual tensile stresses across a crack plane for a modified or mixed INRM (the latter with a fixed sequence of non-updating/updating of iterations, of which only the former are shown in the figure).

stresses. Finally, any state of stress inside the envelope is corrected simply in accordance with the constitutive relations as outlined in Section 3.1.

The subroutine for the implementation of cracking may be described by reference to the case when a first crack is setup at a concrete Gauss point. Second and third cracks (or, in principle, compressive failures) are setup in a similar way, consistent with the D-matrices defined in Section 4.4.3. First, the variable that defines the cracking situation at that location is updated. Then, there follows the calculation of principal stresses and directions, and of the relative magnitudes of the principal stresses. This ordering of stress magnitudes is used so as to determine which column of the matrix defining the axes change relates to which principal stress. Plotting information is then calculated by a subroutine which determines the intersection angle of the crack plane with the plotting plane and also checks whether or not both planes form an angle smaller than 45°. Finally, and consistent with the notion of cracking, the maximum tensile stress is set to zero, local stresses are transformed to global directions and residual forces are computed. It should be noted that the D-matrices are not updated here, for that operation depends on subsequent updating decisions.

The second part of the STRS module is concerned with convergence and updating decisions. First, a check is made on whether or not the convergence criteria discussed in Section 4.3.2.6 are fulfilled. This requires convergence of residual magnitudes (i.e., forces, deformations, stresses, strains), and it also requires 'yield' and 'cracking' convergence, that is, that no steel Gauss point has changed from one linear branch to another and that there are no prospective crack formations or closures left. In practice, the cracking criterion is usually the most restrictive, for residual forces decrease very rapidly when there are no new cracks to form or open. In this respect, Figure 4.24 shows the number of prospective cracks and residual forces for a typical converged load step (this is taken from Gonzalez Vidosa 1989). If all convergence criteria are fulfilled, the next iteration follows a new load increment. It is then that a decision is made on the type of updating for the next iteration. Three NR strategies may be considered: the initial-stiffness method; the mixed NR method; and a selective-updating method (recommended in Section 4.4.3) that updates only newly cracked concrete locations as well as steel locations where yielding takes place, that is, it updates only D-matrices relating to 'strong' non-linear changes, while it does not update matrices relating to 'mild' non-linear changes (this selective updating, which was implemented initially to improve efficiency, was also found to improve numerical stability with respect to the

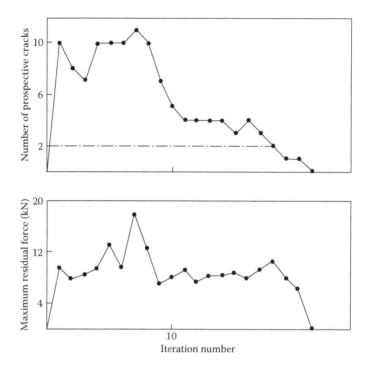

Figure 4.24 Typical curves of number of prospective crack locations and residual forces for a converged step (load step = 25 kN, two cracks allowed per iteration). (From Gonzalez Vidosa F., 1989, Three-dimensional finite element analysis of structural concrete under static loading, University of London, PhD thesis.)

mixed NR method). Finally, D-matrices are updated: this may be done either at all Gauss points or selectively, depending on the updating subroutine chosen.

The above overview of the smeared FE package encompasses the key features of material and structural modelling outlined in this and the previous chapter. The application of the non-linear FE model to 3-D problems forms the subject of Chapter 5.

REFERENCES

ASCE Task Committee on Finite Element Analysis of Reinforced Concrete Structures, 1981, *Constitutive Relations and Failure Theories*, CE – STR – 81–21, New York

ASCE Task Committee on Finite Element Analysis of Reinforced Concrete Structures, 1982, *Finite Element Analysis of Reinforced Concrete*, ASCE Special Publication, New York

Bazant Z. P. and Oh B. H., 1963, Crack band theory for fracture of concrete, *Materials and Structures*, 16, 155–177.

Bedard C., 1983, Non-linear finite element analysis of concrete structures, PhD thesis, University of London.

Bedard C. and Kotsovos M. D., 1986, Fracture processes of concrete for non-linear finite element analysis, *Journal of Structural Engineering, ASCE*, 112, (3), 373–387.

Bergan P. G. and Clough R. W., 1972, Convergence criteria for iterative processes, *AIAA Journal*, 10, 1107–1108.

British Standards Institution, 1972, *British Standards, Structural Use of Concrete, CP110 (Part 1, Design, Materials and Workmanship)*, British Standards Institution, London.

British Standards Institution, 1985a, *British Standards, Structural Use of Concrete, BS 8110 (Part 1, Code of Practice for the Design and Construction)*, British Standards Institution, London.

British Standards Institution, 1985b, *British Standards, Structural Use of Concrete, BS 8110 (Part 2, Code of Practice for Special Circumstances)*, British Standards Institution, London.

Burns N. H. and Siess C. P., 1966, Plastic hinging in reinforced concrete, *Journal of the Structural Division, Proceedings of ASCE*, 92, 45–64.

Cedolin L. and Dei Poli S., 1977, Finite element studies of shear-critical R/C beams, *Journal of the Engineering Mechanics Division, Proceedings of the ASCE*, 103, 395–410.

Cedolin L. and Nilson A. H., 1978, A convergence study of iterative methods applied to finite element analysis of reinforced concrete, *International Journal of Numerical methods in Engineering*, 12, 437–451.

Clough R. W., 1980, The finite element method after twenty-five years: A personal view, *Computers and Structures*, 12, 361–370.

Cook R. D., 1981, *Concepts and Application of Finite Element Analysis*, 2nd edition, John Wiley & Sons, Inc., New York.

Cook R. D., Malkus D. S. and Plesha M. E., 1989, *Concepts and Application of Finite Element Analysis*, 3rd edition. John Wiley & Sons, Inc., New York.

Cristfield M. A., 1982, Local instabilities in the non-linear analysis of reinforced concrete beams and slabs, *Proceedings of the Institution of Civil Engineers*, Part 2, 73, 135–145.

Edwards A. D. and Yannopoulos P. J., 1978, Local bond-stress–slip relationships under repeated loading, *Magazine of Concrete Research*, 30, 62–72.

Fried I., 1971, Basic computational problems in the finite element analysis of shells, *International Journal of Solids and Structures*, 7, 1705–1715.

Gonzalez Vidosa F., 1989, Three-dimensional finite element analysis of structural concrete under static loading, PhD thesis, University of London.

Gonzalez Vidosa F., Kotsovos M. D. and Pavlovic M. N., 1988, On the Numerical instability of the smeared-crack approach in the nonlinear modelling of concrete structures, *Communications in Applied Numerical Methods*, 4, 799–806.

Gonzalez Vidosa F., Kotsovos M. D. and Pavlovic M. N., 1989, Discussion (de Borst R.) and authors' closure of: On the numerical instability of the smeared-crack approach in the nonlinear modelling of concrete structures, *Communications in Applied Numerical Methods*, 1988, 4, 799–806, *Communications in Applied Numerical Methods*, 5, 489–493.

Gonzalez Vidosa F., Kotsovos M. D. and Pavlovic M. N. 1991, A three-dimensional nonlinear finite-element model for structural concrete, Parts 1 and 2, *Proceedings of the Institution of Civil Engineers, Research and Theory*, 91, 517–560.

Gonzalez Vidosa F., Kotsovos M. D. and Pavlovic M. N., 1992, Discussion (Pakianathan L. J., Philips D. V.) and authors' closure of: A three-dimensional nonlinear finite-element model for structural concrete, Parts 1 and 2, *Proceedings of the Institution of Civil Engineers, Research and Theory*, 91, September 1991, 517–560, *Proceedings of the Institution of Civil Engineers*, London, 94, 365–374.

Greunen J., 1979, Nonlinear geometric, material and time dependent analysis of reinforced and prestressed concrete slabs and panels, Report No. UC—STR _ 81–21, University of California, Berkeley.

Hand F. R., Pecknold D. A. and Schnobrich W. C., 1973, Nonlinear layered analysis of reinforced concrete plates and shells, *Journal of the Structural Division, Proceedings of the ASCE*, 99, 1491–1505.

Hitchings D., 1972, *FINEL user's Manual*, Imperial College, London.

Hitchings D., 1980, *FINEL Programming Manual*, Imperial College, London.

Irons B. and Ahmad S., 1980, *Techniques of Finite Elements*, Ellis Horwood, Chichester.

Jelic I., Kotsovos M. D., and Pavlovic M. N., 1999, A study of dowel action in reinforced concrete beams, *Magazine of Concrete Research*, 51(2), 131–141.

Kollegger J. and Mehlhorn G., 1987, Material model for cracked reinforced concrete, *Proceedings IABSE Colloquium on Computational Mechanics of Concrete Structures—Advances and Applications*, Delft, pp. 63–74.

Kotsovos M. D., Bobrowski J. and Eibl J., 1987, Behaviour of RC T-beams in Shear, *The Structural Engineer*, 65B(1), 1–9.

Kotsovos M. D. and Pavlovic M. N., 1986, Non-linear finite element modelling of concrete structures: Basic analysis, phenomenological insight, and design implications, *Engineering Computations*, 3(3), 243–250.

Kotsovos M. D. and Pavlovic M. N., 1995, *Structural Concrete: Finite-Element Analysis for Limit-State Design*, Thomas Telford, London.

Kotsovos M. D. and Spiliopoulos K. V., 1998a, Modelling of crack closure for finite-element analysis of structural concrete, *Computers and Structures*, 69, 383–398.

Kotsovos M. D. and Spiliopoulos K. V., 1998b, Evaluation of structural-concrete design-concepts based on finite-element analysis, *Computational Mechanics*, 21, 330–338.

Labib F., 1976, Non-linear analysis of the bond and crack distribution in reinforced concrete members, PhD thesis, University of London.

Labib F. and Edwards A. D., 1978, An analytical investigation of cracking in concentric and eccentric concrete tension members, *Proceedings of the Institution of Civil Engineers, Part 2*, 65, 53–70.

Lin C. S. and Scordelis A. C., 1975, Nonlinear analysis of reinforced concrete shells of general form, *Journal of the Structural Division, Proceedings of ASCE*, 101, 523–538.

Ngo D. and Scordelis A. C., 1967, Finite element analysis of reinforced concrete beams, *ACI Journal*, 64, 152–163.

Nilson A. H., 1968, Nonlinear analysis of reinforced concrete by the finite element method, *ACI Journal*, 65, 757–766.

Owen D. R. J. and Hinton E., 1980, *Finite Element in Plasticity*, Pineridge Press, Swansea.

Pavlovic M. N. and Stevens L. K., 1981, The effect of prior flexural prestrain on the stability of structural steel columns, *Engineering Structures*, 3, 66–70.

Philips D. V. and Zienkiewicz O. C., 1976, Finite element analysis of concrete structures, *Proceedings of the Institution of Civil Engineers, Part 2*, 61, 59–88.

Rashid Y. M., 1968, Ultimate strength analysis of prestressed concrete pressure vessels, *Nuclear Engineering and Design*, 7, 334–344.

Saouma V. E. and Ingraffea A. R., 1981, Fracture mechanics analysis of discrete cracking, *Proceedings IABSE Colloquium on Advanced Mechanics of Reinforced Concrete*, Delft, pp. 413–436.

Shenglin D., Qigen S. and Bingzi S. A., 1987, A finite element simulation model for cracks in reinforced concrete, *Proceedings IABSE Colloquium on Computational Mechanics of Concrete Structures—Advances and Applications*, Delft, pp. 209–214.

Suidan M. S. and Schnobrich W. C., 1973, Finite element analysis of reinforced concrete, *Journal of the Structural Division, Proceedings of ASCE*, 99, 2109–2122.

Tassios T. P. and Scarpas A, 1987, A model for local crack behaviour, *Proceedings IABSE Colloquium on Computational Mechanics of Concrete Structures—Advances and Applications*, Delft, pp. 35–42.

Timoshenko S. P. and Goodier J. N., 1970, *Theory of Elasticity*, 3rd edition, McGraw-Hill, New York.

Zienkiewicz O. C., 1977, *The Finite Element Method*, 3rd edition, McGraw-Hill, London.

Zienkiewicz O. C., Taylor R. L. and Too J. M., 1971, Reduced integration technique in general analysis of plates and shells, *International Journal of Numerical Methods in Engineering*, 3, 275–290.

Chapter 5

Finite-element solutions
of static problems

As pointed out in Section 4.4, while the study of the objectivity of a given numerical model deals with the sensitivity of its analytical predictions to the factors incorporated into the computational strategy, a model's generality implies that it can successfully predict the behaviour of arbitrary structural configurations. It is, therefore, in order to test its generality that the 3-D FE model, incorporating the material and procedural factors recommended in Section 4.4.3, is now applied to a wide range of problems. In what follows, emphasis is placed mainly on problems concerned with the behaviour of reinforced concrete (RC) structural configurations under load reversals. However, although the model's generality for the case of monotonic loading has already been demonstrated (Kotsovos and Pavlovic 1995), a limited number of such cases are also considered in order to investigate the effect of crack closure on the predicted behaviour, since this effect was initially assumed to be negligible.

Besides the all-important structural strength (i.e., the most relevant characteristic from the viewpoint of limit-state design philosophy), consideration is also usually given to the predicted crack patterns and deformed shapes. With regard to the plotting convention for the latter two aspects of structural response, it should be mentioned that, in all cases, the plots are superimposed onto the mesh lying on the plane parallel to the reference plane xz, as will be illustrated by subsequent examples. The symbols used for the various cracks at the relevant Gauss points are to be interpreted as follows: oriented dashes represent the intersection of a crack plane with the plotting plane xz whenever the angle subtended between these two planes exceeds 45°. However, if these planes form an angle smaller than 45°, then the crack plane is simply indicated by a circle (hence, two cracks both forming an angle less than 45° to xz at one location cannot be distinguished one from the other). Three cracks at the same Gauss point are indicated by an asterisk symbol. The presence of steel bars is marked by dashed lines either superimposed onto the edge(s) of the 20-node isoparametric or extended between successive Gauss points in the x or z directions for the 27-node Lagrangian concrete element(s); displacement shapes at the various load levels are suitably magnified, and the appearance of very large and/or distorted shapes helps to identify quickly the onset of a mechanism (whether real or numerical in nature); finally, load step '777' plots relate to the last performed iteration in an analysis (at the load step that follows the maximum sustained load [MSL]), while, if at all specified, D.M. = N indicates that displacements have been magnified N times.

5.1 EFFECT OF CRACK CLOSURE ON PREDICTIONS OF STRUCTURAL-CONCRETE BEHAVIOUR UNDER MONOTONIC LOADING

5.1.1 Background

For the investigation of the generality of the proposed model when applied for the case of structural-concrete under monotonic loading the assumption has been implicit that the predicted behaviour is essentially independent of crack closure occurring on account of internal stress/strain distributions due to cracking (Kotsovos and Pavlovic 1995). In what follows, the validity of the above assumption is investigated through a comparative study of two sets of results regarding the behaviour of three typical types of RC structural elements (beam, slab, wall) obtained by analysis in which the effect of crack closure was either allowed for or ignored (Kotsovos and Spiliopoulos 1998a).

5.1.2 RC beams in shear

Figure 5.1 summarises the design details of RC beam A-1 for which the numerically predicted behaviour is also reported in Kotsovos and Pavlovic (1995); full design and experimental details are given in Bresler and Scordelis (1963). The compressive (cylinder) strength of concrete was 24 MPa while the yield stress was 556 MPa for the tension bars, 345 MPa for the compression bars and 325 MPa for the stirrups.

The FE mesh adopted for the analysis is shown in Figure 5.2. Only one-fourth of the beam was analysed on account of symmetry and the FE mesh comprised 30 serendipity 20-node

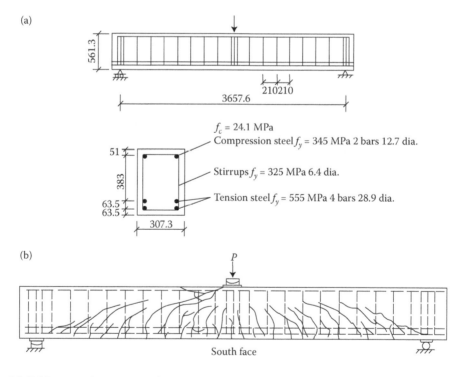

Figure 5.1 RC beam with stirrups under central point load: (a) member characteristics (all dimensions in mm); (b) observed crack pattern at ultimate load (all dimensions in millimetre).

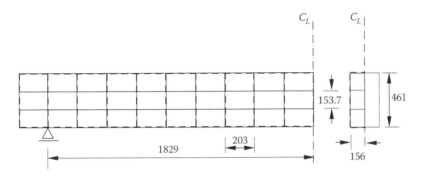

Figure 5.2 RC beam with stirrups under central point load. FE mesh for one-quarter of the structure consisting of 30 brick elements with superimposed dashed lines indicating the position of reinforcing bars (all dimensions in millimetre).

elements (with a $2 \times 2 \times 2$ integration rule) for the concrete and 95 bar elements for the steel indicated by the dashed lines in the figure. As the figure indicates, the mesh adopted provides only a crude representation of the actual beam with the cover of the tension reinforcement being neglected and the compression reinforcement being placed at the top face of the beam, while the spacing of the stirrups is almost equal to that of the actual beam (203 instead of 210 mm) and, hence, the adjustment required to keep the area per unit length the same was minimal.

Figure 5.3 shows the crack patterns at the various load levels up to failure predicted by both methods of analysis. Although for load levels beyond a value of approximately 50% of the MSL internal stress redistribution due to cracking led to some crack closure, the crack patterns were found to be essentially unaffected by such effects. Inclined cracking occurs gradually as load increases (see Figure 5.3 for 175, 225, 300, 375 and 450 kN). It is interesting to note that loss of load-carrying capacity is preceded by splitting of the compressive zone which initiates at 375 kN, but does not lead to immediate collapse since the stirrups sustain the tensile stresses that cannot be sustained by concrete after splitting. As it can be observed by reference to the crack pattern of the last performed iteration (load step '777' [475 kN] in Figure 5.3), failure occurs as a results of extensive splitting of the compressive zone which is in good agreement with the experimental mode of failure (see Figure 5.1b).

The negligible effect of the use of the numerical procedure allowing for crack closure in cases where structural elements are subjected to monotonic loading is also apparent from Figure 5.4. The figure shows the experimentally established load–deflection curve together with those predicted by analysis, the latter being softer than the former. Nevertheless, it should be recalled that the cover to the tension reinforcement has been neglected in the FE discretisation. Now, while this has a negligible effect on the collapse-load predictions, it does play a noticeable role when beam deflections are estimated, as pointed out in Kotsovos and Pavlovic (1995).

5.1.3 RC slabs in punching

The slab studied here is an RC square slab, denoted as B14 by Elstner and Hognestad (1956), and reported fully in this reference in respect to the relevant experimental details, some of which are summarised in Figure 5.5. The percentage of the tension steel was very high (3%) and the slab lacked shear reinforcement. The compressive (cylinder) strength of concrete and the yield stress of the steel were 50.5 and 325 MPa, respectively. The RC plate was simply

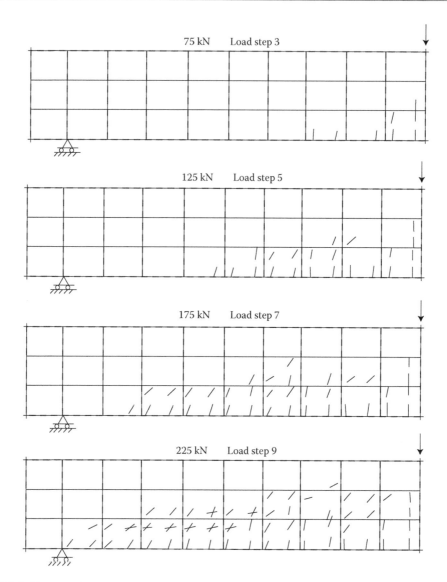

Figure 5.3 RC beam with stirrups under central point load. Crack patterns and deformed shapes at various stages. *(Continued)*

supported at the edges and corners and subjected to a concentrated square load at mid-span monotonically increasing to failure. Failure occurred as a result of punching at a load of 578 kN, before any yielding of the tension bars.

The details of the adopted FE mesh are shown in Figure 5.6. The mesh comprised 25 Lagrangian FEs (363 nodes and 1089 DOF) for concrete and 110 bar elements for the tension steel indicated in the figure by the dashed lines. Although only one-fourth of the slab is analysed on account of symmetry, it is clear that additional savings would be achieved by implementing 'edge' (i.e., triangular-based) brick elements, so as to also make use of the symmetry with respect to the diagonals and hence to analyse one-eighth of the whole slab. With regard to the choice of the Lagrangian FE mesh, it should be noted that, before its choice, three other serendipity meshes were considered (all incorporating under-integrated

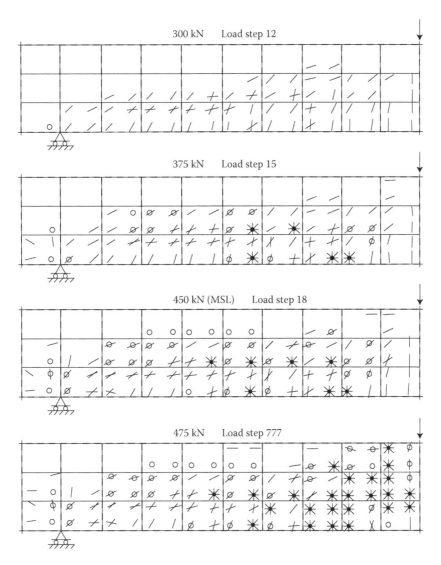

Figure 5.3 (Continued) RC beam with stirrups under central point load. Crack patterns and deformed shapes at various stages.

serendipity meshes); these were based on $5 \times 5 \times 1$ 20-node elements (228 nodes, 684 DOF), $5 \times 5 \times 2$ 20-node elements (360 nodes, 1080 DOF) and $9 \times 9 \times 2$ 20-node elements (1040 nodes, 3120 DOF). However, the Lagrangian mesh was preferred for the following reasons. The first alternative mesh had only two Gauss points across the thickness of the slab, and did not allow a good description of the tension steel in plan. Therefore, this discretisation was considered too rough. The second alternative mesh had four Gauss points across the thickness, but still involved a poor modelling of the steel in plan. Finally, the third alternative mesh was discarded on account of the steep rise in computer resources inherent in this scheme.

The load–deflection curves obtained from the analysis together with measured values are presented in Figure 5.7, while the crack patterns at the section through the mid-span of the slab, both just before and during failure, are shown in Figure 5.8. Although for load levels beyond a value of approximately 50% of the failure load internal stress redistribution due to

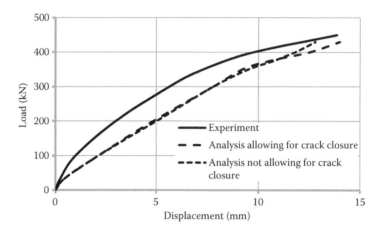

Figure 5.4 RC beam with stirrups under central point load. Load–deflection curves.

cracking led to some crack closure due to unloading in localised regions of the slab, allowing for such effects only marginally improved the predicted load–deflection curve (see Figure 5.7). It is also interesting to note that as for the case of the beam in the preceding section, in spite of the crude FE mesh used, the analyses yielded a close prediction of the load-carrying capacity and the mode of failure, while the predictions of defection near the MSL deviate from the measured values. This deviation coincides with the occurrence of a large number horizontal cracks (splitting) within the compressive zone at load step 13, which causes a loss of overall stiffness of the slab.

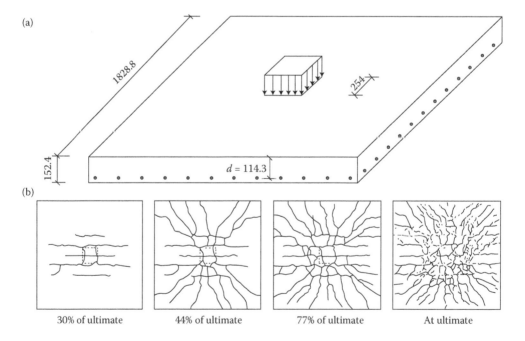

Figure 5.5 RC slab exhibiting punching failure. Experimental data: (a) dimensions and detailing of reinforcement (all dimensions in millimetre); (b) crack pattern on the tensile face at various load stages (all dimensions in millimetre).

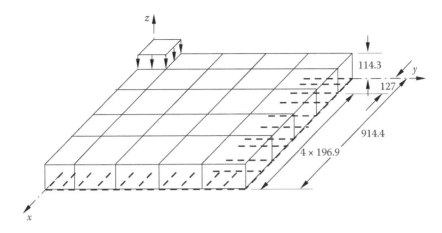

Figure 5.6 RC slab exhibiting punching failure. FE mesh for one-quarter of the structure, consisting of 25 Lagrangian elements with superimposed dashed lines indicating the position of the reinforcing bars (all dimensions in millimetre).

On the other hand, the crack pattern remained virtually unaffected by such effects for all load steps considered in Figure 5.8. The crack pattern at the last performed iteration (load step '777') shows that divergence takes place as a result of a local mechanism in the zone of the concentrated load, which occurs before yielding of the reinforcement. (For a more detailed explanation of the punching-failure mechanism in slabs, see Kotsovos and Kotsovos 2009, 2014, Kotsovos and Kotsovos 2010).

5.1.4 RC structural walls under combined edge compressive and shear stresses

Conclusions similar to those drawn from the preceding case studies may also be drawn for the case of the structural wall studied here, denoted as S4 in Maier and Thurlimann (1985),

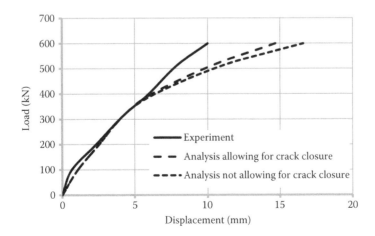

Figure 5.7 RC slab exhibiting punching failure. Load–deflection curves.

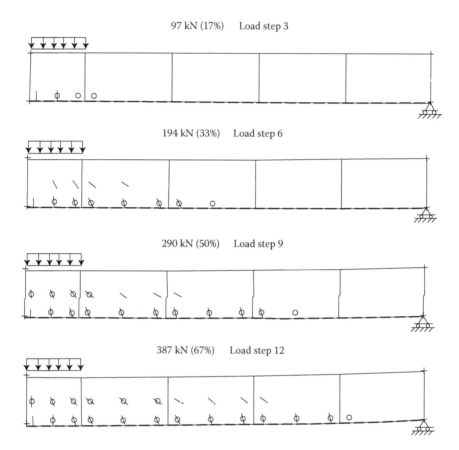

97 kN (17%)   Load step 3

194 kN (33%)   Load step 6

290 kN (50%)   Load step 9

387 kN (67%)   Load step 12

Figure 5.8 RC slab exhibiting punching failure. Crack patterns and deformed shapes at various stages (Note: the percentages refer to the ratio of the analytical to experimental failure loads to facilitate comparison with Figure 5.5b).   (*Continued*)

and fully reported in this reference as regards to its experimental details. The geometric characteristics of the wall are summarised in Figure 5.9, while the compressive (cylinder) strength of concrete and the yield stress of the steel were 30 and 575 MPa, respectively. The wall was fully fixed at the bottom edge and subjected to uniformly distributed normal and shear stresses at the other edge. Failure occurred under a horizontal load of 397 kN which led the diagonal web cracking to penetrate deeply into the left bottom corner of the wall and cause vertical splitting of the concrete.

The FE mesh adopted for the analyses comprised 16 serendipity 20-node (with $2 \times 2 \times 2$ integration rule) brick elements for the concrete, and 80 three-node bar elements for the steel placed along the boundaries of the brick elements as indicated in Figure 5.10. Figure 5.11 depicts both the analytical and the experimental load–deflection curves, while Figures 5.12 and 5.13 present the experimentally established crack pattern of the wall at failure together with its numerically predicted counterpart. It becomes apparent from the figures that, as for the preceding case studies, the analyses yielded close predictions of the behaviour established experimentally. Similarly, the results obtained also indicate that the effect of crack closure on the analysis predictions is insignificant as this particular case of the normal stress applied to the wall before the application of the shear stress was insufficient to cause the formation of cracks.

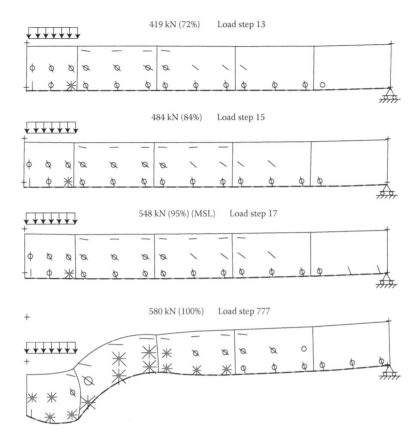

Figure 5.8 (Continued) RC slab exhibiting punching failure. Crack patterns and deformed shapes at various stages (Note: the percentages refer to the ratio of the analytical to experimental failure loads to facilitate comparison with Figure 5.5b).

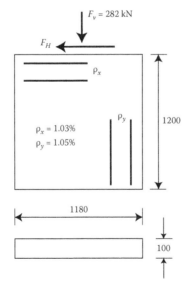

Figure 5.9 RC structural wall: design details (dimensions in millimetre).

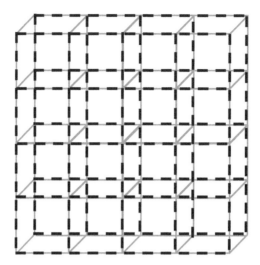

Figure 5.10 RC structural wall: FE mesh comprising 16 brick elements with superimposed dashed lines indicating the position of the reinforcing bars.

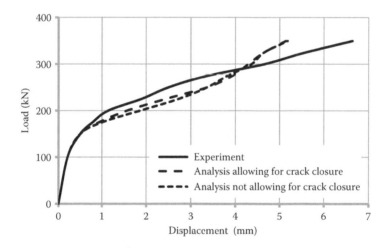

Figure 5.11 RC structural wall: Load–deflection curves.

5.2 PERFORMANCE OF STRUCTURAL-CONCRETE MEMBERS EXHIBITING POINTS OF CONTRA-FLEXURE UNDER SEQUENTIAL LOADING

5.2.1 Introduction

Most available experimental data used for calibrating numerical models for structural concrete are based on tests carried out on statically determinate (and, usually, simply supported) beam elements. Furthermore, the loading employed in such tests is almost invariably monotonic. In practice, however, actual structures are multi-element systems of a (sometimes highly) statically indeterminate nature; and, in addition, sequential loading conditions often need to be taken into account besides the 'neater' (especially in limit-state

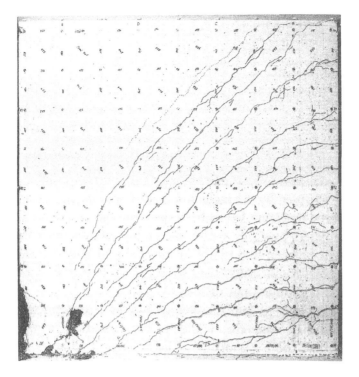

Figure 5.12 RC structural wall: Experimental crack pattern at failure.

collapse analysis) monotonic (and, in the case of multi-loads, proportional) loading assumption. What the above simple laboratory tests rarely provide are points of contra-flexure, in sharp contrast to real structures under practical loading conditions. Moreover, it has been reported that many earthquake failures occur precisely in regions of the structure associated with contra-flexure points, suggesting possible design weaknesses in current codes

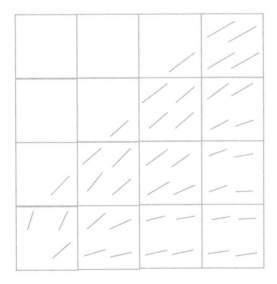

Figure 5.13 RC structural wall: Analytical crack pattern at failure.

of practice, where no special provision is made for these regions (Kotsovos and Pavlovic 2001). This weakness has been confirmed from the results of tests on simply supported RC beams with an overhang under both proportional and sequential loading (Kotsovos and Michelis 1996, Jelic et al. 2003); the beams were found to fail prematurely in a brittle manner in the region of the point of contra-flexure which characterises structural behaviour under the loading regimes adopted for the tests. In an attempt to test the ability of the FE model to provide realistic predictions of the behaviour of structures exhibiting points of contra-flexure, the results obtained from the above experiments were reproduced numerically by Kotsovos and Spiliopoulos (1998b) who also investigated the causes of premature failure. In what follows, the information presented has been extracted from the above work where full details can be found.

The simply supported RC beams investigated are shown in the simplified form adopted for the analysis in Figure 5.14. The beam dimensions, position of longitudinal reinforcement, boundary conditions and loading arrangement were the same for all beams. In order to establish the effect of the amount of longitudinal reinforcement on structural behaviour, two types of longitudinal reinforcement were used; 14 mm diameter bars (A_s = 154 mm^2) with yield stress f_y = 600 MPa and strength f_u = 750 MPa, resulting in an under-reinforced section, and 16 mm diameter bars (A_s = 201 mm^2) with f_y = 545 MPa and f_u = 680 MPa, resulting in a section with a reinforcement percentage just higher than that corresponding to the balanced section. The amount and arrangements of transverse reinforcement used varied depending on whether the beams were designed in compliance with either the code provisions (Eurocode 2 (EC2), 1991, Technical Chamber of Greece 1991) or the compressive force path (CFP) method (Kotsovos and Pavlovic 1999). The cover to the reinforcement was ignored in all cases in order to simplify the FE modelling of the beams. The details of the FE mesh used for the analysis are given in Figure 5.15. The mesh consisted of 24 Lagrangian 27-node brick elements for concrete and 198 3-node bar elements for steel, the latter indicated in the figure by the dashed lines. It is important to note in the figure that the FE mesh adopted imposes a constant spacing of the transverse reinforcement, equal to 100 mm, in all cases investigated. However, the amount of this reinforcement was adjusted so as to conform to the specifications of the design method employed.

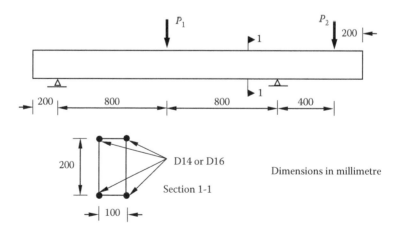

Figure 5.14 RC beams with overhang. Simplified representation adopted for the analysis (dimensions in millimetre).

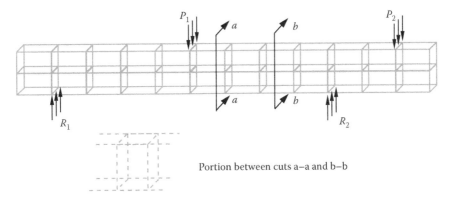

Figure 5.15 RC beams with overhang. FE mesh with the insert indicating the position of the reinforcement for a typical portion of the beam between sections a–a and b–b.

5.2.2 Beam designed in compliance with the Greek code

The design details of the beam are summarised in Figure 5.16, whereas the experimental information presented in the following has been extracted from Kotsovos and Michelis (1996) where full details can be found. The beam, denoted as B1GRC, was designed in compliance with the Greek code of practice (Technical Chamber of Greece 1991) so as to eventually fail in flexure. And yet, in contrast with the code predictions, it was found to fail in a brittle manner under a total load of 130 kN significantly smaller than the code predicted value of 153.5 kN.

The numerically predicted and experimental curves describing the relationships between the applied load and the deflections of the points of application of loads P_1 and P_2 are shown in Figure 5.17 which also shows the loading history of the beam. It is interesting to note that, in spite of the crude FE modelling, the analysis provides a very close prediction of the load-carrying capacity and an adequate prediction of the deformational response of the beam. It is important to note that the prediction of the load-carrying capacity correlates closely with the experimental findings which indicate that compliance with the code of practice specifications could not safeguard the intended structural behaviour (Kotsovos et al. 1994).

The close prediction of the beam load-carrying capacity is attributed to the brittle model of concrete behaviour adopted for the analysis. The use of such a brittle model implies that the beam cannot function as a truss at its ultimate limit state as assumed by the code

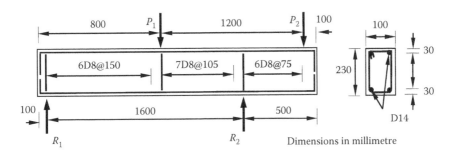

Figure 5.16 RC beam B1GRC under sequential loading. Design details.

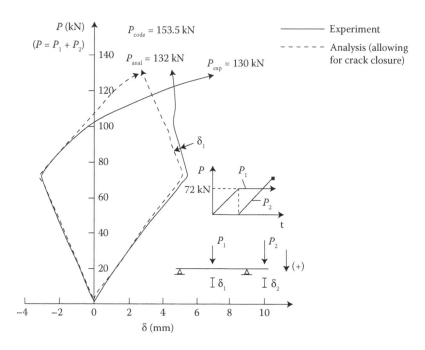

Figure 5.17 RC beam B1GRC under sequential loading. Load–deflection curves.

adopted method of design. This is because the negligible load-carrying capacity of cracked concrete precludes the formation of inclined struts within the web of the beam.

5.2.3 Beam designed in partial compliance with the CFP method

The present beam, denoted as B2CFP, is similar to that in Section 5.2.2 except for the transverse reinforcement which was designed in accordance with the CFP concept (Kotsovos and Bobrowski 1993, Kotsovos and Lefas 1990). Full design details are given in Kotsovos and Michelis (1996) and summarised in Figure 5.18. (Although the above concept also specifies transverse reinforcement within the compressive zone of the beam in the region of point loads, such reinforcement was not used in the present case; its significance is studied in the following section, where the results obtained from the present analysis are compared with

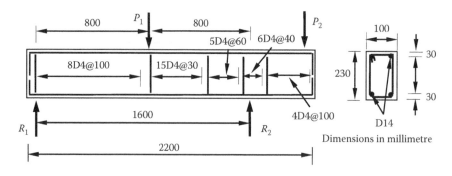

Figure 5.18 RC beam B2CFP under sequential loading. Design details.

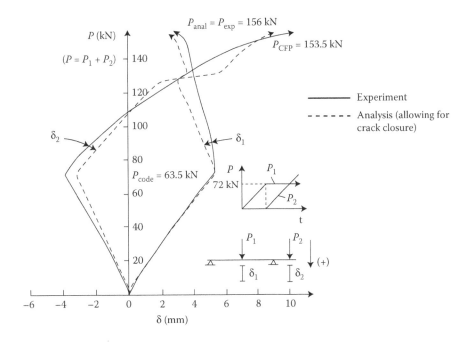

Figure 5.19 RC beam B2CFP under sequential loading. Load–deflection curves.

those obtained by the analysis of a beam designed in full compliance with the CFP concept.) It is important to note that the total amount of links specified for the beam was less than half that of the beam designed to the code, with the links within the overhang being merely 25% of the amount specified by the current code provisions for the design of earthquake-resistant RC structures.

And yet, the beam was found experimentally to fail in a ductile manner under a total load (156 kN) slightly higher than the expected value stemming from the CFP prediction (153.5 kN). The numerically predicted and experimental load–deflection curves are depicted in Figure 5.19 which also shows the values of the load-carrying capacity predicted by the CFP concept and the code. As for the case of the beam discussed in the preceding section, the analysis provides a very close prediction of the load-carrying capacity and a satisfactory description of the deformational response, in spite of the crude FE mesh used to model the beams. Such a prediction sharply contrasts the code prediction that the beam is only capable of sustaining approximately 85% of load P_1 acting alone.

5.2.4 Beam designed in full compliance with the CFP method

The beam, denoted as B3CFP, is the same with that discussed in the preceding section in all respects except for the provision of additional transverse reinforcement within the compressive zone as indicated in Figure 5.20. The figure depicts only the additional reinforcement together with the physical models which underlie the CFP method (Kotsovos and Bobrowski 1993, Kotsovos and Lefas 1990, Kotsovos and Pavlovic 1999). As indicated in Figure 5.20b, the beam model consists of two 'frames' with inclined legs tied by the flexural reinforcement and interacting through the provision of a transverse tie at the cross section including the point of contra-flexure; the additional transverse reinforcement was placed so as to enclose the horizontal members of the 'frames' of the model.

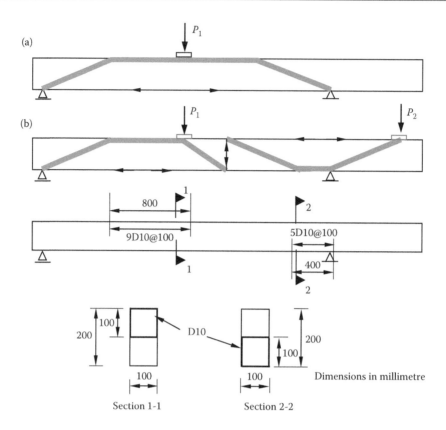

Figure 5.20 RC beam B3CFP under sequential loading: (a) physical model of beam during the first stage of loading sequence; (b) physical model of beam during second stage of loading sequence; (c) details of transverse reinforcement enclosing the horizontal members of 'frames' of physical models (dimensions in millimetre).

Figure 5.21 shows the predicted load–deflection curves describing the relationship between the applied load and the deflections at the points of application of loads P_1 and P_2. The figure also includes the corresponding load–deflection curves predicted for beam B2CFP (already shown in Figure 5.19), together with the load-carrying capacity corresponding to flexural capacity. It is interesting to note in the figure that the provision of the additional transverse reinforcement enclosing the compressive zone resulted in a considerable increase in ductility which was followed by a short strain-hardening branch leading to an increase in load-carrying capacity. In fact, the analysis predicted a value of load-carrying capacity which essentially coincides with the value corresponding to flexural capacity as the latter is calculated in compliance with current design concepts assuming all safety factors equal to 1.

5.2.5 Beams designed in accordance with the European code

The present case study includes three RC beams designed in accordance with Eurocode 2 (1991) and Eurocode 8 (EC8) (1994) for high ductility (beam B4ECH), moderate ductility (beam B4ECM) and low ductility (beam B4ECL). Their transverse reinforcement details are shown in Figure 5.22. The beams also differ from those discussed in the preceding sections in that their longitudinal reinforcement comprises 16 mm, rather than 14 mm, diameter

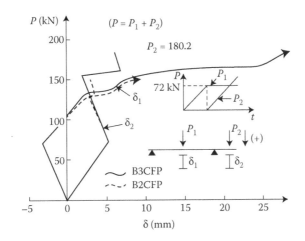

Figure 5.21 RC beam B3CFP under sequential loading. Load–deflection curves (the curves for B2CFP are also included for purposes of comparison).

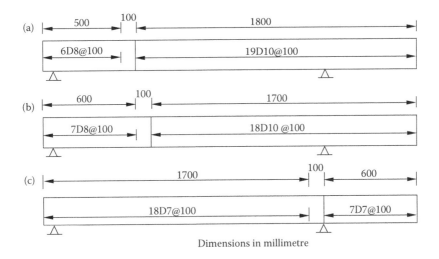

Dimensions in millimetre

Figure 5.22 RC beams under sequential loading. Details of transverse reinforcement designed in accordance with the current code provisions for (a) high ductility (beam B4ECH), (b) moderate ductility (beam B4ECM) and (c) low ductility (beam B4ECL).

bars. As discussed earlier, the FE modelling of the beams (indicated in Figure 5.15) does not allow for the varying spacing of the transverse reinforcement specified by the code for the different levels of ductility; however, the amount of this reinforcement was adjusted so as to conform to the code specifications.

Figure 5.23 depicts the load–deflection curves predicted by the analysis of the beams together with the load-carrying capacity predicted by the code. From the figure, it can be seen that, although the amount of transverse reinforcement (for all three beams investigated) was higher than that required to safeguard against shear types of failure, the beams are predicted to fail prematurely, well before their flexural capacity was exhausted. Such behaviour is similar to that predicted for the beam in Section 5.2.2 and it appears to be independent of

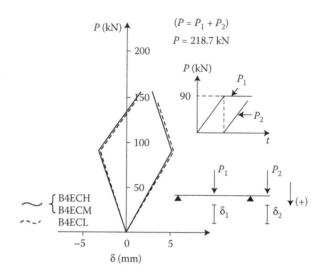

Figure 5.23 RC beams B4ECH, B4ECM and B4ECL under sequential loading. Load–defection curves.

the amount of the longitudinal reinforcement. (It is reminded that for the beams discussed in the present section the longitudinal reinforcement comprises 16 mm diameter bars, as opposed to the 14 mm diameter bars used for the beams discussed in Section 5.2.2.) The causes of the premature failure are discussed in Section 5.2.7.

5.2.6 Safeguarding against shear types of failure

In the present section, it is attempted to safeguard against shear types of failure by adopting two different design approaches: the first is based on the CFP method which was used to design beam B5CFP, and the second on the concepts underlying the current code methods used for designing beam B5COR. Both beams are similar to those discussed in the preceding section except for the transverse reinforcement which is shown in Figures 5.24 and 5.25 for beams B5CFP and B5COR, respectively. Figure 5.24 indicates that for beam B5CFP the transverse reinforcement includes not only links extending throughout the beam depth – with the total amount of such links being significantly lower than that specified by current codes – but, also, links enclosing the compressive zones of the beam – such links not being deemed essential in current design practice. For beam B5COR, Figure 5.25 indicates that the arrangement of transverse reinforcement is similar to that indicated in Figure 5.22 for the case of the beams designed in accordance with the European code, but its amount was nearly double that specified by the code for high ductility. On the basis of the thinking underlying the code provisions, such an amount and arrangement of transverse reinforcement should be considered as 'over-reinforcing' the beam against a shear type of failure.

Figure 5.26 shows the load–deflection curves predicted by analysis for the above beams together with the values of the load-carrying capacity corresponding to the flexural capacity of the beams. The figure indicates that, as for the case of beam B3CFP, designing in compliance with the concept of the CFP achieves the design objectives for load-carrying capacity corresponding to flexural capacity and adequate ductility. On the other hand, the premature failure of beam B5COR in shear indicates that even a significant increase of the code specified amount of transverse reinforcement is insufficient to improve structural behaviour. It appears, therefore, that there is a link between shear failure and the arrangement

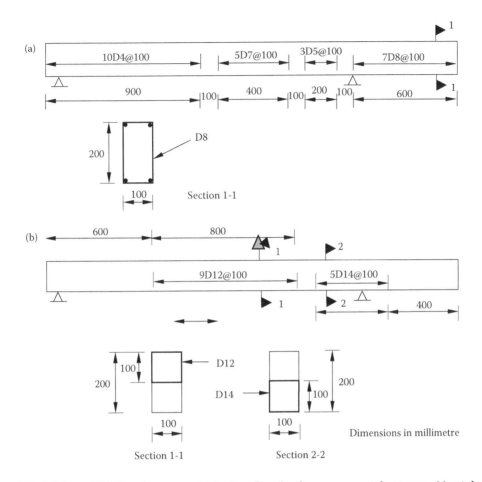

Figure 5.24 RC beam B5CFP under sequential loading. Details of transverse reinforcement: (a) reinforcement extending throughout the beam height; (b) reinforcement enclosing compressive zone. (dimensions in millimetre).

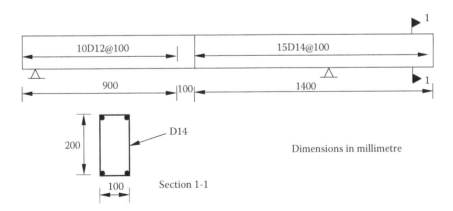

Figure 5.25 RC beam B5COR under sequential loading. Transverse reinforcement details (dimensions in millimetre).

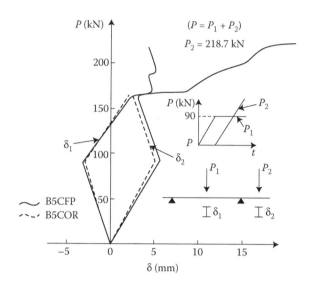

Figure 5.26 RC beams B5CFP and B5COR under sequential loading. Load–deflection curves.

of transverse reinforcement, since the different arrangement of transverse reinforcement specified by the CFP method led to structural behaviour which satisfied the performance requirements of current codes.

5.2.7 Causes of premature failure of beams designed to the European code

The inability of current code-adopted methods for the design of concrete structures to always produce solutions that safeguard the code requirements for structural performance has been attributed to the lack of a sound underlying theory (Kotsovos and Pavlovic 1999, Kotsovos 2014). More specifically, as discussed in Chapter 1, the development of these methods invariably relies on the use of truss or strut-and-tie models for the description of the physical state of structural concrete at its ultimate limit state, the implication of this being that, in contrast with the experimental evidence presented in Chapter 2 which indicates that concrete is brittle in nature – and therefore suffers a complete and immediate loss of strength as soon as visible cracking occurs – cracked concrete is assumed to be characterised by strain-softening behaviour, which is a prerequisite for the formation of inclined struts. It is, therefore, this conflict between assumed and true concrete behaviour that is the underlying cause for the premature failure of the beams designed in accordance with the European code (see Section 5.2.5).

5.3 RC BEAM–COLUMN JOINTS UNDER CYCLIC LOADING

5.3.1 Background

Practical structural analysis of frame-type structures is normally based on the assumption that the common portion (joint) of intersecting beam–column elements behaves as a rigid body. Thus the displacements and rotations of the joint are directly transferred to the ends of the linear elements which intersect at the joint. For the case of RC structures, however, the 'rigid joint' assumption is rather crude, since concrete is weak in tension and therefore all structural elements, joints inclusive, may suffer cracking from early load stages. As a result, the displacements and rotations transferred by a joint to the adjacent linear element ends may be affected by the cracking of the joint, and this effect, which may also affect overall structural behaviour, is not reflected on the results obtained by practical FE analyses employing linear elements.

Amongst the work published to date on the investigation of the effect of cracking of the joint on overall structural behaviour is that of Cotsovos and Kotsovos (2008) which was based on the use of the proposed FE package. Indicative results of this work are presented in the following, since not only was the package found capable to produce realistic predictions of the behaviour of beam–column joints, but also proved to be invaluable as a research tool. The results presented are concerned with the behaviour of one of the RC beam–column sub-assemblages denoted as A2 in Shiohara and Kusuhara (2006) where it was investigated experimentally as part of a benchmark test programme for the validation of mathematical models of RC beam–column joints. It is first shown that the proposed FE package is capable of producing results that correlate closely with the experimental findings and, then, the effect of cracking in the joint on structural behaviour is investigated through a comparative study of the predicted behaviour of the beam–column sub-assemblage with that of the same structure but without allowing cracking in the joint.

5.3.2 Structural form investigated

A schematic representation of the RC beam–column sub-assemblage investigated is shown in Figure 5.27, whereas its cross-sectional characteristics and reinforcement details are provided in Figure 5.28. The longitudinal reinforcement in both the beams and the columns comprises 13 mm diameter bars (D13) with a nominal cross-sectional area of 139 mm² and a yield stress (f_y) of 456 MPa, in the beams, and 357 MPa in the columns. For both beams and columns, the transverse reinforcement comprises 6 mm diameter stirrups (D6) (nominal cross-sectional area of 32 mm²) with yield stress of 326 MPa and a spacing of 50 mm. The mean compressive strength (f_c) of concrete was 28 MPa. It should be noted that the design was based on the Japanese Code of Practice (AIJ 1999), the requirements of which for earthquake-resistant structures are similar to those of ACI 318 (2002) and partly of EC2 (1991), although the latter code specifies a larger amount of stirrups within the joint. However, there is experimental evidence (Ehsani and Wight 1985, Hwang et al. 2005) which shows

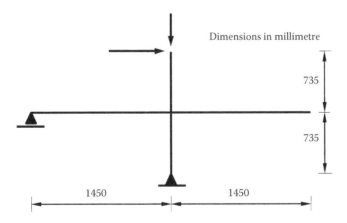

Figure 5.27 RC beam–column sub-assemblage. Schematic representation.

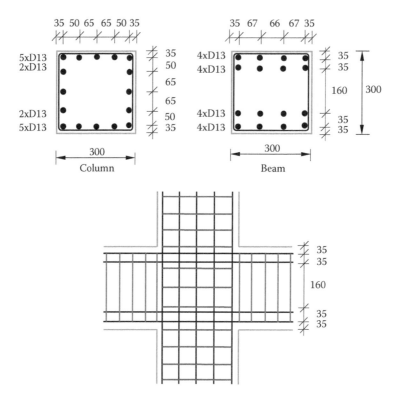

Figure 5.28 RC beam–column sub-assemblage. Design details.

that a larger amount of stirrups than that recommended by ACI-318 may be detrimental to the structural element behaviour and this is, in fact, confirmed in the following.

The structure was subjected to the combined action of a constant axial load (equal to 216 kN) and a lateral displacement-controlled load imposed at the location shown in Figure 5.27 in the manner indicated in Figure 5.29. In the latter figure, the imposed lateral displacement is

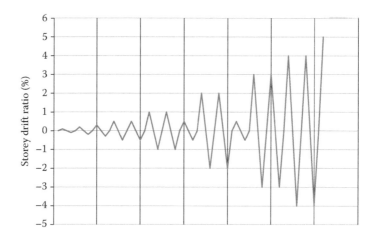

Figure 5.29 RC beam–column sub-assemblage. Transverse loading history.

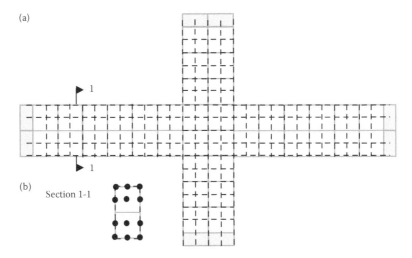

Figure 5.30 RC beam-column sub-assemblage. Finite-element mesh with brick elements (modelling concrete and shaded ends of beam-column members) indicated by continuous lines and truss elements (modelling steel reinforcement) indicated by dashed lines in (a) and thick dots in (b).

expressed in the form of storey drift ratio, that is, the value of lateral displacement of the top end of the column element divided by the vertical distance between the column ends.

5.3.3 FE discretisation

The FE mesh adopted to model the beam–column joint sub-assemblage is shown in Figure 5.30. The structure is subdivided into 48 brick elements as shown in Figure 5.30a, with the load being applied through prismatic members (shaded end portions of the FE mesh) monolithically connected to the end faces of the beam and column elements; these members were subdivided into $2 \times 1 \times 1 = 2$ brick elements, as indicated also in Figure 5.30a. The line elements representing the steel reinforcement were placed along successive series of nodal points in both vertical and horizontal directions as indicated in the figure, with Figure 5.30b showing a typical cross section of the beam–column joint sub-assemblage. Since the spacing of these line elements was predefined by the location of the brick elements' nodes, their cross-sectional area was adjusted so that the total amount and position of the geometrical centre of the compression and longitudinal tension, as well as the total amount of the transverse reinforcement comply with the design specifications.

As discussed in Section 4.4.2, the size of the 27-node Lagrangian brick FEs used is dictated by the philosophy upon which the FE model adopted in the present work is based, which does not employ small FEs. This is because the material model adopted is based on data obtained from experiments in which concrete cylindrical specimens were subjected to various triaxial loading conditions. Consequently these cylinders may be assumed to constitute a 'material unit' for which average material properties are obtained and hence the volume of these specimens provides a guideline to the order of magnitude of the size of the FE that should be used for the modelling of concrete structures.

5.3.4 Results of analysis and discussion

From Figures 5.31 and 5.32, it appears that the correlation between numerical and experimental results is realistic; not only are the predicted values of both load-carrying capacity

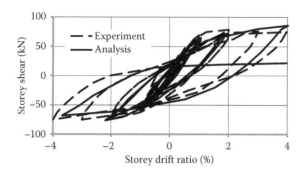

Figure 5.31 RC beam–column sub-assemblage. Numerical (continuous lines) and experimental (dashed lines) relationships between applied load (storey shear) and storey drift ratio under cyclic loading.

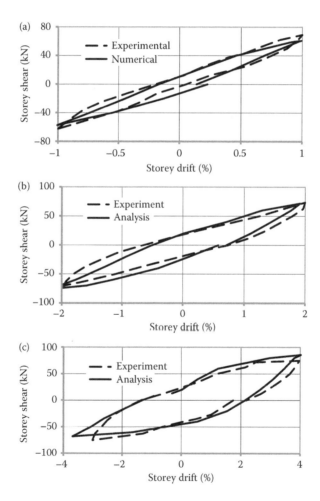

Figure 5.32 RC beam–column sub-assemblage. Numerical (continuous lines) and experimental (dashed lines) curves extracted from Figure 3.31 indicating the relationship between applied load (storey shear) and storey drift ratio for individual hysteresis loops at values of the storey drift ratio equal to: (a) 1%, (b) 2% and (c) 4%.

and maximum horizontal storey drift similar to their experimental counterparts, but also the area enclosed by the hysteretic loops of the numerically predicted storey shear–storey drift ratio curve is similar in size to the area enclosed by the experimentally established hysteretic loops, with the deviations of the predicted from the experimentally established values being of the order of 10%. In fact, the close correlation between analytical and experimental results is more clearly shown in Figure 5.32, which depicts the storey shear–storey drift ratio curves of individual hysteretic loops extracted from Figure 5.31 and corresponding to maximum values of the storey drift ratio equal to 1%, 2% and 4%.

In order to investigate the effect of crack formation within the joint on structural behaviour, the structure was also analysed without allowing crack formation to occur within the joints. The resulting storey shear–storey drift ratio relationship (designated as '1') under monotonic loading is shown in Figure 5.33. The figure also includes the storey shear–storey drift ratio relationship (designated as '2') obtained by analysis which did not allow crack formation within the joints, together with the levels of load-carrying capacity (designated as '3') assessed by hand calculation based on the assumption that these levels correspond to the flexural capacity of the beam components of the structural forms investigated. It should also be noted that the suffixes ' + ' and '−' to the designations '1', '2' and '3' indicate the direction of loading: ' + ' from left to right and '−' from right to left. From the figure, it is interesting to note that, for the structural element analysed without allowing cracking to occur within the joint, the analysis yields values of load-carrying capacity larger than those assessed through hand calculation by an average value of around 10%. Such a difference in the predicted values of load-carrying capacity should be attributed to the approximate nature of the assumptions underlying the hand calculations.

From the same figures, the comparison of curves '1' and '2' indicates that cracking within the joint reduced not only the load-carrying capacity of the structural forms investigated, but also their stiffness, and hence led to a nearly 50% increase of the storey drift ratio when the load exceeded a value of around 25% of the structure's load-carrying capacity; in fact, this load level appears to mark the start of cracking within the joint (see Figure 5.34).

The significant cracking suffered by the joint can be observed in the numerical predictions presented in Figures 5.34 and 5.35 which depict the predicted crack pattern and the deformed shape (with the deflections being magnified by a factor of 10) of the beam–column

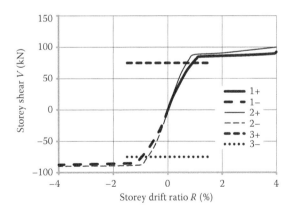

Figure 5.33 RC beam–column sub-assemblage. Numerically established relationships between storey shear and storey drift ratio under monotonic loading (1: cracking of joint allowed; 2: cracking of joint not allowed; 3: hand-calculated value of load-carrying capacity; ' + ' indicates load applied from left to right; '−' indicates load applied from right to left).

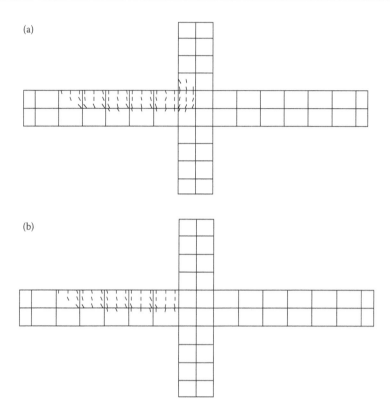

Figure 5.34 RC beam–column sub-assemblage. Predicted crack patterns under a load equal to 25% of its predicted load-carrying capacity under static monotonic loading when cracking of the joint area (a) was and (b) was not allowed to occur. (Short lines within finite elements represent the trace of cracks intersecting the plane of the specimen.)

sub-assemblage at two critical levels of the applied load increasing monotonically to failure: at the level of crack formation within the joint (see Figure 5.34), and at load-carrying capacity (see Figure 5.35). For purposes of comparison the figures also includes the corresponding crack patterns predicted from the analysis of the same structure, but, this time, without allowing crack formation within the joint. From the figures, it can be seen that, except for the joint area, the predicted crack patterns were similar: cracking first initiated at the left-hand side beam; for the case where it was allowed also to occur within the joint, cracking quickly extended to the joint area at a load (storey shear) level of approximately 25% of the predicted load-carrying capacity, and progressively covered a large part of the upper column of the specimen, also penetrating within the lower column in its region adjacent to the joint. Although allowing cracking to occur within the joint appears to have caused a larger horizontal storey drift of the top end of the column component of the specimen, in this form of representation, the effect of joint cracking is not as pronounced as it is when the results are expressed in the form of the storey shear–storey drift ratio curves shown in Figure 5.33, since in the former case this effect is overshadowed by the significant distortion of the specimen's shape caused by the magnification of the deflections.

The effect of crack formation within the joint on the overall deformational response becomes even more apparent when the structure investigated is subjected to cyclic loading. Figure 5.36 shows the storey shear–storey drift ratio relationship established by analysis

(a)

(b)

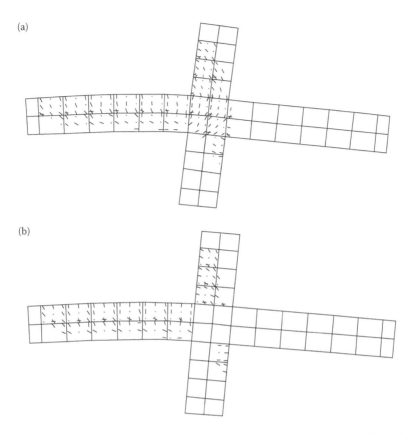

Figure 5.35 RC beam–column sub-assemblage. Predicted crack patterns at the predicted load-carrying capacity under static monotonic loading when cracking of the joint area (a) was and (b) was not allowed to occur. (Short lines within finite elements represent the trace of cracks intersecting the plane of the specimen.)

which allowed crack formation within the joint (curve denoted as '1') together with its counterpart (curve denoted as '2') also established by analysis, but without allowing such cracking to occur. The figure also provides an indication of the values of load-carrying capacity assessed by hand calculation assuming that these values correspond to the flexural capacity of the beam components of the elements. The storey shear–storey drift ratio curves of typical individual hysteresis loops extracted from Figure 5.36 are shown in Figures 5.37 and 5.38. From the figures, it can be seen that, for the case where cracks were not allowed to form within the joint core, the structure exhibited a load-carrying capacity higher than that corresponding to the flexural capacity of the beam component as assessed by hand calculation.

However, the most interesting feature of structural behaviour under cyclic loading is the significant loss of stiffness that characterises the structure when cracking of the joint is allowed to occur: at any given level of the storey shear, the storey drift ratio appears to be larger than that of its counterpart element, for which cracking was not allowed to occur in the joint, by a factor exceeding 2 at load levels close to the structure's load-carrying capacity. This significant loss of stiffness, which characterises RC structures comprising linear elements even under service loading conditions (as indicated in Figure 5.37 extracted from Figure 5.36) is not allowed for in practical structural analysis of RC skeletal structures. As a result, the use of such analysis methods as the means to check whether the (often stringent)

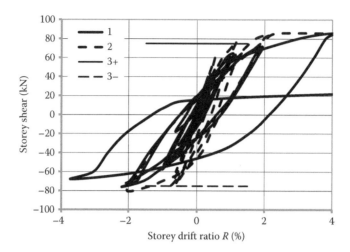

Figure 5.36 RC beam–column sub-assemblage. Numerically established relationships between storey shear and storey-drift ratio under cyclic loading (For notation see Figure 5.33).

performance requirements of current codes for earthquake-resistant design are satisfied appears to be ineffective.

Recent attempts to provide a solution to the above problem have been predominantly directed towards developing, and incorporating into analysis packages suitable for use in practice, purpose-developed constitutive models of RC joint behaviour (see, e.g., Morikawa 2007). The development of such constitutive models usually places emphasis on the description of the post-peak characteristics of joint behaviour and relies on the use of experimental data for the calibration of the models. However, such data are dependent on the interaction between specimen and testing device, with the effect of this interaction, although often significant, being rarely, if ever, considered in the development of the constitutive models. As a

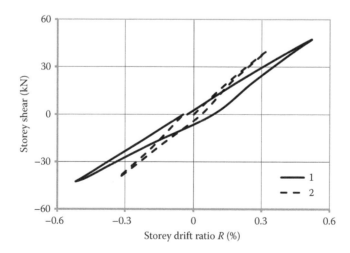

Figure 5.37 RC beam–column sub-assemblage. Numerically established relationships between storey shear and storey drift ratio for the individual load cycle at maximum storey drift ratio 0.5% extracted from Figure 5.36 (For notation see Figure 5.33).

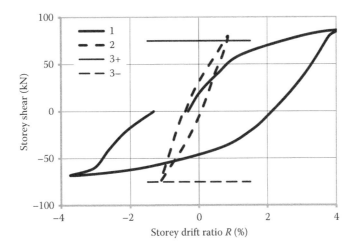

Figure 5.38 RC beam–column sub-assemblage. Numerically established relationships between storey shear and storey drift ratio for the individual load cycle at maximum storey drift ratio 4% extracted from Figure 5.36. (For notation see Figure 5.33.)

result, in the authors' opinion, attempts to improve analysis results through the use of constitutive models of joint behaviour are unlikely to be successful in the not too distant future. On the other hand, analysis models, such as the one used in the present work, though free of the problems described above, are essentially research tools, not user friendly and costly for practical applications.

It would appear from the above, therefore, that an alternative approach would be to aim for an improvement of current design methods so as to safeguard the condition for rigid joint behaviour that underlies current methods for frame analysis. However, such an improvement is unlikely to be achieved through an increase of the amount of transverse reinforcement within the joint, in excess of the amount specified by current codes, since it has already been established by experiment that this approach is ineffective (Ehsani and Wight 1985, Hwang et al. 2005). This was, in fact, also established in the present work by comparing the storey shear–storey drift ratio curve for the structure with its counterpart obtained by analysing the same structure with three times the amount of transverse stirrups within the joint (see Figure 5.39). From the figure, it can be seen that structural behaviour deteriorated, rather than improved. The causes for such behaviour appear to be linked with the findings of an earlier work which demonstrated that providing artificial confinement to concrete when such confinement is naturally provided by the surrounding concrete is ineffective (Kotsovos and Pavlovic 1995). In the case of the structural element investigated, such natural confinement to concrete in the joint is naturally provided by the beam and column elements intersecting at the joint; the provision of transverse stirrups, beyond the amount specified by current codes, as the means to artificially safeguard confinement to the joint appears, not only to be ineffective, but also to reduce the effectiveness of the intersecting beam–column elements to provide confinement.

5.3.5 Concluding remarks

The proposed FE model is not only found to produce realistic predictions of the behaviour of beam–column elements under cyclic loading, but also to demonstrate that crack formation

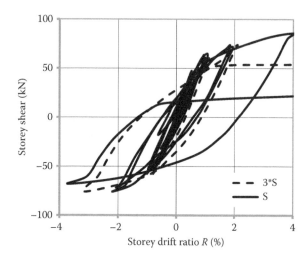

Figure 5.39 RC beam–column sub-assemblage. Effect of an amount of transverse stirrups, in excess of that specified by ACI, on the behaviour of beam–column joint elements under cyclic loading.

within the joint of the beam–column elements has a significant effect on the overall structural behaviour; as a result, structural analysis based on the assumption of rigid joints yields results of dubious validity when applied to RC structures.

In view of the above, it appears that there is an urgent need for improving current design methods so as to safeguard the condition for rigid joint behaviour that underlies current methods for frame analysis.

5.4 STRUCTURAL WALLS UNDER CYCLIC LOADING

The present case study is concerned with a numerical investigation of the behaviour of structural elements for which, as for the case discussed in preceding section, there is published information obtained from cyclic tests. The structural elements considered are three of the RC structural walls denoted as walls SW31, SW32 and SW33 in Lefas (1988) where full details of the experimental investigation are provided. In what follows, only the main results of the numerical investigation are presented; full details can be found in Cotsovos and Pavlovic (2005).

5.4.1 Wall details

All walls were 650 mm wide, 1,300 mm high and 70 mm thick. The values of the uniaxial cylinder compressive strength (f_c) of the concrete used was equal to approximately 35, 53 and 49 MPa for walls SW31, SW32 and SW33, respectively, whereas the values of the yield stress (f_y) of the steel was 420, 520 and 470 MPa, for the 4, 6 and 8 mm diameter bars used, respectively. An identical arrangement of reinforcement was used for all three specimens and this is shown in Figure 5.40. During the experiment each wall was monolithically connected to a rigid RC prism at both top and bottom. The bottom prism was firmly bolted to the laboratory strong floor in order to resemble a rigid type of foundation, while at the end faces of the top prism the external loading was imposed through two 50-ton jacks. The prisms were designed so as to remain undamaged throughout the course of the experiment.

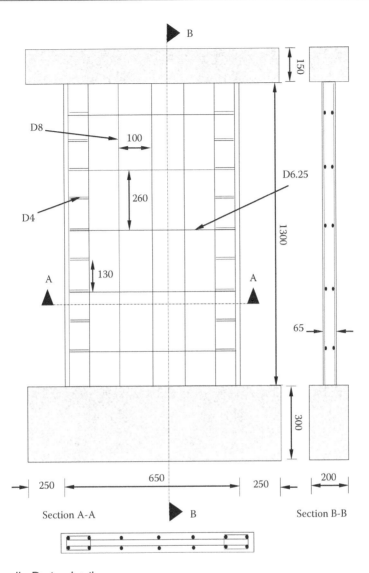

Figure 5.40 RC walls. Design details.

Each wall was subjected to different cyclic loading conditions. After a prescribed number of load cycles, the load increased monotonically up to the full loss of load-carrying capacity. The loading histories used for the tests are presented in Figure 5.41. On the right-hand side, the loading history is presented in the form of imposed displacements, whereas, on the left-hand side, it is presented in the form of imposed forces.

5.4.2 FE modelling

Concrete was modelled by using 27-node Lagrangian brick elements whereas 3-node iso-parametric truss elements were used for the modelling of the reinforcement. The structural walls discussed above were discretised as shown in Figure 5.42. The steel elements were placed along successive rows and columns of nodes in the longitudinal and transverse

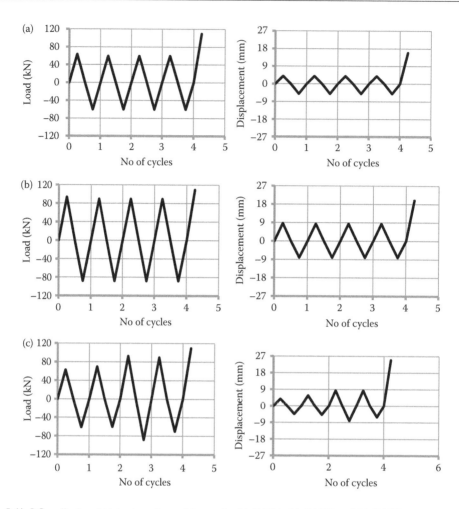

Figure 5.41 RC walls. Load histories adopted for walls: (a) SW31, (b) SW32 and (c) SW33.

directions in a manner that matched, as closely as possible, the amount and location of the actual reinforcement. On the top, a layer of rigid brick elements was added to represent the rigid RC prism monolithically connected to the top of each wall. Because of the symmetry of the specimen's cross section with respect to the vertical plane in the direction of the applied load, only one half of the actual specimen was modelled. As a result, a smaller number of FEs were used, thus reducing the computational cost. The above symmetry in the FE model was effected by restraining the displacements normal to the plane of symmetry. Finally, the bottom face of the FE mesh was fixed in order to represent the rigid type of foundation imposed in the experiment. It should be noted that, although the experimental load was applied in the form of imposed displacements, in the actual analysis the state of development of the software at the time the work was carried out allowed loading to be applied only in the form of imposed forces. Although this might have led to some deviation of the numerical results from their experimental counterparts at load levels near the wall's load-carrying capacity, such deviation was minimised through the use of the smallest possible load increments.

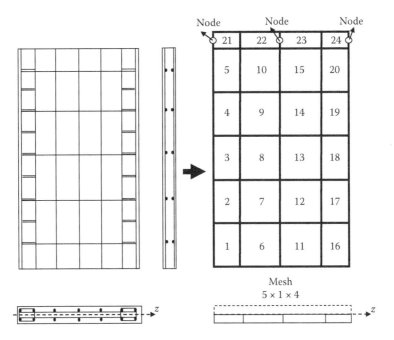

Figure 5.42 RC walls. FE model.

5.4.3 Numerical results

The results of the analysis are presented in Figures 5.43 through 5.45 in the form of load–displacement curves, whereas typical results describing the cracking process that specimen SW31 undergoes during the first loading cycle are presented in Figure 5.46. The cracks begin to appear in the bottom left area of the specimen (where the tensile stresses are most critical) at the value of the external load of 10 kN. As the imposed load increases (from 0 through 64 kN), new cracks form. During unloading (from 64 through 0 kN) many cracks gradually start to close. When the external load is imposed in the opposite direction (from 0 to −64 kN) the crack-closure procedure continues (although the number of cracks that need to close now are significantly less), but at the same time is followed by the opening of new cracks due to the change in direction of the imposed load.

5.4.4 Discussion of numerical results and comparison with experimental data

A comparison between the numerical predictions (cyclic and monotonic case studies) and the experimental data for specimen SW31 is presented in Figure 5.43. It can be seen that under monotonic loading specimen SW31 had a load-carrying capacity of 130 kN, whereas under cyclic loading its residual load-carrying capacity was found to be 104 kN. It is also evident that the ductility attained under monotonic loading was twice that reached under cyclic-loading conditions. This should mainly be attributed to the fact that the specimen suffered excessive cracking under successive load cycles, thus lowering its ultimate strength and ductility.

The maximum load applied to specimen SW31 at each loading cycle was 64 kN, which is approximately 50% of the (numerically predicted) ultimate strength under monotonic

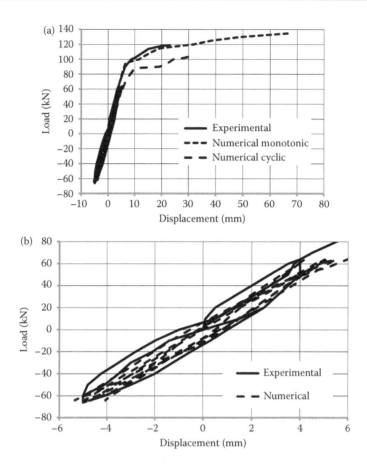

Figure 5.43 RC walls. Comparison between numerical predictions obtained for the monotonic and cyclic case studies with experimental data for specimen SW31: (a) full loading range; (b) region of cyclic loops magnified.

loading and 62% of the (numerically predicted) residual load-carrying capacity when it was loaded to failure after it had been subjected to the prescribed number of loading cycles. At these levels of loading the program did not show any signs of numerical instability during the solution process.

The numerical results obtained from the cyclic-loading case study are similar to the experimental data (this can be seen more clearly in the magnified Figure 5.43b), indicating that the program is able to realistically predict the behaviour of RC structures when subjected to this medium level of cyclic loading. The maximum deflection at each load cycle and the corresponding applied load are nearly identical for test data and numerical predictions, the deviation becoming somewhat larger mainly when the load was eventually increased monotonically to failure. The experimental results show that specimen SW31 had a load-carrying capacity of 119 kN whereas the numerically predicted load-carrying capacity of the same specimen subjected to the same loading conditions was 104 kN, that is, 87% of the value of the ultimate strength established experimentally. Similarly good agreement was found for the values of deflections at failure established by experiment (23 mm) and analysis (30 mm). Furthermore, the areas enclosed by the hysteretic loops formed by the external load–displacement curves established by experiment and predicted analysis are similar. This area expresses the energy dissipated by the RC wall due to the

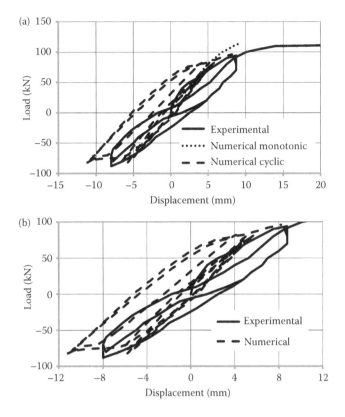

Figure 5.44 RC walls. Comparison between numerical predictions obtained for the monotonic and cyclic case studies with experimental data for specimen SW32: (a) full loading range; (b) region of cyclic loops magnified.

non-linear behaviour of concrete as a result of cracking and yielding of the reinforcement. The fact that in both cases – experiment and analysis – this area is approximately the same shows that the proposed FE model is able to realistically model the crack opening and closure procedures that the RC wall undergoes during each load cycle.

For the case of specimen SW32 the (numerically) predicted ultimate strength was 115 kN under monotonic loading and 96 kN under cyclic loading (see Figure 5.44a). As for the case of specimen SW31, specimen SW32 under cyclic loading failed at a lower load and exhibited smaller maximum deflection when compared to its monotonically loaded counterpart. The maximum load applied to specimen SW32 at each loading cycle was 84 kN, which is approximately 73% of the specimen's (numerically predicted) ultimate strength when subjected to monotonic loading but 87% of the specimen's (numerically predicted) residual load-carrying capacity when loaded to failure after it had been subjected to the prescribed number of loading cycles. Even at these high levels of loading the program was found to be numerically stable.

By comparing successive cyclic loops of the horizontal load–displacement curve, it appears that the size of these loops increases with the number of cycles. The area of the cyclic loop corresponds to the amount of energy lost due to the non-linear behaviour of concrete caused by the cracking of concrete and the yielding of steel. The higher cyclic-load level (84 kN) imposed on specimen SW32 appears to have caused more significant cracking and hence a loss of energy larger than that suffered by specimen SW31 which was subjected to a lower level of cyclic load (64 kN).

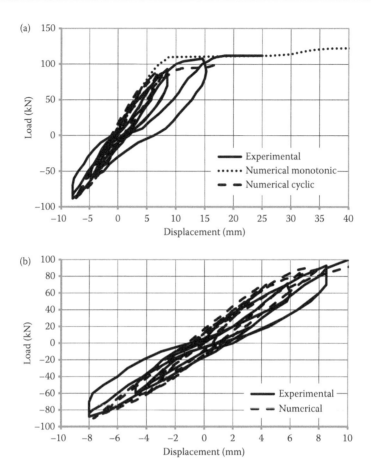

Figure 5.45 RC walls. Comparison between numerical predictions obtained for the monotonic and cyclic case studies with experimental data for specimen SW33: (a) full loading range; (b) region of cyclic loops magnified.

As for the case of specimen SW31, the numerical results obtained for specimen SW32 under cyclic loading are similar to their experimental counterparts (see Figure 5.44), indicating that the program used is capable of realistically predicting the behaviour of RC structures even when these are subjected to successive levels of loading close to the ultimate strength of the specimen. The maximum displacement at each load cycle and the corresponding load are similar for both experiment and analysis. Furthermore, as in the case of SW31, the area enclosed by the hysteretic loops formed by the external load–displacement curves predicted by both experiment and analysis are similar, thus proving once again that the FE model adopted in the current investigation is efficient in modelling the cracking procedure that concrete undergoes. Under cyclic loading, the experimental results indicate that specimen SW32 had a load-carrying capacity of 112 kN, whereas the numerically predicted load-carrying capacity of the specimen under the same loading conditions was 96 kN or approximately 86% of the ultimate strength established experimentally.

However, a discrepancy in the maximum attained displacement is noticeable in the last stage when the load increased monotonically to failure. Experimentally, this maximum displacement was recorded to be around 20–25 mm, whereas the numerical prediction is

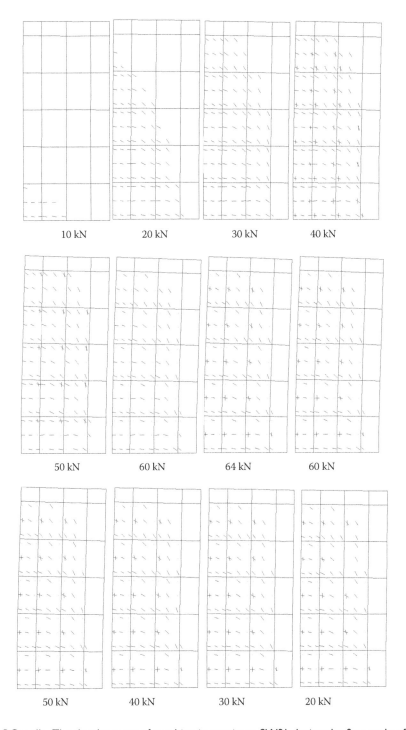

Figure 5.46 RC walls. The development of cracking in specimen SW31 during the first cycle of the cyclic-loading regime.   *(Continued)*

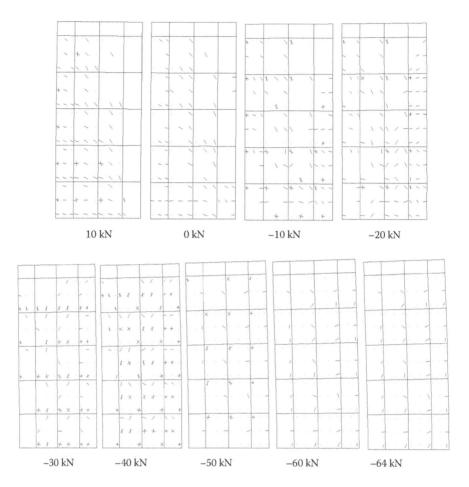

10 kN   0 kN   −10 kN   −20 kN

−30 kN   −40 kN   −50 kN   −60 kN   −64 kN

Figure 5.46 (*Continued*) RC walls. The development of cracking in specimen SW31 during the first cycle of the cyclic-loading regime.

9 mm. The main reason for such differences is the fact that in the numerical investigation failure is considered to occur when the structure's stiffness matrix becomes non-positive definite whereas in the actual experiment failure occurs when the RC structural form under investigation suffers total loss of its load-carrying capacity. In the present case, this occurred due to the disruption of the continuity of concrete caused by excessive crack formation in the lower area of the RC wall. After the degradation of the concrete within the lower regions of the wall, the actual structure used in the test may still have been capable of maintaining its load-carrying capacity through alternative resistance mechanisms such as, for example, dowel action. However, this type of behaviour cannot be described numerically, as the development of such alternative resistance mechanisms are not allowed for by the analysis procedure developed. It is significant to note that, when observing the experimental results, it is clear that in the final stages of the loading procedure the load–displacement curve becomes practically horizontal beyond a displacement of about 13 mm (comparable with the numerical prediction of 9 mm) which means, that at this point, the RC wall suffered extensive cracking (especially in its lower area) which led its stiffness to drop dramatically and that the RC wall resorted to alternative resistance mechanisms. Evidently, such alternative mechanisms (not catered for by the numerical model) are highly unstable, of short

duration, and cannot be relied upon in safe design: they are really manifestations of post-failure phenomena and, as such, are of no relevance to designers.

In the last case study, the ultimate strength of specimen SW33 predicted by the program was 120 kN when subjected to monotonic loading but, under cyclic loading, the program was unable to complete all the prescribed loading cycles: namely, in the experiment there had been five cycles followed by monotonic loading to failure, while in the FE analysis failure occurred during the fifth load cycle due to premature failure of the specimen during the fifth load cycle (Figure 5.45). The maximum load applied to specimen SW33 at the two initial loading cycles was 64 kN whereas for the next two it was 90 kN. After concluding these loading cycles the specimen failed at the last prescribed loading cycle at a load level of 100 kN which is close to the specimen's (experimentally determined) load-carrying capacity (112 kN).

The experimental maximum displacement at each load cycle and the corresponding applied load are similar to the displacements predicted by the program and the corresponding imposed load (see Figure 5.45). The maximum load applied to specimen SW33 during the first two loading cycles was 70 kN, which is approximately 53% of the (numerically predicted) ultimate strength under monotonic loading and 70% of the (numerically predicted) residual load-carrying capacity when it was loaded to failure after it had been subjected to the prescribed number of loading cycles. Then, the maximum load applied to specimen SW33 during loading cycles 3 and 4 was increased to 90 kN, which is approximately 75% of the (numerically predicted) ultimate strength under monotonic loading and 90% of the (numerically predicted) residual load-carrying capacity in the cyclic case study. At these levels of loading the FE scheme did not show signs of numerical instability during the solution process, proving again that the program used is capable of realistically predicting the behaviour of RC structures when subjected to levels of cyclic loading close to the ultimate strength of the specimen.

The experimental results indicate that specimen SW33 had a load-carrying capacity of around 112 kN, whereas the numerically predicted load-carrying capacity of the specimen under the same loading conditions was 100 kN, or approximately 89% of the ultimate strength established experimentally. Furthermore, the maximum displacement established experimentally was found to be about 25 mm whereas in the numerical investigation it was found to be 17 mm. Moreover, as in the case of SW31 and SW32, the area enclosed by the hysteretic loops formed by the external load–displacement curves predicted by both experiment and analysis are similar, thus proving that the FE model adopted in the current investigation is efficient in modelling the cracking procedure that concrete undergoes even during load cycles in which the level of loading reaches the load-carrying capacity of the specimen.

The premature failure of the specimen (in terms of the number of cycles attained, namely four and a half instead of five and a half) of the specimen in the last load cycle and the small deviation between experimental and numerical predictions could be attributed to the fact that, as in the case of RC wall SW32, failure, during the numerical investigation, is considered to occur when the structure's stiffness matrix becomes non-positive definite and that at this stage the RC structure is resorting to alternative resistance mechanisms such as, for example, dowel action, in order to extend (albeit very briefly) its load-carrying capacity.

5.4.5 Conclusions

Through the comparative study between the numerical predictions and the experimental data, it is established that the non-linear strategy adopted is able to describe the hysteretic behaviour of concrete even when the RC specimens are subjected to high levels of loading, that is, close to their load-carrying capacity. Any deviation between experimental results and numerical predictions are within the order of accuracy of structural engineering design and are due to a number of causes. These include post-failure phenomena, applied load control instead of displacement control (especially relevant when large increases in displacement under constant load are experienced, as in the case of specimen SW32), and time-dependent effects (Lefas 1988) which are noticeable in static cyclic tests (as those presently modelled) where cracking in the concrete medium continues to evolve even when the external load remains constant (such effects becoming more apparent as the value of the applied load increases and approaches the load-carrying capacity of the specimen), whereas these effects are much less prominent in dynamic analysis.

It may be concluded, therefore, that, despite the small divergence between experimental and numerical results, the non-linear procedure adopted in the FE model used in the present work was found to perform satisfactorily, providing a realistic description of the hysteretic behaviour of RC structural elements under cyclic loading even when this was of a severe and static nature. The use of this model to describe the behaviour of RC structures under dynamic cyclic loading (Cotsovos 2004) yielded equally satisfactory results, as is reported in Chapters 7 and 8.

5.5 NUMERICAL EXPERIMENTS ON FLAT SLABS

The FE model has often been used to conduct numerical experiments for the verification of new concepts which are considered to provide a better understanding of structural concrete and their implementation in practical structural design may lead to an improvement of the behaviour of RC structures. One such concept has led to the development of a unified ultimate limit-state method – the CFP method (Kotsovos and Pavlovic 1999, Kotsovos 2014) – for the design of concrete structures. The effectiveness of the reinforcement designed in accordance with the CFP method for safeguarding against punching of flat slabs was verified by carrying out numerical tests, through the use of the FE model, with the results obtained showing a significant improvement of flat slab behaviour when compared with that of similar slabs reinforced in compliance with the methods implemented in current codes such as ACI 318 and EC2. A concise description of the above numerical-test programme is presented in what follows, with full details being provided elsewhere (Kotsovos and Kotsovos 2010).

5.5.1 Slabs investigated

The slabs investigated are square in shape with 2000 mm side, 200 mm depth and flexural reinforcement ratio (ρ) of 0.77%. The uniaxial (cylinder) compressive strength of concrete and the yield stress of the longitudinal and shear reinforcement are taken equal to $f_c = 30$ MPa, $f_y = 500$ MPa and $f_{yv} = 220$ MPa, respectively. The slab is considered to be monolithically connected to a column with a square cross section of 400 mm side, and subjected to a monotonically increasing uniform displacement imposed on points arranged symmetrically about the slab's axes of symmetry so as to form a near circular curve with its centre coinciding with the geometric centre of the slab and its radius being approximately equal to 900 mm (see Figures 5.47 through 5.49).

The figures also provide an indication of the reinforcement arrangements resulting from the methods investigated. It should be noted that, in contrast with the codes which specify only transverse reinforcement within a predefined distance from the column–slab interface, the CFP method specifies both horizontal and vertical reinforcement for sustaining the tensile

Total amount of reinforcement against punching: 4 * 2834 = 11,336 mm²

——— Location of longitudinal reinforcement

⊗   Locations of induced displacement

●   Locations of shear reinforcement (2); 174 mm²@100

●   Locations of shear reinforcement (1); 28 mm²@100

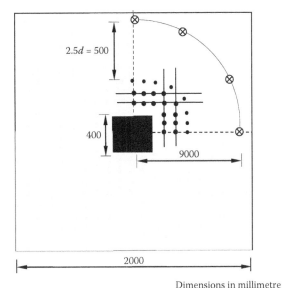

Dimensions in millimetre

Figure 5.47 RC flat slabs. Design details in accordance with the CFP method.

Total amount of reinforcement against punching: 4 * 1608 = 6432 mm²

⊗ Locations of induced displacement

● Locations of shear reinforcement; 67 mm²@100

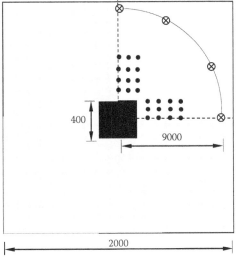

Dimensions in millimetre

Figure 5.48 RC flat slabs. Design details in accordance with ACI318.

Total amount of reinforcement against punching: $4 * 1272 = 5088$ mm^2

⊗ Locations of induced displacement

● Locations of shear reinforcement; 53 mm^2@100

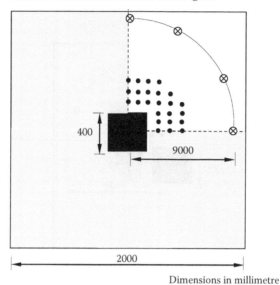

Dimensions in millimetre

Figure 5.49 RC flat slabs. Design details in accordance with EC2.

stresses developing within the compressive zone. The CFP specified reinforcement is distributed within the strips with width $w_c + d$ (where w_c is the side of the column's cross section and d the effective depth of the slab) intersecting at the column head and extending in parallel to the slab sides from the column–slab intersection to a distance of $2.5d$ from the location of the imposed displacement. (As discussed later, distributing the vertical reinforcement within a strip with the aforementioned width is found to produce the most effective design solution.) Moreover, the CFP method also specifies transverse reinforcement along the perimeter of the region extending radially from the column's centre to a distance of $2.5d$ from the geometric locus of the imposed displacement. It should also be noted that, unlike the codes, the CFP method ignores any contribution other than that of the reinforcement to the slab's resistance to punching once the tensile strength of concrete in the compressive zone is exhausted.

In all cases the stirrup spacing was taken constant and equal to 100 mm, with the amount of stirrups being expressed in terms of the stirrup cross-sectional area rather than diameter for purposes of easier comparison. Depending on the method of design, the slabs are referred to as CFP, EC2 and ACI, whereas those without shear reinforcement as CON (control). It is interesting to note in the figures that the CFP method specifies a significantly larger amount of transverse reinforcement than that specified by current-code methods: moreover, the proposed method also specifies a layer of horizontally placed bars across the slab strips in the region of their compressive zone adjacent to the column–slab interface.

5.5.2 Mesh discretisation adopted

Due to twofold symmetry only one-quarter of the slabs is analysed, with this portion being discretised as shown in Figure 5.50. Concrete is modelled by means of 27-node Lagrangian brick elements. Longitudinal and vertical reinforcement is represented by 3-node line

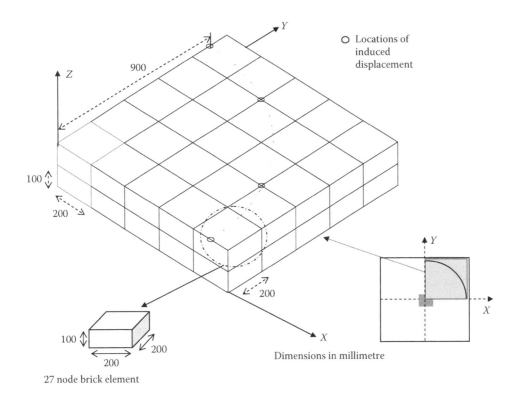

Figure 5.50 RC flat slabs. FE mesh.

elements of appropriate cross-sectional areas possessing axial stiffness only. The line elements used to model the steel reinforcement are not shown in the figure for clarity purposes; these elements are placed along consecutive nodes of the brick elements so as to maintain the amount and arrangement of various types of reinforcement per unit area equal to that of the slabs shown in Figures 5.47 through 5.49.

5.5.3 Results of analysis and discussion

The verification of the analysis package used has formed the subject of the preceding sections; for the purposes of the present work, however, it is considered essential to provide additional evidence of the package's ability to yield realistic predictions of flat slab behaviour. Such evidence is provided in Figure 5.51 which shows the load–displacement curves of two typical flat slabs (with and without transverse reinforcement) with geometric characteristics and boundary conditions similar to those of the slab used as the basis for investigating the validity of the proposed design method. The figure also shows the experimentally established values of load-carrying capacity, with the latter being also shown in Table 5.1 together with the slabs' design details; full design and test details are provided elsewhere (Kinnunen et al. 1978). From both figure and table, it can be seen that the predicted values of load-carrying capacity correlate closely with their experimental counterparts, with the former being smaller than the latter by less than 5%.

Having established the package's ability to yield realistic predictions of slab behaviour, it is first used to establish the most effective spread (w_r) of the vertical reinforcement specified by the CFP method across the slab-strip width. Three cases of w_r are investigated:

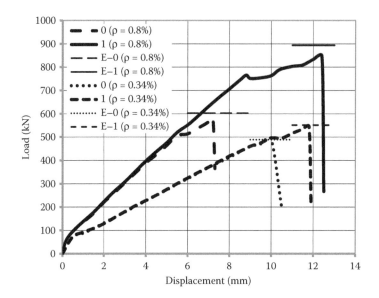

Figure 5.51 RC flat slabs. Analytical relationships between load and displacement for two typical slabs with (denoted as 1) and without (denoted as 0) conventional transverse reinforcement and experimentally established values (indicated with the prefix E) of load-carrying capacity for the cases of flexural reinforcement ratios equal to 0.34% and 0.8%.

$w_r = w_c + 2d - 800$ mm, $w_r = w_c + d = 600$ mm and $w_r = w_c = 400$ mm. The results obtained, expressed in the form of load–deflection relationships, are shown in Figure 5.52. From the figure, it can be seen that distributing the vertical reinforcement over a width $w_r = w_c + d = 600$ mm yields a small, yet distinct, increase in load-carrying capacity. It is proposed, therefore, to distribute the vertical reinforcement specified by the proposed method across the slab strips to a distance $(w_c + d)/2$ on either side of their axes of symmetry. This rule has been followed thereafter when designing in accordance with the proposed method.

An indication of the effect of the amount and arrangement of the transverse reinforcement specified by the proposed and the code methods investigated herein is given in Figure 5.53. The figure includes (a) the load–displacement relationship predicted for the control slab (slab without transverse reinforcement), (b) the value of load-carrying capacity corresponding to

Table 5.1 RC flat slabs. Design details and comparison between experimental and analysis predicted ultimate loads (P_u)

Slab ref.	Dimensions (mm)			Material properties (mm)		Main reinf. $\rho \times 10^{-3}$	Stirrups (mm^2)	P_u (kN)		Test/ analysis
	d	c	a	f_c	f_y			Test	Analysis	
S2.1	200	250	2,400	24.2	657	8	–	603	583	1.03
S2.1s	195	250	2,400	24.9	501	8.2	2513	894	848	1.05
S2.3/S2.4	200	250	2,400	25.4	668	3.4	–	489	490	1.00
S2.3s/S2.4s	198	250	2,400	24.7	671	3.4	1256	552	545	1.01

Note: *d* is the slab's effective depth; *c* is side of the square loaded area; *a* is the slab span; ρ is the flexural steel-reinforcement ratio.

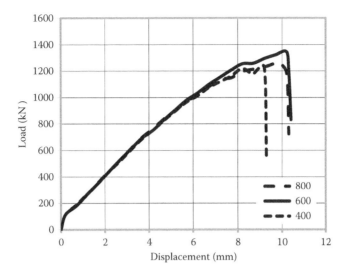

Figure 5.52 RC flat slabs. Analytical relationships between load and displacement for a typical slab with the vertical reinforcement designed in accordance with the CFP method distributed at various widths (800, 600 and 400 mm) across the slab strips specified by the design method adopted.

flexural capacity assessed through the use of the yield line theory and used as the basis for the design of the transverse reinforcement and (c) the value of the load corresponding to punching failure as assessed in accordance with the CFP method. From the figure, it can be seen that designing in accordance with the proposed method leads to a realistic prediction of the load-carrying capacity, since the predicted value is smaller than the value corresponding to flexure capacity by less than 9%. In fact, this margin may even be smaller, as the latter

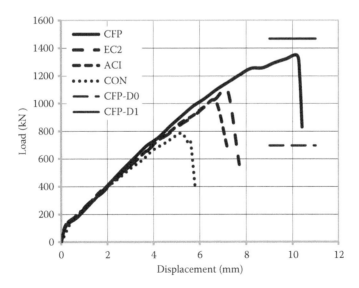

Figure 5.53 RC flat slabs. Analytical load–displacement curves for a typical slab without (curve CON) and with the transverse reinforcement designed in accordance with the CFP, ACI and EC2 methods. Load-carrying capacities assessed by the CFP method for the slab with and without transverse reinforcement are indicated as CFP-D1 and CFP-D0, respectively.

value, as discussed above, has been assessed by yield line theory which is known to produce values that tend to overestimate load-carrying capacity. The value of load-carrying capacity predicted by analysis for the control slab appears also to correlate closely with the value assessed in accordance with the CFP method. On the other hand, designing the transverse reinforcement in accordance with the code adopted methods leads to analytical predictions of the load-carrying capacity which are considerably smaller than the design value.

It could be argued that such a difference in behaviour may be attributed to the considerably larger amount of transverse reinforcement specified by the proposed method, which is approximately twice as large as that specified by the code methods. And yet, Figure 5.54 indicates that an increase of the amount of transverse reinforcement designed in compliance to code provisions to the level specified by the proposed method, but without changing its arrangement, is essentially ineffective. It appears from the above, therefore, that the superior performance of the slab designed in accordance with the proposed method should be predominantly attributed to the specified arrangement, rather than amount, of the transverse reinforcement.

The beneficial effect of the transverse reinforcement arrangement specified by the proposed method is also demonstrated in Figure 5.55. The figure shows the load–deflection curves of the slabs designed in compliance with the code specifications together with that of the slab designed in accordance with the CFP method modified so as to allow for the contribution of concrete to the punching resistance of the slab. Although this modification results in a significant reduction of the amount of the transverse reinforcement to a level comparable to the code specified amount, the figure indicates that it leads to a considerable improvement of the slab load-carrying capacity when compared with the load-carrying capacity of the slabs designed to the code specifications. However, the results also indicate that the contribution of concrete is not as effective as the contribution of the additional transverse reinforcement specified when the contribution of concrete is ignored.

An indication of the validity of the concepts underlying the proposed method may be obtained by investigating whether the causes of punching do indeed relate with the

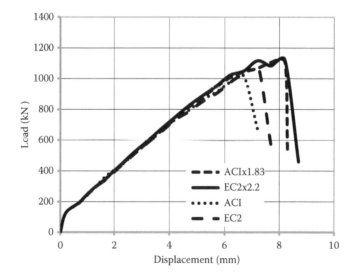

Figure 5.54 RC flat slabs. Numerically established relationships load–displacement curves for a typical slab with the transverse reinforcement arranged in accordance with the ACI and EC2 methods in the amounts specified by the proposed (curves ACI × 1.83 and EC2 × 2.2) and the code methods (curves ACI and EC2).

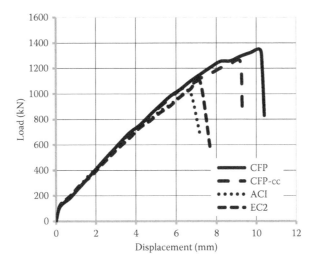

Figure 5.55 RC flat slabs. Analytical relationships between load and displacement for a typical slab with the vertical reinforcement designed in accordance with the ACI, EC2 and CFP methods, as well as the CFP method modified so as to allow for the contribution of concrete to the slab resistance to punching (CFP-cc).

transverse tensile stresses developing within the compressive zone of the slab strips. This may be achieved by comparing the load–deflection curve obtained for a slab with the vertical reinforcement within the slab strips extending throughout the slab depth with that of the same slab with the same vertical reinforcement but this time extending to half the slab depth, the latter being slightly larger than the compressive zone depth. Such a comparison is made in Figure 5.56 which shows that placing vertical reinforcement within the compressive zone only is sufficient for the slab to attain its design load-carrying capacity, thus confirming that the development of such tensile stresses is one of the underlying causes of punching.

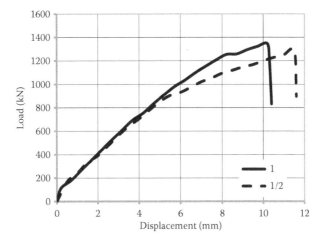

Figure 5.56 RC flat slabs. Analytical relationships between load and displacement for a typical slab with the vertical reinforcement designed in accordance with the CFP method extending to either half (1/2) or the full (1) slab effective depth.

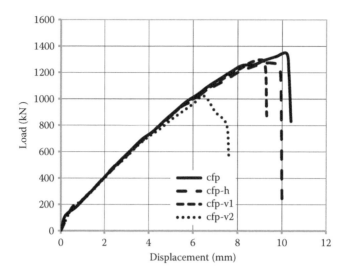

Figure 5.57 RC flat slabs. Effect of the various types of transverse reinforcement specified by the CFP method on the numerically established load–displacement relationships. (Notation: 'cfp' indicates full compliance with design method; 'cfp-h' slab without compression reinforcement; 'cfp-v1' without the vertical reinforcement specified for the region of abrupt change in the direction of the CFP; 'cfp-v2' with the amount of vertical reinforcement specified for slab strips reduced to a nominal value.)

An indication of the significance of the various types of transverse reinforcement specified by the proposed method may be obtained through a comparison of the load–deflection curves shown in Figure 5.57. From the figure, it can be seen that omitting either the horizontal compressive reinforcement (curve cfp-h) or the vertical reinforcement along the perimeter of the region extending from the column's centre to a distance of 2.5*d* from the geometric locus of the induced displacements (curve cfp-v1) results in a small reduction of both the load-carrying capacity and the maximum deflection of the slab. The loss of load-carrying capacity becomes significant when reducing the vertical reinforcement of the slab strips (curve cfp-v2) to a nominal amount, and this is a further indication of the significance of such reinforcement in preventing failure due to the development of transverse tensile stresses within the compressive zone. In fact, the slab behaviour without such reinforcement is similar to that of the slabs designed in compliance with the code provisions, as indicated by comparing the relevant load–deflection curves in Figures 5.53 and 5.57.

5.5.4 Concluding remarks

The CFP method specifies transverse reinforcement which differs from that resulting from the methods adopted by current codes in both arrangement and quantity, and, in contrast with code methods, ignores the contribution of concrete to a slab's resistance to punching. The application of the CFP method for safeguarding against punching is verified by numerical testing through the use of the NLFEA package presented in the preceding chapter which is found to be capable of yielding realistic predictions of the behaviour of concrete structures, flat slabs included.

Through such numerical testing, it is found that in contrast with the code methods the CFP method achieves the design aim for load-carrying capacity, with the transverse reinforcement arrangement underlying the different structural behaviour exhibited by the slab when designed by different methods. In fact, it is found that the effect of increasing the code specified amount of vertical reinforcement to the amount specified by the CFP method is ineffective.

The effect of the various types of transverse reinforcement specified by the CFP method on slab behaviour has also been established by numerical testing, which also provided evidence in support of the validity of the concepts underlying the proposed method.

REFERENCES

ACI 318, 2002, *Building Code Requirements for Structural Concrete (ACI 318-02) and Commentary (ACI 318R-02)*, American Concrete Institute, Detroit, MI.

AIJ 1999, *Design Guidelines for Earthquake resistant Reinforced Concrete Buildings based on Inelastic Displacement Concept*, Architectural Institute of Japan, Tokyo (in Japanese).

Bresler B. and Scordelis A., 1963, Shear strength of reinforced concrete beams, *Journal of the ACI*, 60(1), 51–74.

Cotsovos D. M., 2004, Numerical investigation of structural concrete under dynamic (earthquake and impact) Loading. PhD thesis, University of London, UK

Cotsovos D. M. and Kotsovos M. D., 2008, Cracking of RC beam/column joints: Implications for practical structural analysis and design, *The Structural Engineer*, 86(12), 33–39.

Cotsovos D. M. and Pavlovic M. N, 2005, Numerical investigation of RC structural walls under cyclic loading, *Computers and Concrete*, 2, 215–235.

Ehsani M. R. and Wight J. K., 1985, Exterior reinforced concrete beam-to-column connections subjected to earthquake-type loading, *ACI Journal*, 82(4), 492–499.

Elstner R. C. and Hognestad E., 1956, Shearing strength of reinforced concrete slabs, *ACI Journal*, 53, 29–58.

Eurocode 2 (EC2), 1991, *Design of Concrete Structures – Part 1: General Rules and Rules for Buildings*, European Committee for Standardization, CEN.

Eurocode 8 (EC8), 1994, *Design Provisions for Earthquake Resistance of Structures – Part 1-1: General Rules, Seismic Actions and General Requirements for Structures*, European Committee for Standardization, CEN.

Hwang S.-J., Lee H.-J., Liao T.-F., Wang K.-C. and Tsai H.-H., 2005, Role of hoops on shear strength of reinforced concrete beam-column joints, *ACI Structural Journal*, 102(3), 445–453.

Jelic I., Pavlovic M. N. and Kotsovos M. D., 2003, Towards the development of a method suitable for the strength assessment of existing reinforced concrete structures, *Proceedings of the FIB Symposium Concrete Structures in Seismic Regions*, Athens, May, Paper #246 (6 pages/on CD-ROM).

Kotsovos M. D., 2014, *Compressive Force-Path Method: Unified Ultimate Limit-State Design of Concrete Structures*, Springer, Cham.

Kotsovos M. D., Bazes S., and Lefas I. D., 1994, A contribution into the investigation of the validity of the new Hellenic code of practice for the design of concrete structures, *Proceedings of the 11th Hellenic Congress on Concrete Structures*, Corfu, Greece, *III*, pp. 136–150 (in Greek)

Kotsovos M. D. and Bobrowski J., 1993, Design model for structural concrete based on the concept of the compressive force path, *ACI Structural Journal*, 90, 12–20.

Kotsovos G. M. and Kotsovos M. D., 2009, Flat slabs without shear reinforcement: Criteria for punching, *The Structural Engineer*, 87(23/24), 32–38.

Kotsovos G. M. and Kotsovos M. D., 2010, A new design method for punching of RC slabs: Verifications by non-linear finite-element analysis, *The Structural Engineer*, 88(8), 20–25.

Kotsovos M. D. and Lefas I. D., 1990, Behavior of reinforced concrete beams designed in compliance with the concept of the compressive-force path, *ACI Structural Journal*, 87, 127–139.

Kotsovos M. D. and Michelis P., 1996, Behavior of structural concrete elements designed to the concept of the compressive force path, *ACI Structural Journal*, 93, 428–437.

Kotsovos M. D. and Pavlovic M. N., 1995, *Structural Concrete: Finite-Element Analysis for Limit-State Design*, Thomas Telford, London.

Kotsovos M. D. and Pavlovic, M. N., 1999, *Ultimate Limit-State Design of Concrete Structures: A New Approach*, Thomas Telford, London.

Kotsovos M. D. and Pavlovic, M. N., 2001, The 1999 Athens earthquake: Causes of damage not predicted by structural concrete design methods, *The Structural Engineer*, 79(15), 23–29.

Kotsovos M. D. and Spiliopoulos, K. V., 1998a, Modelling of crack closure for finite-element analysis of structural concrete, *Computers & Structures*, 69, 383–398.

Kotsovos M. D. and Spiliopoulos K. V., 1998b, Evaluation of structural-design concepts based on finite element analysis, *Computational Mechanics*, 21, 330–338.

Kinnunen S., Nylander H. and Tolf P., 1978, *Undersokningar rorande genemostansning vid Institutionen for Byggnadsstatik, KTH*, Nordisk Betong (Stockholm), No. 3, 1978, pp. 25–27. (cited in Task Group 3.1/4.10, *Punching of structural concrete slabs*, Technical report, Bulletin 12, fib (CEB-FIP), April 2001, 307 pp.)

Lefas I., 1988, Behaviour of reinforced concrete walls and its implementation for ultimate limit state design, Ph.D. thesis, University of London.

Maier J. and Thurlimann B., 1985, *Bruchversuche an Stahlbetonscheiben*, Bericht Nr. 8003-1, Institut fur Baustatik und Konstuktion (Eidgenossische Technische Hochschule), Zurich.

Morikawa H., 2007, Finite element analysis of benchmark test on P/C beam–column joints under cyclic loading, *ACI Technical Session on Blind Prediction, ACI Conference*, Atlanta, GA.

Shiohara, H. and Kusuhara, F., 2006, *Benchmark Test for Validation of Mathematical Models for Nonlinear and Cyclic Behaviour of R/C Beam–Column Joints*, Department of Architecture, School of Engineering, University of Tokyo, http://www.rcs.arch.t.u-tokyo.ac.jp/shiohara/benchmark/

Technical Chamber of Greece, 1991, *Code for the Design and Construction of Concrete Structures* (in Greek)

Chapter 6

Extension of finite element modelling to dynamic problems

6.1 BACKGROUND

Dynamic analysis is by its nature a non-linear problem since the equations of motion are functions of not only the displacements, but also of the first and second derivatives of the displacements. This chapter presents a widely adopted method through which the dynamic problem can be converted into an equivalent static one, the solution of which is achieved as described in Chapter 4, thus extending the application of the finite-element model presented in the preceding chapter to dynamic problems.

6.2 EQUATION OF MOTION

When subjected to a dynamic external force, a structural element is set into motion. This causes deformation of the element and an internal force (F_{int}) field to develop resisting the element's deformation (u). The internal force is given by the product of the element stiffness (K) and the imposed deformation (u), that is, $F_{int} = K\,u$. Because of the element motion, forces due to inertia and damping also develop. Inertia forces (F_I) oppose the change of velocity (acceleration) and their value is defined as the product of mass (M) and acceleration ($\ddot{u}$), that is, $F_I = M\,\ddot{u}$. Damping forces (F_d), on the other hand, oppose the change of displacement with time (velocity) and their value is given as the product of the damping constant (C) and velocity ($\dot{u}$), that is $F_d = C\dot{u}$. The work of the damping forces F_d represents a percentage of the energy of motion that is lost. The damping constant C is difficult to quantify. It is usually assessed experimentally but a high factor of uncertainty remains since the value of C depends on multiple parameters, the effect of which is not clear.

In view of the above, when forming the equation of equilibrium, the forces due to inertia ($F_I = M\ddot{u}$) and damping ($F_d = C\dot{u}$) cannot be overlooked. The equation of equilibrium is no longer a simple algebraic equation but a second-order differential equation of motion:

$$M\,\ddot{u}(t) + C\dot{u}(t) + K(t)\,u(t) = F_{ext}(t) \tag{6.1}$$

6.3 NUMERICAL SOLUTION OF THE EQUATION OF MOTION

Because of the non-linear material behaviour present in the problem under consideration, Equation 6.1 is solved numerically (Hinton and Owen 1980, Karabalis and Beskos 1990, Bathe 1996). This can be achieved by using a number of methods. All of these methods are based on the notion that, for each time step Δt, a solution may be achieved by transforming the second-order differential equation of motion into a simpler algebraic equation that can be easily solved. To accomplish this, the acceleration $\ddot{u}$ and the velocity $\dot{u}$ of the structure are expressed as functions of the change in displacement Δu. Some of the most common methods used to accomplish this are those developed by Houmbolt, Newmark, Wilson and the a-method. A full description of these numerical methods is given elsewhere (Hinton and Owen 1980, Cook et al. 1989, Karabalis and Beskos 1990, Bathe 1996).

By using one of the above methods the equation of motion within a given time step can be transformed into an equivalent static problem, which can be expressed by the following simple algebraic equation:

$$K^*\Delta u = \Delta F^* \tag{6.2}$$

where K^* is the effective stiffness matrix, and Δu and ΔF^* are the vectors of the increments of displacement and effective force, respectively.

As will be seen in Section 6.4.1, the effective stiffness matrix K^* and the effective force vector ΔF^* are functions of the structure's stiffness matrix K and the force increment vector ΔF, respectively, as well as the structure's mass M and damping C matrices and the time step used in order to solve the equation of motion numerically. These functions depend on the particular method used for the numerical solution of the equation of motion (the method adopted in the solution technique proposed herein makes use of the Newmark family of approximations and is fully described in Section 6.4). Finally, the numerical solution of the equation of motion (6.2) can be accomplished by using either the *explicit* or the *implicit* method.

6.3.1 Explicit method

When using the explicit method (Hinton and Owen 1980), for each time step, the evaluation of the acceleration and velocity is carried out only once followed by the construction of the effective stiffness and the force increment vectors. Then, Equation 6.2 is solved to evaluate the displacement increment. The time step used in this method must be extremely small for the error to remain small and for the accuracy of the method to be maximised. The advantage of this method is that the formation of the effective stiffness and loading matrices is carried out only once during each time step. As a result, the computational cost of the numerical procedure per time step is low. However, the use of a small time step may increase the overall computational cost of the whole problem, especially if the problem has a long duration. Moreover, the error at every time step is accumulated, as it is added to that resulting at the next time step, thus leading in a continuous increase of the divergence between the numerical predictions and the actual behaviour of the structural element or structure analysed. A possible formulation of the explicit method may comprise the following steps:

1. At the beginning of each time step (Δt), initial values for the velocity $\dot{u}$ and the acceleration $\ddot{u}$ are assessed through the use of methods such as that discussed in Section 6.4.
2. Because of the approximate values selected in step 1, the value on the left-hand side of the equation of equilibrium (6.1) differs from that on the right-hand side. The difference between these two values represents the residual force Ψ. Thus, neglecting damping, Equation 6.1 can be written as

$$^{t+\Delta t}\Psi^{i+1} = M\,^{t+\Delta t}\ddot{u}^{i+1} + \,^{t+\Delta t}F_{\text{int}}^{i+1} - \,^{t+\Delta t}F_{\text{ext}}^{i+1}$$

3. The effective stiffness matrix K^* is assessed through the use of expressions such as those presented in Section 6.4.1.

4. Having assessed the stiffness matrix in step 3 and the residual force vector Ψ in step 2, the algebraic equation $K^* {}^{t+\Delta t}\Delta u^{i+1} = {}^{t+\Delta t}\Psi^{i+1}$ can be solved. The solution yields the incremental change of displacement ${}^{t+\Delta t}\Delta u^{i+1}$ corresponding to iteration $i + 1$. From the values of ${}^{t+\Delta t}\Delta u^{i+1}$, the values of the increments of strain ${}^{t+\Delta t}\Delta \varepsilon^{i+1}$ and stress ${}^{t+\Delta t}\Delta \sigma^{i+1}$ are determined, with the values of the increments ${}^{t+\Delta t}\Delta u^{i+1}$ and ${}^{t+\Delta t}\Delta \sigma^{i+1}$ being added to the total displacements ${}^{t+\Delta t}u^i$ and stresses ${}^{t+\Delta t}\sigma^i$ to produce ${}^{t+\Delta t}u^{i+1}$ and ${}^{t+\Delta t}\sigma^{i+1}$, respectively.

5. After evaluating the total stresses, the internal forces are obtained by integration of the total stress vector.

$$^{t+\Delta t}F_{int}^{i+1} = \int_V B \, {}^{t+\Delta t}\sigma^{i+1} dV$$

The new internal force vector will be used to define the residual forces of the next time step.

6. Using the values of the incremental change of displacement determined in step 4 it is easy to determine new approximate values of the velocity and the acceleration of the structural element by using an approximation method such as that described in Section 6.4.

7. The values of displacement u, velocity $\dot{u}$ and acceleration $\ddot{u}$ represent the final values for the time step $t + \Delta t$ and they are stored:

$$^{t+\Delta t}u = {}^{t+\Delta t}u^{i+1}$$

$$^{t+\Delta t}\dot{u} = {}^{t+\Delta t}\dot{u}^{i+1}$$

$$^{t+\Delta t}\ddot{u} = {}^{t+\Delta t}\ddot{u}^{i+1}$$

8. The process is repeated in the next time step.

6.3.2 Implicit method

In the implicit method (Hinton and Owen 1980), the solution process used in the explicit method described in the preceding section is repeated until convergence is accomplished, that is, until the difference between the calculated and true values is small. For each iteration, the values of velocity and acceleration are assessed, the effective stiffness and force matrices are constructed, and Equation 6.2 is solved in order to evaluate the displacement increment. If the difference between internal and external forces (residual force) is too high, then, the residual force is reapplied to the system as an external load and the whole procedure is repeated. When the difference becomes smaller than a predefined value (i.e., the difference satisfies the convergence criteria) the solution procedure moves on to the next time step. A possible formulation for the implicit method is described below:

1. As for the case of the explicit method, initial values for the velocity $\dot{u}$ and the acceleration $\ddot{u}$ are assessed at the beginning of each time step (Δt) through the use of expressions such as those presented in Section 6.4.1.

2. Owing to the approximate values selected in step 1, the value on the left-hand side of the equation of equilibrium (6.1) differs from that on the right-hand side. The difference between these two values represents the residual force Ψ. Thus, neglecting damping, Equation 6.1 can be written as

$$^{t+\Delta t}\Psi^{i+1} = M \, ^{t+\Delta t}\ddot{u}^{i+1} + \, ^{t+\Delta t}F_{\text{int}}^{i+1} - \, ^{t+\Delta t}F_{\text{ext}}^{i+1}$$

3. The effective stiffness matrix K^* is assessed in a manner similar to that adopted for the case of the explicit method.
4. The assessment of the stiffness matrix in step 3 and the value of the residual forces Ψ in step 2 is followed by the solution of the algebraic equation $K^* \, ^{t+\Delta t}\Delta u^{i+1} = \, ^{t+\Delta t}\Psi^{i+1}$. This produces the incremental change of displacement $^{t+\Delta t}\Delta u^{i+1}$ corresponding to iteration $i+1$. From the values of $^{t+\Delta t}\Delta u^{i+1}$ the values of strain $^{t+\Delta t}\Delta\varepsilon^{i+1}$ and stress $^{t+\Delta t}\Delta\sigma^{i+1}$ are determined, with the values of the increments $^{t+\Delta t}\Delta u^{i+1}$ and $^{t+\Delta t}\Delta\sigma^{i+1}$ being added to the total displacements $^{t+\Delta t}u^i$ and stresses $^{t+\Delta t}\sigma^i$ to produce $^{t+\Delta t}u^{i+1}$ and $^{t+\Delta t}\sigma^{i+1}$, respectively.
5. After evaluating the stress, the internal forces are obtained by integration of the stress vector

$$^{t+\Delta t}F_{\text{int}}^{i+1} = \int_V B \, ^{t+\Delta t}\sigma^{i+1}dV$$

The new internal force vector will be used to define the residual forces of the next iteration.
6. Using the values of the incremental change of displacement determined in step 4 it is easy to determine new approximate values of the velocity and the acceleration of the structural element by using an approximation method such as that presented in Section 6.4.
7. Convergence is accomplished when the maximum absolute value of the residual forces is less than a small predefined positive value e, that is, if $\left|^{t+\Delta t}\Psi^{i+1}\right| \leq e$, convergence has been accomplished and therefore the process continues to step 8, whereas if $\left|^{t+\Delta t}\Psi^{i+1}\right| > e$, convergence has not been accomplished and the process returns to step 2.
8. The values of displacement u, velocity $\dot{u}$ and acceleration $\ddot{u}$ are considered as final for time step $t + \Delta t$ and are stored:

$$^{t+\Delta t}u = \, ^{t+\Delta t}u^{i+1}$$

$$^{t+\Delta t}\dot{u} = \, ^{t+\Delta t}\dot{u}^{i+1}$$

$$^{t+\Delta t}\ddot{u} = \, ^{t+\Delta t}\ddot{u}^{i+1}$$

9. The process is repeated in the next time step.

In contrast with the explicit method, the computational cost of the implicit method during each time step is much higher since the solution process is repeated as many times as necessary to satisfy the convergence criteria. However, because of the iterative procedure used, the time step may be much larger than that used in the explicit method which results in a reduction of the computational cost otherwise incurred.

6.4 NUMERICAL PROCEDURE ADOPTED FOR STRUCTURAL CONCRETE

An extensive literature survey has shown that the Newmark family of approximations is the most commonly used approach for the numerical solution of the equation of motion (Cotsovos 2004). An explicit procedure is usually adopted when dealing with problems involving high rates of loading occurring in a small period of time and the numerical solution makes use of a very small time step (Belytschko 1976). For the numerical investigation of problems involving events of longer duration, such as earthquake problems, the use of a very small time step would result in a large increase of the computational cost; therefore the use of an implicit scheme is preferable (Belytschko 1976). However, as regards structural concrete it has been suggested that an implicit scheme can also be adopted for problems making use of a very small time step – problems involving high rates of loading such as impact and blast (Cotsovos 2004); this is because the use of an explicit procedure often leads to predictions exhibiting excessive deviation from the true structural behaviour as a result of the processes of cracking (both opening and closure of cracks) which produce large residual forces not accounted for in an explicit procedure.

6.4.1 Newmark family of approximations

In these approximations, the acceleration $\ddot{u}$ and the velocity $\dot{u}$ of the structure are represented as functions of the change of the displacement Δu. This is accomplished through the use of the following expressions:

$$^{t+\Delta t}\dot{u}^{i+1} = {}^{t}\dot{u} + \Delta t[(1-\gamma)^{t}\ddot{u} + \gamma\,{}^{t+\Delta t}\ddot{u}^{i}] \tag{6.3}$$

$$^{t+\Delta t}\ddot{u}^{i+1} = \frac{1}{\beta\Delta t^2}\left[{}^{t+\Delta t}u^{i} - {}^{t}u - \Delta t\,{}^{t}\dot{u} - \Delta t^2\left(\frac{1}{2}-\beta\right){}^{t}\ddot{u}\right] \tag{6.4}$$

where
  i is the number of the iteration
  γ, β are constants of the particular Newmark approximations; their values are always smaller than 1 and dependent on the Newmark method adopted
  $^{t+\Delta t}u^{i+1}$, $^{t+\Delta t}\dot{u}^{i+1}$, $^{t+\Delta t}\ddot{u}^{i+1}$ the values of displacement, velocity and acceleration at $t+\Delta t$ of iteration $i+1$
  ^{t}u, $^{t}\dot{u}$, $^{t}\ddot{u}$ are the final values of displacement, velocity and acceleration at t

For two sequential iterations i and $i+1$, Equation 6.1 takes the following forms:

$$M\,{}^{t+\Delta t}\ddot{u}^{i+1} + C\,{}^{t+\Delta t}\dot{u}^{i+1} + {}^{t+\Delta t}F_{int}^{i+1} = {}^{t+\Delta t}F_{ext} \tag{6.5}$$

$$M\,{}^{t+\Delta t}\ddot{u}^{i} + C\,{}^{t+\Delta t}\dot{u}^{i} + {}^{t+\Delta t}F_{int}^{i} = {}^{t+\Delta t}F_{ext} \tag{6.6}$$

Replacing velocity and acceleration in the above equations with their expressions (6.3) and (6.4), respectively, and then by subtracting Equation 6.6 from Equation 6.5, the equation of motion for iteration $i+1$ becomes

$$K^*\Delta u = \Delta F^* \tag{6.7}$$

with $\Delta u = {}^{t+\Delta t}u^{i+1} - {}^{t}u$ being the incremental change of displacement at $t + \Delta t$ at iteration $i + 1$, whereas K^* is the effective stiffness expressed in the form

$$K^* = C\frac{1}{\beta\Delta t} + M\frac{1}{\beta\Delta t^2} + K \tag{6.8}$$

and ΔF^*, the effective load vector, which may be expressed either as a function of the values of displacement (${}^{t}u$), velocity (${}^{t}\dot{u}$) and acceleration (${}^{t}\ddot{u}$) predicted in the previous time step

$$\Delta F^* = \Delta F + \frac{1}{\beta\Delta t^2} M[{}^{t}u + \Delta t\,{}^{t}\dot{u} + \Delta t^2(0.5 - \beta){}^{t}\ddot{u}]$$

$$+ C\left[\frac{\gamma}{\beta\Delta t}{}^{t}u + \left(\frac{\gamma}{\beta} - 1\right){}^{t}\dot{u} + \Delta t\left(\frac{\gamma}{2\beta} - 1\right){}^{t}\ddot{u}\right] \tag{6.9}$$

or as a function of the values of displacement (${}^{t}u$ and ${}^{t-\Delta t}u$), velocity (${}^{t}\dot{u}$ and ${}^{t-\Delta t}\dot{u}$) and acceleration (${}^{t}\ddot{u}$ and ${}^{t-\Delta t}\ddot{u}$) predicted in two previous time steps (Bathe 1996):

$$\Delta F^* = \Delta F + C\left[\frac{\gamma}{\beta\Delta t^2}({}^{t}u - {}^{t-\Delta t}u) + \left(\frac{\gamma}{\beta} - 1\right)({}^{t}\dot{u} - {}^{t-\Delta t}\dot{u}) + \Delta t\left(\frac{\gamma}{2\beta} - 1\right)({}^{t}\ddot{u} - {}^{t-\Delta t}\ddot{u})\right]$$

$$+ M\left[\frac{1}{\beta\Delta t^2}({}^{t}u - {}^{t-\Delta t}u) + \frac{1}{\beta\Delta t}({}^{t}\dot{u} - {}^{t-\Delta t}\dot{u}) + \left(\frac{1}{2\beta} - 1\right)({}^{t}\ddot{u} - {}^{t-\Delta t}\ddot{u})\right] \tag{6.10}$$

Expression (6.7), although an algebraic equation that can be easily solved, is equivalent to the second-order differential equation of relation (6.1). Its solution produces the change in displacement that occurs during iteration $i + 1$ from which the values of strain and stress increments are readily obtained.

The work of the forces due to damping represents a percentage of energy that is lost during the motion of the structural element. In RC structures the energy loss during motion and deformation is primarily caused by the non-linear behaviour of the materials involved. As the computer program used for the numerical analysis of the RC structural elements incorporates constitutive models describing the non-linear behaviour of steel and concrete, it is considered that the action of the damping forces ($F_d = C\dot{u}$) is taken into consideration by these constitutive models. Consequently, the damping forces F_d will not be explicitly used, and so the second-order differential equation of motion (6.1) and its equivalent algebraic equation (6.7) are simplified as follows:

$$M\ddot{u}(t) + K(t)u(t) = F_{\text{ext}}(t) \tag{6.11}$$

$$K^*\Delta u = \Delta F^* \tag{6.12}$$

whereas Equations 6.8 and 6.10 become

$$K^* = M\frac{1}{\beta\Delta t^2} + K \tag{6.13}$$

$$\Delta F^* = \Delta F + M\left[\frac{1}{\beta \Delta t^2}({}^{t}u - {}^{t-\Delta t}u) + \frac{1}{\beta \Delta t}({}^{t}\dot{u} - {}^{t-\Delta t}\dot{u}) + \left(\frac{1}{2\beta} - 1\right)({}^{t}\ddot{u} - {}^{t-\Delta t}\ddot{u})\right] \quad (6.14)$$

However, as it will be seen in Chapter 7, there are cases in which the effect of damping cannot be overlooked. Such is the case of structural elements subjected to shake-table excitation; in such cases the development of additional damping is mainly attributed to the interaction between the test specimen and the experimental set-up used to investigate the structural element response and, therefore, cannot be ignored. It should be stressed, however, that allowing for such damping is not linked in any way to the properties of the concrete and steel reinforcement but accounts for energy loss directly reflecting the testing procedure effects.

6.4.2 Stability conditions

The values of β and γ in the Newmark approximations are summarised in Table 6.1. For each pair of values, there is a suitable time step that must be selected in order to obtain a stable solution for the dynamic problem. The critical time step Δt_{cr} (stability condition) is chosen by using the largest (frequency) eigenvalue ω_n of an n- degree-of-freedom system which is linked to the corresponding period $T = 2\pi/\omega$. The values of the critical time step are also shown in Table 6.1. The use of a time step larger than the critical time step may result in instability of the numerical procedure, whereas the use of a much smaller time step may provide a more accurate and detailed solution but at a higher computational cost.

All approximations presented in Table 6.1 are linked to a stability condition, with the exception of the average acceleration approximation. This latter approximation is unconditionally stable, which means that, by using any time step, the procedure remains stable and always provides a solution. However, when using a large time step the solution provided by this method may prove inaccurate (Karabalis and Beskos 1990, Bathe 1996). In such cases, the use of a smaller time step may give a different result from that obtained for a larger time step. Therefore, it is evident that, although the stability of this particular approximation is independent of the time step used, the accuracy of the solution obtained is not. To ensure the accuracy of the solution, a number of test runs must be made, each time using a smaller time step, until the solutions obtained converge. An initial time step that can be used in the average-acceleration approximation is one-tenth of the period that corresponds to the largest eigenvalue (frequency) of the structure investigated.

6.5 IMPLEMENTATION OF THE DYNAMIC SCHEME

The scheme adopted for the numerical solution of the dynamic problem is based on the Newmark approximation functions of velocity and acceleration introduced in the preceding

Table 6.1 Newmark's family of approximations

Method	β	γ	Stability condition
Average acceleration	1/4	1/2	Unconditionally stable
Linear acceleration	1/6	1/2	$\Delta t_{cr} = 2 \cdot (\sqrt{3}/\omega_n)$
Fox–Goodwin	1/12	1/2	$\Delta t_{cr} = \sqrt{6}/\omega_n$
Central difference	0	1/2	$\Delta t_{cr} = \sqrt{2}/\omega_n$

section. This scheme has formed the basis of the modifications implemented to the original FE model in order to extend its use to the solution of non-linear dynamic problems. The main steps of the proposed scheme are as follows:

1. Through the use of the values of displacement ($^t u$ and $^{t-\Delta t}u$), velocity ($^t\dot{u}$ and $^{t-\Delta t}\dot{u}$) and acceleration ($^t\ddot{u}$ and $^{t-\Delta t}\ddot{u}$) predicted in previous time steps, the effective force-increment matrix ($^{t+\Delta t}\Delta F^*$) is assessed from Equation 6.10, that is,

$$
^{t+\Delta t}\Delta F^* = {}^{t+\Delta t}\Delta F + C\left[\frac{\gamma}{\beta \Delta t^2}(^t u - {}^{t-\Delta t}u) + \left(\frac{\gamma}{\beta}-1\right)(^t\dot{u}-{}^{t-\Delta t}\dot{u}) + \Delta t\left(\frac{\gamma}{2\beta}-1\right)(^t\ddot{u}-{}^{t-\Delta t}\ddot{u})\right]
$$
$$
+ M\left[\frac{1}{\beta \Delta t^2}(^t u - {}^{t-\Delta t}u) + \frac{1}{\beta \Delta t}(^t\dot{u}-{}^{t-\Delta t}\dot{u}) + \left(\frac{1}{2\beta}-1\right)(^t\ddot{u}-{}^{t-\Delta t}\ddot{u})\right]
$$

whereas the effective stiffness matrix is assessed from Equation 6.8, that is,
$$
K^* = C\frac{1}{\beta \Delta t} + M\frac{1}{\beta \Delta t^2} + K
$$

2. Following the assessment of the effective force increment and stiffness matrices in step 1, the algebraic equation $K^* \, {}^{t+\Delta t}\Delta u^1 = \Delta F^*$ can be solved. The solution produces the incremental change of displacement $^{t+\Delta t}\Delta u^1$ corresponding to the initial iteration ($i = 0$). From the values of $^{t+\Delta t}\Delta u^1$, the incremental values of strain $^{t+\Delta t}\Delta\varepsilon^1$ and stress $^{t+\Delta t}\Delta\sigma^1$ are determined, with the values of the increments $^{t+\Delta t}\Delta u^1$ and $^{t+\Delta t}\Delta\sigma^1$ being added to the total displacements $^{t+\Delta t}u$ and stresses $^{t+\Delta t}\sigma$ to produce $^{t+\Delta t}u^1$ and $^{t+\Delta t}\sigma^1$, respectively.
3. Using the values of the incremental change of displacement determined in the previous step, it is easy to determine new approximate values of the velocity and acceleration of the structural element from the Newmark approximation functions as follows:

$$
^{t+\Delta t}\dot{u}^{i+1} = {}^t\dot{u} + \Delta t[(1-\gamma)^t\ddot{u} + \gamma \, {}^{t+\Delta t}\ddot{u}^i]
$$

$$
^{t+\Delta t}\ddot{u}^{i+1} = \frac{1}{\beta \Delta t^2}\left[{}^{t+\Delta t}u^i - {}^t u - \Delta t^t\dot{u} - \Delta t^2\left(\frac{1}{2}-\beta\right)^t\ddot{u}\right]
$$

4. After evaluating the stress, the internal forces are obtained by integration of the stress vector:

$$
^{t+\Delta t}F_{\text{int}}^{i+1} = \int_V B \, {}^{t+\Delta t}\sigma^{i+1}dV
$$

The new internal force vector will be used to define the residual forces of the next iteration.
5. Because of the approximate values selected in step 3, the value on the left-hand side of the equation of equilibrium (6.1) may differ from that on the right-hand side. The difference between these two values represents the residual forces Ψ. Thus, Equation 6.1 can be written as

$$
^{t+\Delta t}\Psi^{i+1} = C \, {}^{t+\Delta t}\dot{u}^{i+1} + M \, {}^{t+\Delta t}\ddot{u}^{i+1} + {}^{t+\Delta t}F_{\text{int}}^{i+1} - {}^{t+\Delta t}F_{\text{ext}}^{i+1}
$$

6. Convergence is accomplished when the maximum absolute value of the residual forces is less than a small predefined positive value e, that is, if $\left|^{t+\Delta t}\Psi^{i+1}\right| \leq e$ convergence has been accomplished and therefore step 7 is bypassed and the process continues to step 8; if $\left|^{t+\Delta t}\Psi^{i+1}\right| > e$ convergence has not been accomplished, and the process continues to step 7.

7. Having assessed the values of the residual forces Ψ in step 5 and by reassembling the effective stiffness matrix (as in step 1), the algebraic equation $K^* {}^{t+\Delta t}\Delta u^{i+1} = {}^{t+\Delta t}\psi^{i+1}$ is solved and a new value of the incremental change of displacement $^{t+\Delta t}\Delta u^{i+1}$ corresponding to iteration $i+1$ is obtained. From the values of $^{t+\Delta t}\Delta u^{i+1}$ the incremental values of strain $^{t+\Delta t}\Delta\varepsilon^{i+1}$ and stress $^{t+\Delta t}\Delta\sigma^{i+1}$ are determined, with the values of the increments $^{t+\Delta t}\Delta u^{i+1}$ and $^{t+\Delta t}\Delta\sigma^{i+1}$ being added to the total displacements $^{t+\Delta t}u^i$ and stresses $^{t+\Delta t}\sigma^i$ to produce $^{t+\Delta t}u^{i+1}$ and $^{t+\Delta t}\sigma^{i+1}$, respectively. Then, the process returns to step 4.

8. If convergence has been accomplished, the values of displacement u, velocity $\dot{u}$ and acceleration $\ddot{u}$ are considered as final for the time step $t + \Delta t$ and they are stored:

$$^{t+\Delta t}u = {}^{t+\Delta t}u^{i+1}$$

$$^{t+\Delta t}\dot{u} = {}^{t+\Delta t}\dot{u}^{i+1}$$

$$^{t+\Delta t}\ddot{u} = {}^{t+\Delta t}\ddot{u}^{i+1}$$

9. Step 1 through to step 8 is repeated in the next time step.

The implicit method used for solving numerically the dynamic problem is essentially an iterative procedure, the convergence of which is checked at the end of each iteration. During each iteration, the evaluation of the stiffness matrices (K and K^*) and the solution of the equation $\psi = K^*\Delta u$ by inverting the effective stiffness matrix K^* are carried out at least once. To accomplish convergence within any one given time step, several iterations must be executed and it is, therefore, obvious that the computation time for this method is very high. However, because of the repeated checks for convergence, the time step does not need to be extremely small in order to avoid numerical problems and this moderates the computational cost of the implicit method. The length of the time step used can be defined as a percentage of the period that corresponds to the largest eigenvalue of the structure, as discussed in Section 6.4.2.

The modifications implemented to the static program in order to extend its use for the solution of dynamic problems are indicated in flowchart form in Figure 6.1. From the figure, it can be seen that the part of the flowchart encompassed by the dotted lines is the flowchart of the original static problem (see in Figure 4.20), which forms the basic unit of the dynamic program. The effective stiffness matrix K^* is used within the equation $F = K^*\Delta u$ which yields the incremental change of displacements Δu, strains $\Delta\varepsilon$ and stresses $\Delta\sigma$ during each iteration.

6.6 VERIFICATION STUDIES FOR THE DYNAMIC SCHEME

The verification of the proposed dynamic scheme has been based on two linearly elastic case studies (Cotsovos 2004). The first case study is concerned with the dynamic response of the column shown in Figure 6.2. The column is fixed at its lower end and supports a mass monolithically connected to its upper free end (vertical cantilever). It is 3000 mm long with

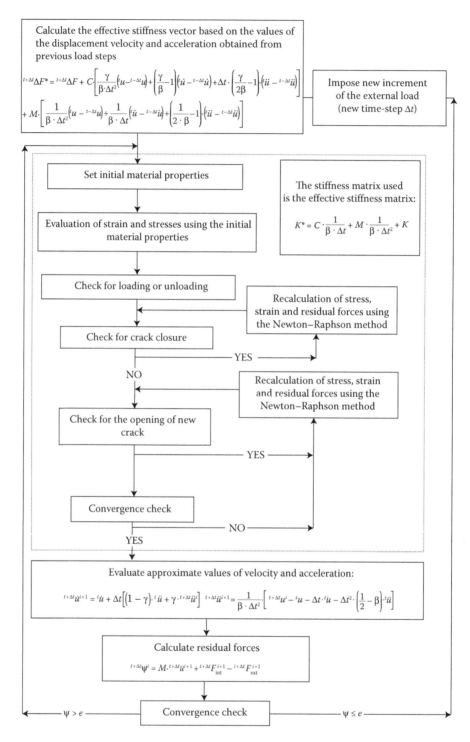

Figure 6.1 Proposed scheme for the dynamic non-linear program.

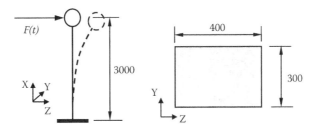

Figure 6.2 Case study 1.

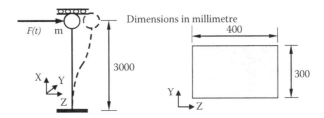

Figure 6.3 Case study 2.

a rectangular cross section 400 mm high × 300 mm wide. A time-dependent external force $F(t)$ acts at the mass centre as shown in Figure 6.2.

The second case study, shown in Figure 6.3, differs from the first one in that the upper end of the column is no longer free; it is allowed to translate freely only on the Y–Z plane (i.e., rotation about the X, Y and Z axes and translation along the X axis are constrained).

Both the above case studies involve simple structures for which analytical solutions describing their motion under various loading conditions are readily available. The input information required for obtaining the analytical solutions is provided in Table 6.2. Such solutions were used to verify the validity of the dynamic scheme introduced in this chapter by reference, for now, to linear elastic problems only. Thus, the solutions obtained from the proposed FE program are compared with their analytical counterparts and the corresponding deviations are considered to provide an indication of the validity of the dynamic scheme developed.

Two different FE meshes, shown in Figures 6.4a and b, were used to represent the column in each of the case studies considered. For both FE meshes, 27-node brick Lagrangian elements with a 3 × 3 × 3 integration rule were used. Moreover, the results of the numerical analysis depend on the selection of an appropriate time step. In general, numerical instabilities associated with the linear-acceleration method adopted for the analysis are avoided by using a time step smaller than a 'critical' value. However, in order to obtain a solution as accurate

Table 6.2 Input information required for the analytical solution of the case studies considered

Case study	Modulus of elasticity E (MPa)	Moment of inertia I (m⁴)	Stiffness K (N/m)	Mass M (kg)	Eigen value ω	Period T (s)	Critical time step (s)	Time step used (s)
1	56,808	0.0016	10,099,200	203,874	7.038	0.892	0.492	0.02
2	56,808	0.0016	40,396,800	203,874	14.076	0.446	0.248	0.02

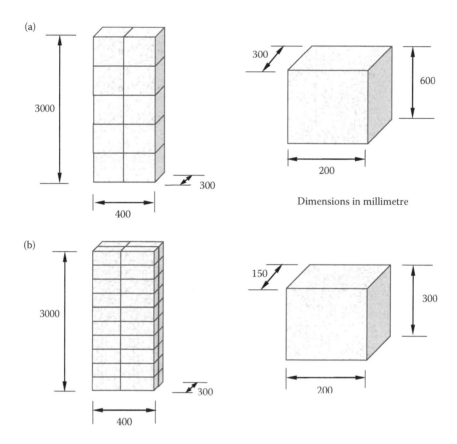

Figure 6.4 Finite-element mesh for the two study cases: (a) mesh 1, (b) mesh 2.

as possible, the time step must be much smaller, usually less than a tenth of the period of the structure, and this was ensured through the use of the values shown in Table 6.2.

For each case study, two different loading histories are considered. The results obtained include the values of displacement, velocity and acceleration of the mass, in the direction of the imposed external force. The loading histories, the analytical (closed-form) solution and the numerical (proposed dynamic scheme) results of the above case studies are shown in graphical form in Figures 6.5 through 6.8.

6.7 GENERAL REMARKS

By comparing the numerical and the analytical results obtained, it can be seen that the divergence is small for both case studies. The main sources of divergence are the concentrated lumping of both stiffness and mass in the single degree-of-freedom analytical solution, whereas the FE mesh distributes both these characteristics throughout the structure. In addition, the analytical solution relies on the Bernoulli assumption of (one dimensional) beam behaviour while the FE modelling is based on the more formal 3D elasticity approach. Within these discrepancies (in the fundamental sense but, clearly, minor for practical purposes as the beam theory is, for the problems presently considered, quite accurate), the results of both case studies suggest that the dynamic scheme proposed in this chapter within the modified FE package appears to be adequate and robust, leading to accurate and reliable results.

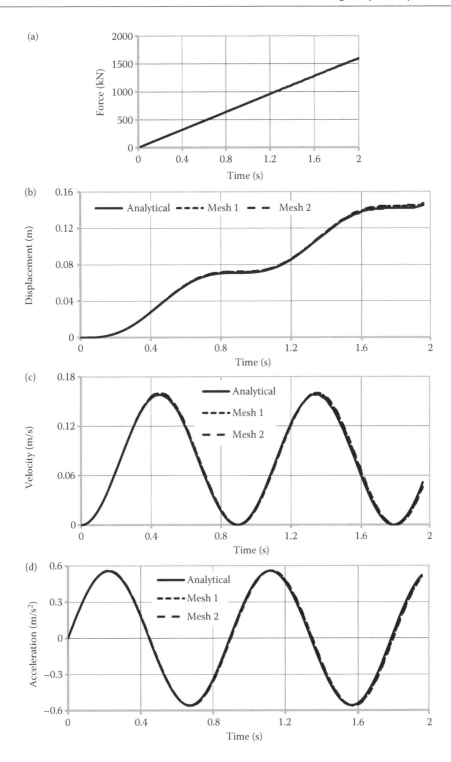

Figure 6.5 Case study 1: (a) loading history 1, (b) displacement response of mass, (c) velocity response of mass, (d) acceleration response of mass.

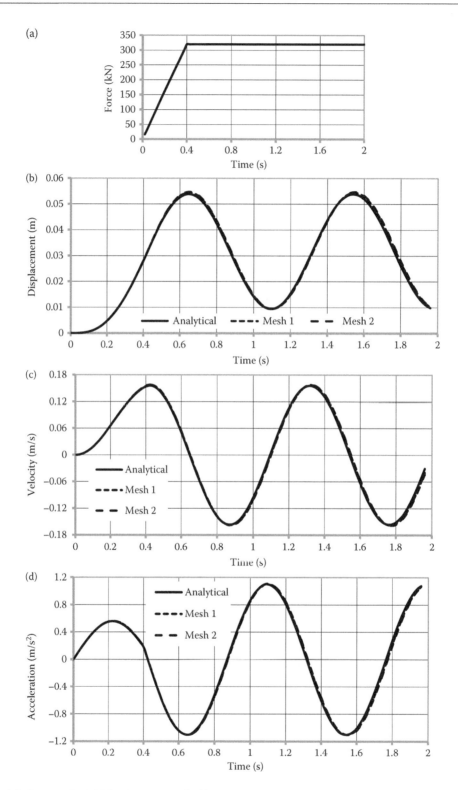

Figure 6.6 Case study 1: (a) loading history 2, (b) displacement response of mass, (c) velocity response of mass, (d) acceleration response of mass.

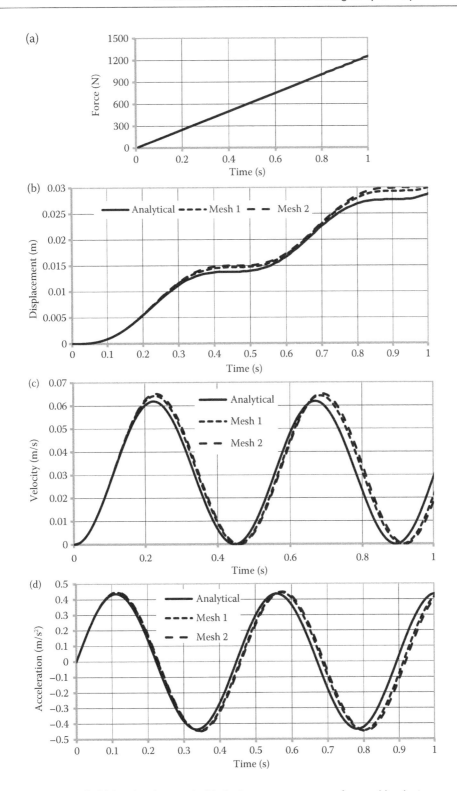

Figure 6.7 Case study 2: (a) Loading history 1, (b) displacement response of mass, (c) velocity response of mass, (d) acceleration response of mass.

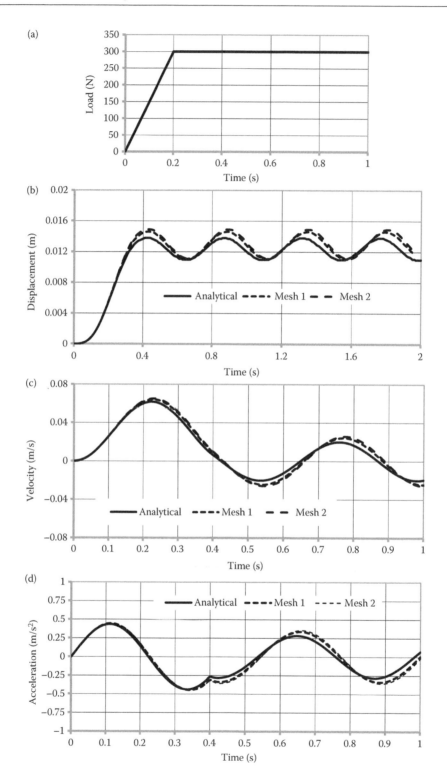

Figure 6.8 Case study 2: (a) loading history 2, (b) displacement response of mass, (c) velocity response of mass, (d) acceleration response of mass.

Admittedly, the dynamic scheme incorporated into the non-linear FE static program has been tested for linear elastic cases only. Next, it is to be tested for the more general problems for which it was intended, namely for the cases where the non-linear material properties of concrete and steel are taken into consideration. It is necessary to investigate how the newly incorporated dynamic scheme interacts with the existing non-linear static FE program and to ensure that the various software components are completely compatible, thus yielding a numerically stable scheme. The most serious source of numerical instability is the cracking procedure that concrete undergoes when subjected to external loading. To estimate the efficiency and effectiveness of the dynamic program in this non-linear range, a number of non-linear case studies are analysed numerically in the chapters to follow, and their results compared with available experimental data. This enables assessment of the capabilities and limitations of the proposed scheme, and points to any necessary corrections needed to increase its effectiveness and robustness.

REFERENCES

Bathe K. J., 1996, *Finite Element Procedures*, Prentice-Hall, Upper Saddle River, NJ.

Belytschko T., 1976, A survey of numerical methods and computer programs for dynamic structural analyses, *Nuclear Engineering & Design*, 37, 23–34.

Cook R. D., Malkus D. S. and Plesha M. E., 1989, *Concepts and Applications of Finite Element Analysis*, 2nd edition, John Wiley & Sons, Inc., New York.

Cotsovos D. M., 2004, Numerical modelling of structural concrete under dynamic (earthquake and impact) loading, PhD thesis, University of London.

Hinton E. and Owen D. R. J., 1980, *Finite Elements in Plasticity: Theory and Application*, Pineridge Press, Swansea.

Karabalis D. L. and Beskos D., 1990, Numerical methods in earthquake engineering in *Computer Analysis and Design of Earthquake Resistant Structures: A Handbook*, edited by D. E. Beskos and S. A. Anagnostopoulos, CMP, Southampton, pp. 1–104.

Chapter 7

Reinforced concrete structural members under earthquake loading

7.1 INTRODUCTION

The ability of the FE model described in Chapter 4 to produce realistic predictions of structural behaviour under short-term static (both monotonic and cyclic) loading has been demonstrated in Chapter 5. Through a comparative study of the model's predictions with the results of shake-table tests on a number of typical RC structural forms (columns, frames, walls, etc.), this chapter is intended to show that the modifications implemented in Chapter 6 resulted in a stable numerical scheme capable of also producing realistic predictions of structural response under seismic excitation.

7.2 APPLICATION OF THE EARTHQUAKE LOAD

In the case studies presented in this chapter, the seismic excitation is applied incrementally as force rather than displacement. Because the RC structural forms analysed are simple and can be modelled by systems with a small number of degrees of freedom, the application of the external load in the form of force, rather displacement, increments do not affect the accuracy of the predictions obtained. This can be shown by making use of a simple one-degree-of-freedom system, such as that adopted for modelling the RC frames discussed in Section 7.4. In Figure 7.1, the frame model is depicted at two different moments in time. Initially, for $t = 0$, no load is applied to the frame model. For $t = \tau$ the model is subjected to earthquake loading in the form of the displacement u_g of its foundation, with u_1 being the displacement (due to the deformation of the frame) of the mass lumped at the geometric centre of the rigid beam connecting the two columns. On the basis of the above, the total displacement of the mass is equal to the sum of u_g and u_1, that is,

$$u = u_g + u_1 \tag{7.1}$$

and, by differentiating Equation 7.1 twice, a similar equation is obtained for the acceleration

$$\ddot{u} = \ddot{u}_g + \ddot{u}_1 \tag{7.2}$$

The current problem is a one-degree-of-freedom problem, since the mass of the frame has been assumed to be lumped at the geometric centre of the rigid beam connecting the two columns. Furthermore, the lumped mass is assumed to move only horizontally considering

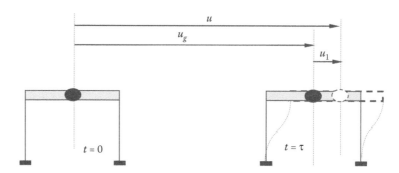

Figure 7.1 One-degree-of-freedom model of frame at times $t = 0$ and τ.

the vertical component of motion to be insignificant. On the basis of the above assumptions, the equation of motion may be written as

$$m\ddot{u} + ku_1 = 0 \qquad (7.3)$$

where
   k is the sum of the stiffness of the two columns and
   m is the concentrated mass situated at the geometric centre of the rigid beam.

Depending on the selection of the method for applying the earthquake load to the frame, Equation 7.3 may take the two forms described in the following:

Form 1. By using Equation 7.2, Equation 7.3 may be transformed as shown below:

$$m(\ddot{u}_g + \ddot{u}_1) + ku_1 = 0 \Rightarrow m\ddot{u}_g + m\ddot{u}_1 + ku_1 = 0 \Rightarrow m\ddot{u}_1 + ku_1 - m\ddot{u}_g \qquad (7.4)$$

The resulting equation of motion takes into account the earthquake loading as a force, the value of which, at each time step, is equal with the product of the value of the acceleration (provided by an accelerogram) and the mass. The force is then applied incrementally at the node at which the concentrated mass is situated.

Form 2. By using Equation 7.1, Equation 7.3 may be transformed as shown below:

$$m\ddot{u} + k(u - u_g) = 0 \Rightarrow m\ddot{u} + ku - ku_g = 0 \Rightarrow m\ddot{u} + ku = ku_g \qquad (7.5)$$

In this case the equation of motion takes into account the earthquake loading as a displacement which is transformed into an equivalent load by multiplying its value by the stiffness of the frame. It can therefore be concluded that, for the case study currently examined, the application of the external load in the form of force, rather than displacement, increments have no significant effect on the predictions obtained because, as already demonstrated, both options (when used for the analysis of systems with a small number of degrees of freedom) are equivalent and yield practically the same results. The same set of equations and the resulting conclusion can be extended to the rest of the RC structural forms investigated. However, it should be pointed out that this conclusion is not valid in the case of structural systems with multiple degrees of freedom. Such structural systems have a more complex response and the application of the seismic excitation either as a displacement at the bottom of the structure or as a force on the masses may give results with significant differences.

7.3 RC COLUMNS

The first case study considered has been extracted from Cotsovos (2004), where full details may be found; it involves an RC column investigated experimentally by Takeda et al. (1970).

7.3.1 Design details

The column was fixed to the shake table used to impose the earthquake excitation. The set-up used in the experimental investigation is presented in Figure 7.2, whereas the specimen dimensions and the arrangement of the reinforcement are shown in Figure 7.3. The yield

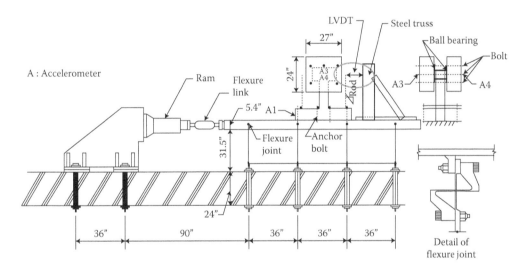

Figure 7.2 RC column under seismic excitation. Experimental set-up. (Adapted from Takeda T., Sozen M. A. and Nielsen N. N., 1970, *Journal of Structural Engineering, ASCE*, 96, 2557–2573.)

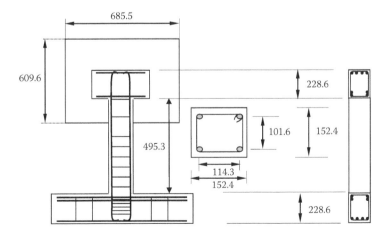

Figure 7.3 RC column under seismic excitation. Design details.

stress (f_y) of the longitudinal reinforcement was 350 MPa, whereas that of the transverse reinforcement was 276 MPa. The uniaxial cylinder compressive strength of the concrete used was $f_c = 30$ MPa. At the column's top end, two steel masses of 914 kg were hung on either side on a 1-in steel shaft resting on ball bearings. To restrain the large rotations of the steel masses, two 0.25 in pre-stressed rods tied the mass to the platform as shown in Figure 7.2.

7.3.2 FE discretisation

The FE model adopted for the RC column is shown in Figure 7.4. It comprises seven layers of two 27-node Lagrangian brick elements with the upper layer providing a simplified, yet effective, representation of the arrangement used to support the mass. The latter is attached at the geometric centre of the layer which was forced to behave as a rigid body by assigning to it a large modulus of elasticity (that of steel). As the seismic excitation was induced along the longitudinal axis of symmetry, as shown in Figures 7.4 and 7.5, and by making use of the symmetry of the column's cross section, only half the cross section of the column was modelled (see Figure 7.5), thus reducing the size of the numerical problem and minimising the computational effort needed for its solution.

As already mentioned, 3-node truss elements are used to model the reinforcement bars. The cross-sectional area of the longitudinal bars was distributed to the nodes of the column cross section so as to be equivalent, in terms of both cross-sectional area and location, to the

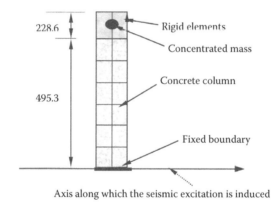

Figure 7.4 RC column under seismic excitation. FE model.

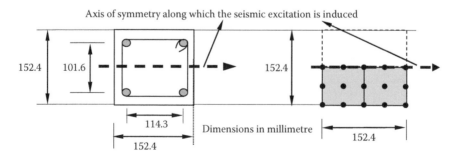

Figure 7.5 RC column under seismic excitation. Modelling of cross section.

four longitudinal bars of the real column (see Figure 7.5). The same reasoning was followed for modelling the transverse reinforcement.

7.3.3 Static loading

The investigation of structural response under earthquake loading was preceded by the assessment of the load-carrying capacity, deformational response and cracking process of the column under static loading. The column under an axial load (N) equivalent to the weight of the mass attached to its upper end was subjected to a horizontal load (P) at the centre of the mass and increased in steps to failure (see Figure 7.6). The column's response, in the form of base shear–displacement curve and crack patterns at various load levels, is shown in Figures 7.7 and 7.8, respectively. From the figures, it is seen that the column's load-carrying capacity reached the code predicted value of 26 kN, whereas cracking first appeared in the lower part of the column and gradually extended upwards with increasing load. Failure eventually occurred at the lower part of the column when the flexural cracks penetrated deeply into the compressive zone.

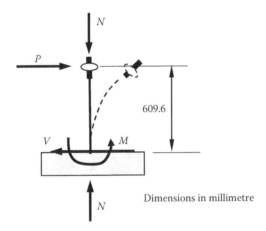

Figure 7.6 RC column. Applied loading and reactions under static loading.

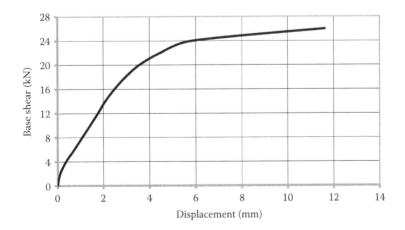

Figure 7.7 RC column. Base shear–displacement curve under static loading.

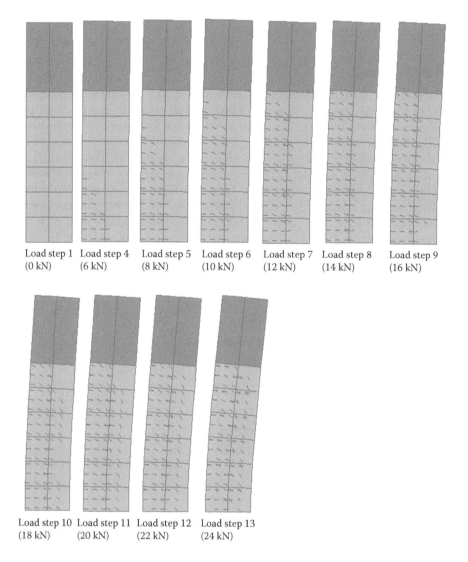

Load step 1   Load step 4   Load step 5   Load step 6   Load step 7   Load step 8   Load step 9
(0 kN)   (6 kN)   (8 kN)   (10 kN)   (12 kN)   (14 kN)   (16 kN)

Load step 10 Load step 11 Load step 12 Load step 13
(18 kN)   (20 kN)   (22 kN)   (24 kN)

Figure 7.8 RC column. Crack patterns at various load levels under static loading.

7.3.4 Dynamic loading

Before carrying out the dynamic analysis, an important factor that needs to be considered is the assessment of a sufficiently small time step in order to minimise the likelihood of numerical instability due to the development of excessive residual forces resulting from the formation or closure of cracks during each iteration. As discussed in Section 6.4.2, such a time step may be a fraction of the period of a simplified elastic model of the column such as the one-degree-of-freedom model shown in Figure 7.9. Through the use of the elastic properties of concrete and assuming cross-sectional characteristics equivalent to those of the gross cross section of the RC column, the period of the model is found to be $T = 0.03$ s, and the time step selected is $\Delta T = 0.001$ s.

The acceleration record to which the RC column was subjected during the shake-table test is similar to the El Centro 1940 acceleration record (Figure 7.10), except that the time length was compressed to a sixth (meaning that, although the original acceleration record

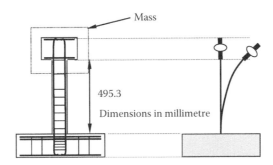

Figure 7.9 RC column under seismic excitation. One-degree-of-freedom elastic model.

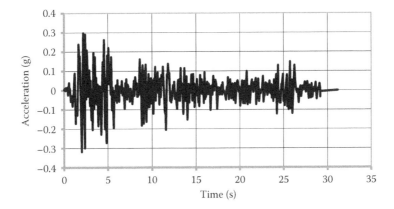

Figure 7.10 RC column under seismic excitation. Original El-Centro acceleration record.

lasted 30 s, the one used in the experimental [and adopted for the numerical] investigation lasted only 5 s), whereas the values of the acceleration record were multiplied by a factor of about 4.3 in order to achieve a maximum acceleration of 1.28 g (instead of 0.3 g of the original record). The variations with time of the imposed base acceleration and the recorded displacement response of the mass are presented in Figure 7.11. The acceleration record used for the analysis was essentially that used in the shake-table experiment (see Figure 7.12), whereas the predicted response of the specimen is presented in Figures 7.13 and 7.14 in the form of displacement–time and base shear–time curves.

7.3.5 Discussion of the numerical results

Although the predicted maximum value of the displacement correlates closely with its experimental counterpart, there are differences between the predicted (Figure 7.13) and recorded (Figure 7.11) variations of the displacement with time. These differences have been attributed to the rotation of the steel masses attached to the top end of the column, which could not be completely prevented by the steel rods used to tie the mass to the shake table (Cotsovos 2004).

From Figure 7.11, it can be seen that the maximum displacement exhibited during the shake-table test is just over 0.4 in (10.16 mm). In the static monotonic case study, this level of displacement corresponds to an applied load close to the load-carrying capacity of the specimen, which, as indicated in Figure 7.7, is characterised by ductile behaviour. The dynamic analysis also predicts ductile behaviour, since the predicted maximum values

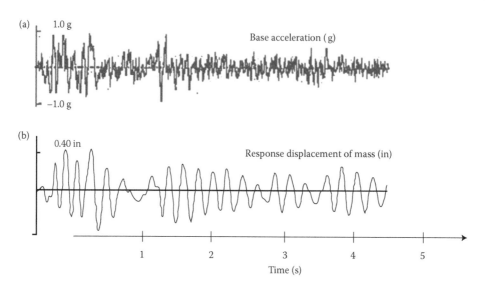

Figure 7.11 RC column under seismic excitation: (a) acceleration input used in shake-table test and (b) resulting displacement response of the RC column. (Adapted from Takeda T., Sozen M. A. and Nielsen N. N., 1970, *Journal of Structural Engineering, ASCE*, 96, 2557–2573.)

of displacement (see Figure 7.13) correspond to near constant values of the applied load (expressed in the form of base shear in Figure 7.14).

Numerically, failure of the RC column occurs when the structure's stiffness matrix becomes non-positive definite due to disruption of the continuity of concrete caused by excessive cracking of the lower part of the column. Although the actual column tested in the shake table suffered similar disruption at its lower part, collapse was prevented by the rods used to tie the mass to the platform which allowed the development of alternative resistance mechanisms. However, this stage of behaviour – which, clearly, is neither stable nor sustainable – cannot be described numerically, as the development of such alternative resistance mechanisms are not allowed for by the numerical procedure adopted. Nevertheless, the numerical model clearly serves the purpose of predicting, with sufficient accuracy, the maximum values of load and displacement. The approximate time of failure of the specimen – an arguably less important parameter – is predicted at around 1 s: it is interesting that it is

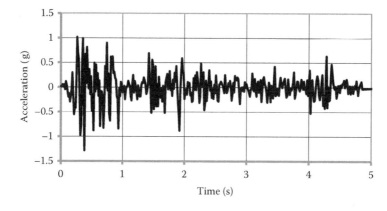

Figure 7.12 RC column under seismic excitation. Acceleration record input adopted for the numerical investigation.

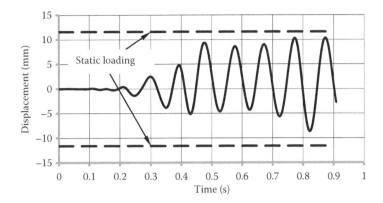

Figure 7.13 RC column under seismic excitation. Numerical predictions of the displacement response compared with the maximum displacement predicted for the case of static loading.

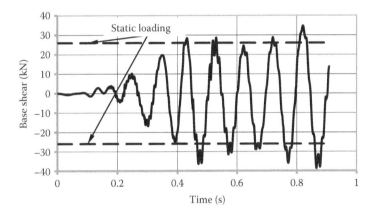

Figure 7.14 RC column under seismic excitation. Numerical predictions of the base shear compared with the maximum base shear predicted for the case of static loading.

around this time (i.e., between 1 and 2 s) that the measured displacement of the column seems to experience a change in behaviour (see Figure 7.11b) which might signal structural failure in a classic sense, followed by some further response under more unstable conditions stemming from other more complex post-failure mechanisms.

7.4 RC FRAMES

The structural form discussed in what follows has also been extracted from Cotsovos (2004); it is one of the four RC space frames (denoted as frame W050) investigated by Minowa et al. (1995) and consists of four RC columns with their lower end fixed on a shake table and their upper end monolithically connected to a thick rigid steel slab (see Figure 7.15). The frame was subjected to one-dimensional (1-D) excitations along its longitudinal axis of symmetry.

7.4.1 Design details

The column had a height of 850 mm and a square cross section with a side of 130 mm as shown in Figure 7.16. A dead weight of 300 kN was uniformly distributed on the steel slab.

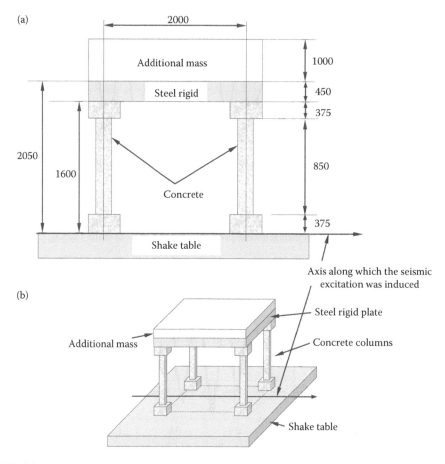

Axis along which the seismic
excitation was induced

Figure 7.15 RC frame under seismic excitation. (a) 2-D and (b) 3-D views.

The slab itself was 450 mm thick and had a span of 2000 mm. The uniaxial cylinder compressive strength of concrete used was $f_c = 26$ MPa for all columns. The column's longitudinal reinforcement consisted of four 16 mm diameter bars with a yield stress $f_y = 380$ MPa and an ultimate strength $f_u = 544$ MPa placed at the four corners of the cross section. The transverse reinforcement consisted of 6 mm diameter stirrups with a yield stress $f_y = 392$ MPa

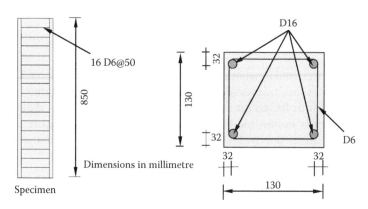

Figure 7.16 RC frame under seismic excitation. Design details.

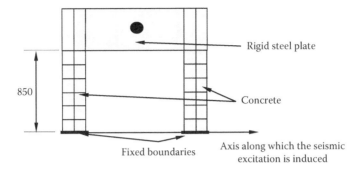

Figure 7.17 RC frame under seismic excitation. FE model.

and a strength $f_u = 490$ MPa. The stirrup spacing for each of the four columns was 50 mm (see Figure 7.16).

7.4.2 FE discretisation

Owing to the symmetry of the induced 1-D excitations, only half the frame represented by the 2-D model shown in Figure 7.17 was analysed. The FE mesh of each column of the frame consisted of $2 \times 6 = 12$ 27-node Lagrangian brick elements. Such elements were also used for modelling both the column–slab joints and the rigid slab, which, however, were forced to behave as rigid bodies by assigning to them a very large modulus of elasticity (that of steel). The mass was lumped at the geometric centre of the slab. Once again, due to the symmetry of the column's cross section, only half the cross section needs to be considered (see Figure 7.18). As for the column discussed in Section 7.3, 3-node truss elements were used to model the reinforcement bars. The cross-sectional area of the longitudinal bars was distributed to all the nodes of the column cross section so as to be equivalent, in terms of both cross-sectional area and location, to the four longitudinal bars of the real columns (see Figure 7.18). The same reasoning was followed for modelling the transverse reinforcement.

7.4.3 Results of the analysis

The load-carrying capacity, non-linear response and mode of failure of the RC frame were initially established under static loading. Then, the RC frame was analysed under seismic

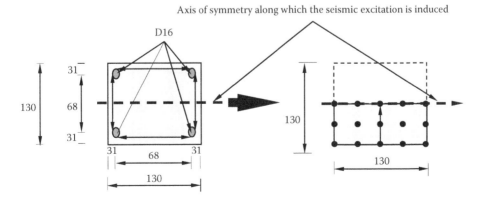

Figure 7.18 RC frame under seismic excitation. Modelling of column cross section.

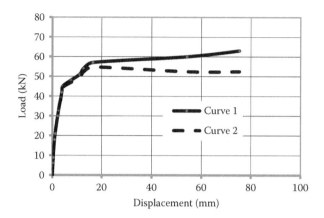

Figure 7.19 RC frame under static loading. Horizontal load–displacement curve.

excitation practically identical to its experimental counterpart (Minowa et al. 1995). The main results presented in what follows were extracted from Cotsovos (2004) where the full results are provided.

7.4.3.1 Static loading

The frame under the weight of the mass supported by its horizontal member was subjected to a horizontal point load applied at the mass centre. The vertical load was also imposed as a point load in one load increment, since it was only a small fraction of the axial load-carrying capacity of the columns and insufficient to cause crack formation. The horizontal load was gradually increased to failure in small load increments in order to obtain a prediction of the non-linear behaviour of the specimen, as accurate as possible.

It should be pointed out that the analysis carried out does not allow for geometric non-linearities. As a result, the predicted load–displacement curve (curve 1 in Figure 7.19) does not take into account second-order (P–Δ) effects which appear to be significant due to the large deflections occurring for load levels beyond 50 kN. The figure also includes the load–displacement curve corrected so as to allow for the P–Δ effects (curve 2). The correction of the horizontal load (V), $\Delta V = M_{P-\Delta}/h$, has been based on the assessment of the additional moment $M_{P-\Delta} = P\Delta$ due to the horizontal displacement (Δ) of the vertical load (P) (see Figure 7.20). This correction, which has been made for every load step, has been subtracted

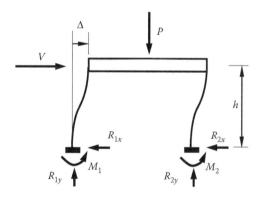

Figure 7.20 RC frame under static loading. Deformed shape and corresponding actions.

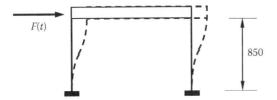

Figure 7.21 RC frame under seismic excitation. One-degree-of-freedom elastic model.

from the applied load and the resulting values plotted against the corresponding horizontal deflections produced curve 2 in Figure 7.19. From the latter curve, the load-carrying capacity of the frame was found to be 54.7 kN.

7.4.3.2 Dynamic loading

As for the case of the RC column discussed in Section 7.3, the selection of an appropriate time step has been based on the use of a simplified representation of the frame through the use of the one-degree-of-freedom model shown in Figure 7.21. Using the elastic properties of concrete and assuming cross-sectional characteristics equivalent to those of the gross cross section of the RC columns of the frame, the period of the model was found to be $T = 0.14$ s, and the time step selected was $\Delta T = 0.00128$ s.

The induced dynamic excitation formed part of the acceleration record presented in Figure 7.22. It corresponded, roughly, to the last part of the excitation actually used for the shake-table experiment (see Figure 7.23). By inspecting the displacement and acceleration response of the mass of the specimen recorded during the shake-table experiments (see Figures 7.24 and 7.25), it is apparent that only small deflections and accelerations were recorded up to approximately the 18th second of the seismic excitation. During this time, the behaviour of the structure was essentially elastic and, therefore, there was nothing of particular significance to monitor. Thus, the acceleration record used in the numerical investigation corresponds to that part of the original record that produced significant values of acceleration and deformation during the experiment. Within this stage of the experiment, the non-linear response of concrete progressively becomes more and more pronounced, with crack formation and extension, eventually leading to failure.

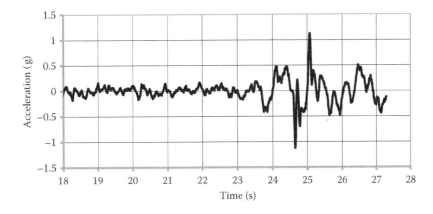

Figure 7.22 RC frame under seismic excitation. Acceleration record used in the numerical investigation indicating the duration of the numerical investigation.

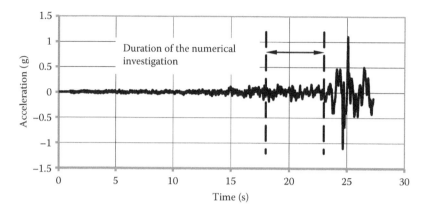

Figure 7.23 RC frame under seismic excitation. Acceleration record used in the experimental investigation.

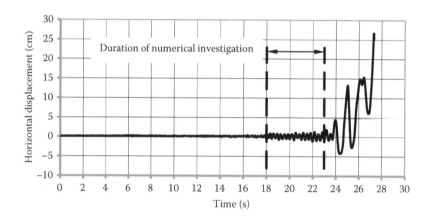

Figure 7.24 RC frame under seismic excitation. Horizontal displacement response recorded during the shake-table experiment.

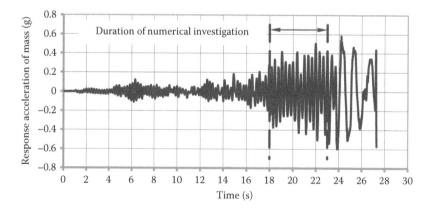

Figure 7.25 RC frame under seismic excitation. Acceleration response recorded during the shake-table experiment.

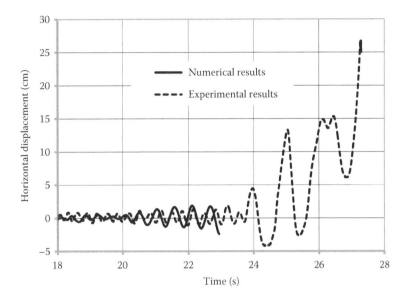

Figure 7.26 RC frame under seismic excitation. Numerical and experimental displacement response.

Adopting only that part of the acceleration record after the 18th second resulted in a more efficient use of the available computer resources, without compromising the aims of the numerical investigation. The numerical results are compared with their experimental counterparts in Figures 7.26 through 7.29. The maximum values of the displacement predicted numerically compare well with their experimental counterparts up to around the 23rd second when the analysis predicted failure, with the predictions slightly overestimating the experimental values (see Figure 7.26). The maximum values of the acceleration (Figure 7.27), the velocity (Figure 7.28) and the base shear (Figure 7.29) also compare well with their experimental counterparts up to this time of 23 s, when the numerical model predicted failure. It should be noted that only displacement and acceleration response values were

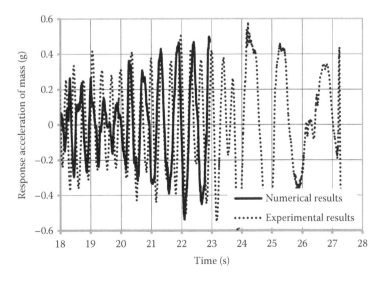

Figure 7.27 RC frame under seismic excitation. Numerical and experimental acceleration response.

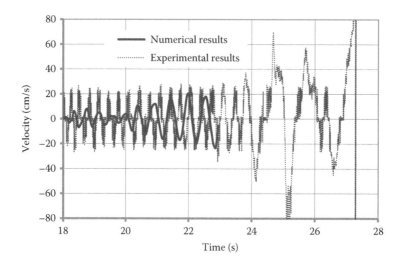

Figure 7.28 RC frame under seismic excitation. Numerical and experimental velocity response.

directly measured during the experiments, with the corresponding values of velocity and base shear shown in the figures being obtained indirectly. The velocity was obtained by differentiating the displacement record, whereas the calculation of the base shear was based on the use of the equation of motion.

7.4.3.3 Discussion of results

Under both static and dynamic loading, the specimen behaviour was ductile. Cracks formed within the top and bottom regions of each column since the bending moment was higher in these areas. As the external load increased, the cracks began to gradually spread throughout the whole column. Before failure, the longitudinal reinforcement yielded in tension in the upper and lower regions of the columns, thus enabling the specimen to exhibit large displacements.

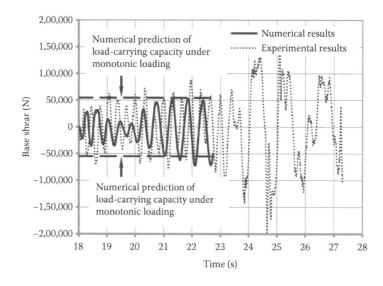

Figure 7.29 RC frame under seismic excitation. Numerical and experimental values of base shear.

Under static loading, ductile behaviour is indicated by the large increase in displacement for small load increments characterising the load–displacement curves of Figure 7.19 for load levels close to the load-carrying capacity. Under dynamic loading, the ductile behaviour of the specimen is indicated in Figures 7.26 and 7.29 by the fact that, after the 21st second, the maximum displacement gradually increased with time (Figure 7.26), whereas the maximum value of the base shear remained essentially constant (Figure 7.29). Such behaviour is considered to indicate that the specimen had formed plastic hinges in the areas of maximum bending moment. Failure occurred at the 23rd second when the load reached the specimen load-carrying capacity (indicated in Figure 7.29 by the horizontal line) under static loading (see Figure 7.19).

It appears from Figures 7.26 through 7.29 that the numerical analysis predicted failure to have occurred at the 23rd second rather than at the 27th second indicated by the experimental values. Although this discrepancy between the predicted and the measured values is not excessive, it is important to draw a distinction between the definition of numerical failure and its experimental counterpart. During the shake-table experiments the loading procedure ended after the specimen had suffered severe destruction of the lower and upper regions of the columns where concrete disintegrated. In fact, concrete disintegration may have been the cause of the abrupt increase of the vertical displacement of the frame slab which started just before the 24th second and increased rapidly thereafter as indicated in Figure 7.30. On the other hand, numerical failure occurs when the structure's stiffness matrix becomes non-positive definite. In the present case, this occurred when the geometry of the columns changed dramatically due to the disruption of the continuity of concrete caused by excessive crack formation.

It would appear from above that, after the severe destruction of concrete within the upper and lower regions of the columns, the real structure may have still been capable of sustaining, for a few seconds, the induced excitation, as indicated in Figures 7.26 through 7.29, by resorting briefly to alternative resistance mechanisms, such as, for example, dowel action. However, this stage of behaviour – which, clearly, is neither stable nor sustainable and as such, of no real significance for design purposes – cannot be described numerically, as the development of alternative resistance mechanisms such as the above are not allowed for by the numerical model, as already explained. Nevertheless, the numerical model clearly serves the purpose of predicting, with sufficient accuracy, the maximum load and displacement values attained and the approximate time of failure of the specimen. The extent of the

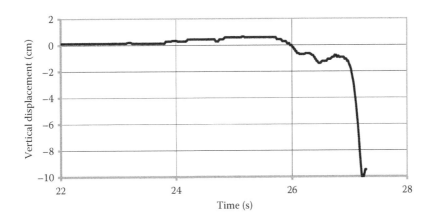

Figure 7.30 RC frame under seismic excitation. Record of the vertical deflection of the slab during the shake-table test from the 22nd to the 27th second.

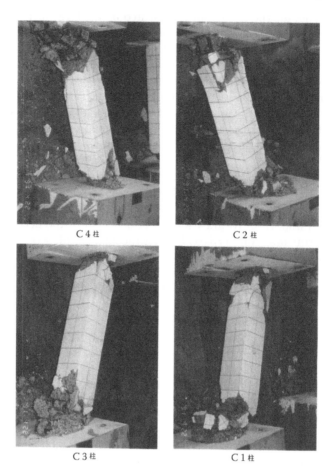

C 4 柱

C 2 柱

C 3 柱

C 1 柱

Figure 7.31 RC frame under seismic excitation. Experimentally established mode of failure. (Adapted from Minowa C. et al., 1995, *Nuclear Engineering and Design*, 156, 269–276.)

damage caused to the four columns of the RC frame in the shake-table test after inducing the seismic excitation can be seen in Figure 7.31. From these figures, it is evident that concrete in the upper and lower areas suffered extensive disintegration and that the specimen (just prior to failure) resorted to alternative resistance mechanisms in what could be considered as a post-failure regime (in a classic structural sense).

7.5 THREE-STOREY RC WALL

The present case study is concerned with the behaviour of an RC wall for which, as for the cases discussed in Sections 7.3 and 7.4, there is published information obtained from shake-table tests (Lestuzzi et al. 1999). The wall, which was considered to represent a scaled down three-storey building, has a 900 × 100 mm cross section and a 3 m height; three 12 ton masses with 1 m spacing along the height are attached to the wall as indicated in Figure 7.32. Each mass, which is considered to be the mass of a floor of the equivalent three-storey building, is supported by a separate rigid three-storey steel frame (see Figure 7.32) and can move only in the horizontal direction. Therefore, it can be assumed that the inertia of the masses affects only the horizontal motion of the wall. The wall is also subjected to

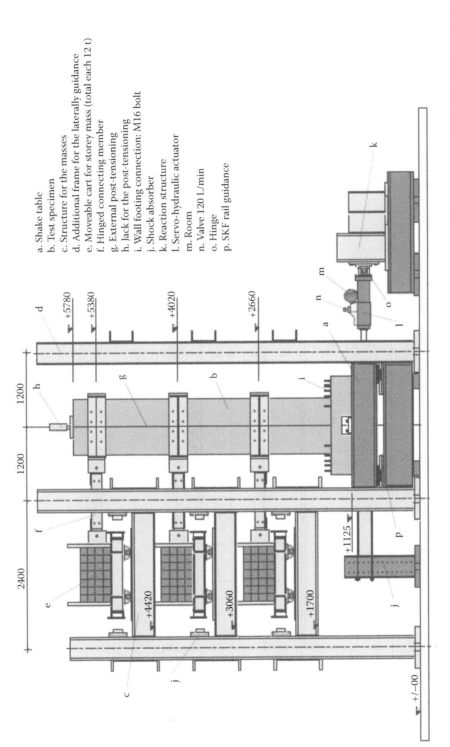

Figure 7.32 Three-storey RC wall under seismic excitation. Experimental set-up. (Adapted from Lestuzzi P., Wenk T. and Bachmann H., 1999, *IBK Report No. 240*, Institut für Baustatik und Konstruktion, ETH, Zurich.)

concentric axial compression equal to approximately 30% of the wall's load-carrying capacity under such loading.

The uniaxial (cylinder) compressive strength of the concrete used is equal to approximately 35 MPa, while the values of the yield stress and ultimate strength of the various reinforcement bars used varied between approximately 480 and 570 MPa and 580 and 670 MPa, respectively, with the arrangement of the reinforcement being depicted in Figure 7.33. Full design details can be found in Lestuzzi et al. (1999).

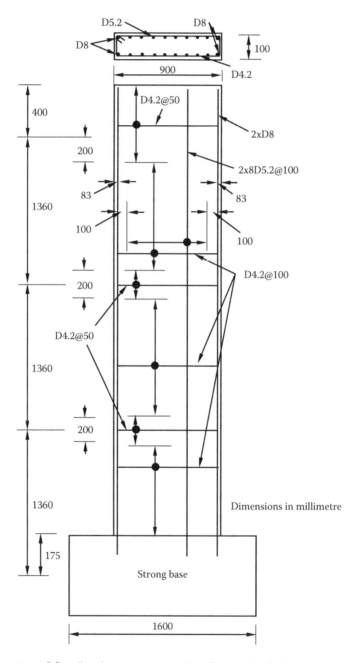

Figure 7.33 Three-storey RC wall under seismic excitation. Design details (dimensions in millimetre).

7.5.1 FE discretisation

As shown in Figure 7.34, the FE mesh adopted for modelling the RC wall consists of $3 \times 15 = 45$ 27-node Lagrangian brick elements. As for the case studies presented in Sections 7.3. and 7.4, the seismic excitation is induced along the longitudinal axis of symmetry of the wall's cross section and, therefore, only half the wall needs to be analysed (see Figure 7.35). The strengthened regions of the wall where the masses were attached (see Figure 7.32) are represented by highlighting with a lighter shade horizontal strips where cracking was not allowed to occur, thus allowing – in an effective, yet simple, manner – for the effect of the

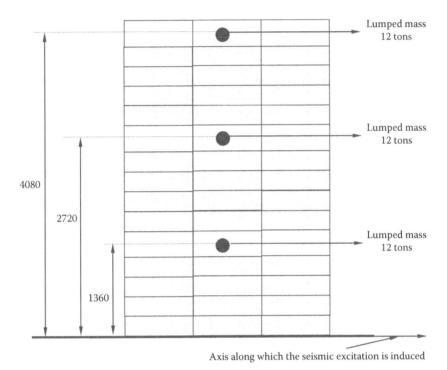

Figure 7.34 Three-storey RC wall under seismic excitation. FE model (dimensions in millimetre).

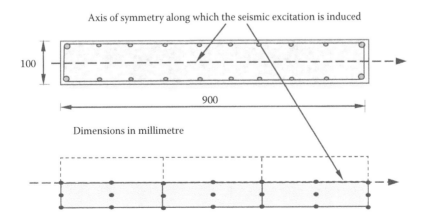

Figure 7.35 Three-storey RC wall under seismic excitation. Model of wall cross section.

local strengthening in structural behaviour. The masses are considered to act at the geometric centre of the above strips.

Three-node truss elements are used to model both the transverse and the longitudinal reinforcement bars. The cross-sectional area of the reinforcement bars is distributed to the nodes of the FE model adopted so as to be equivalent in terms of both cross-sectional area and location with the reinforcement bars of the actual specimen.

7.5.2 Results of analysis

7.5.2.1 Static loading

The wall, under the constant axial load of 25 kN applied at its top end, is subjected to three equal horizontal loads (P) at the locations indicated in Figure 7.36. The latter loads, the sum of which is equal to the base shear (V), are increased in steps to failure and the wall's response is described in Figures 7.37 and 7.38 in the form of base shear–displacement curves and crack patterns at various load steps, respectively. In contrast with the frame discussed in Section 7.4, Figure 7.37 shows that the horizontal displacements are small thus resulting in insignificant second-order effects that can be ignored without any loss of accuracy. The figure also shows that the wall is capable of sustaining a base shear of 64 kN. Figure 7.38 shows that cracking first appears in the bottom region of the wall, and, with increasing load, both spreads upwards and extends towards the side face in compression. Eventually, failure occurs when cracking at the wall base penetrates deeply into the compressive zone, with the upper region of the wall remaining essentially free of cracking.

7.5.2.2 Dynamic loading

The selection of an appropriate time step for the dynamic analysis that ensures a stable numerical procedure has been based on the use of the three-degrees-of freedom model of the

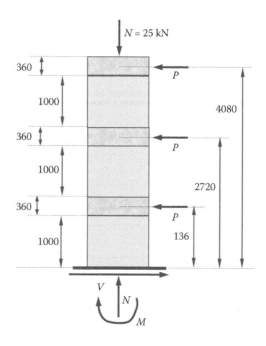

Figure 7.36 Three-storey RC wall under seismic excitation. Applied loads and reactions.

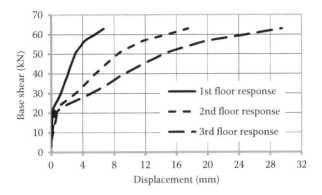

Figure 7.37 Three-storey RC wall under static loading: Base shear–displacement curve.

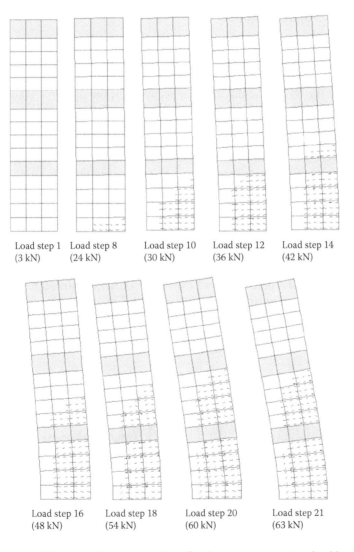

Figure 7.38 Three-storey RC wall under static loading. Crack patterns at various load levels (the figures in brackets denote the values of base shear i.e., the total horizontal load acting on the wall).

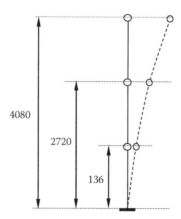

Figure 7.39 Three-storey RC wall under seismic excitation. Three-degree-of-freedom model for selecting time step for dynamic analysis.

wall shown in Figure 7.39. Assuming geometric characteristics and material properties similar to those of the real structure, the smallest of the three periods of the model was found to be $T = 0.057$ s and, on the basis of this value, the time increment selected is $\Delta T = 0.000625$ s.

As in the case of the RC frame discussed in Section 7.4, the dynamic load is applied in the form of an acceleration record, which is presented in Figure 7.40. The full response of the specimen during the experimental investigation is presented in Figure 7.41a–c in the form of variations with time of displacement, acceleration and base shear, respectively. From Figure 7.41c, it can be seen that the maximum value of the base shear under static loading (indicated by the horizontal lines) practically coincides with its shake-table counterparts.

In the case of the RC frame discussed in Section 7.4, the energy lost during seismic excitation was primarily attributed to the non-linear structural response. In the present case study, energy is also lost due to the interaction between the wall and the frame structure supporting the masses which are connected to the wall. This additional loss of energy is accounted for through the use of a damping matrix containing constant values of the damping coefficients C at locations corresponding to the nodes where the masses are attached. A comparative study of the predicted and experimentally established responses of the wall for values of C (kNm/s) equal to 0, 5 and 10 revealed that, although realistic predictions as regards load-carrying capacity and maximum displacement can be obtained for all three values of C, $C = 5$ kNm/s was found to lead to the closest correlation between numerical and experimental results as indicated by the variations of the displacement and base shear with

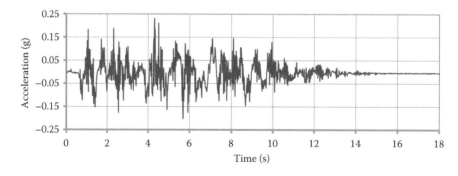

Figure 7.40 Three-storey RC wall under seismic excitation. Acceleration record used in the numerical and experimental investigation.

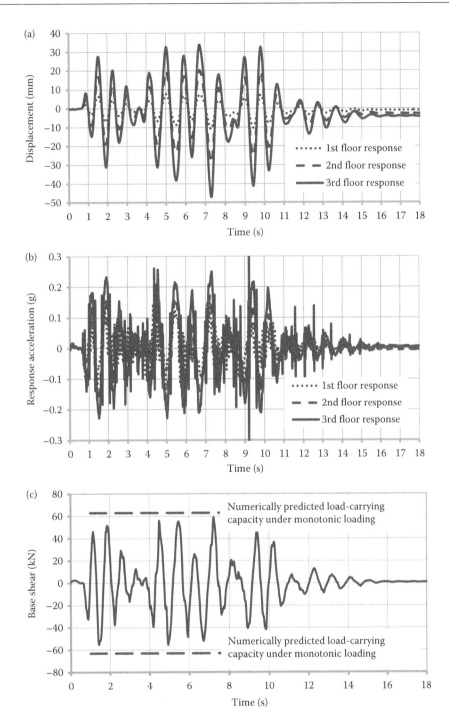

Figure 7.41 Three-storey RC wall under seismic excitation. Response of the RC wall recorded during the shake-table experiment: (a) horizontal displacement; (b) acceleration; (c) base shear.

time shown in Figures 7.42 and 7.43, respectively. As regards failure, it was found that the smaller the value of C, the sooner failure occurs; such behaviour being consistent with the fact that as C decreases a larger part of the seismic energy is dissipated in causing disruption in the form of cracking.

However, in general, failure is predicted to occur earlier than what the experimental results indicate (see Figures 7.42 and 7.43). As explained for the case studies discussed in Sections 7.3 and 7.4, this difference between the predicted and the experimental values of the time of failure reflects secondary tests effects which enable the structure, after disruption

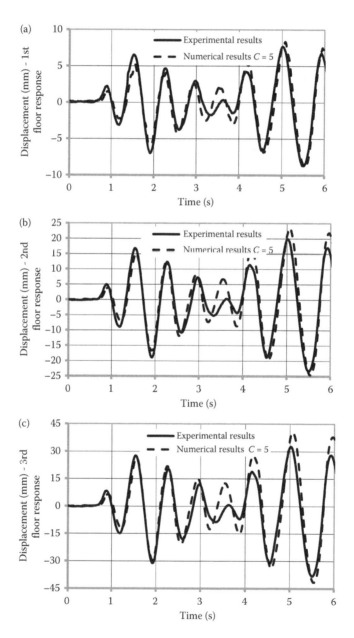

Figure 7.42 Three-storey RC wall under seismic excitation. Numerical (for C = 5 kNm/s) and experimental displacement response: (a) first floor; (b) second floor; (c) third floor.

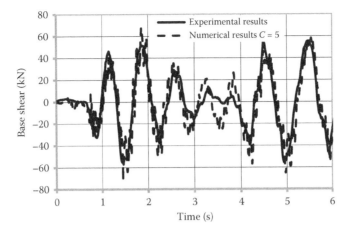

Figure 7.43 Three-storey RC wall under seismic excitation. Variation with time of the predicted (for $C = 5$ kNm/s) and experimentally obtained base shear.

during testing, to resort to alternative mechanisms of load transfer which are unstable in nature and constitute post-failure phenomena.

Under static loading, the ductile behaviour is indicated by the large displacements under small load increments characterising the load–displacement curves of Figure 7.37 for load levels close to the load-carrying capacity. Under dynamic loading, the ductile behaviour is indicated in Figure 7.41 by the fact that, even though the maximum deformation reached a high value at a relatively early stage of the shake-table test, this maximum deformation continued to increase gradually with time (see Figure 7.41a), while the maximum value of the base shear (Figure 7.41c) increased quickly up to a certain level and from then onwards remained practically constant. Such behaviour is considered to indicate an early formation of a plastic hinge in the region of wall base which is the key characteristic of ductile behaviour.

7.6 TWO-LEVEL RC FRAME UNDER SEISMIC ACTION

The work described in the following has been extracted from Cotsovos (2013). It first shows that the proposed FE package is capable of producing realistic predictions of the seismic behaviour of a two-level RC frame, which is a structural configuration more complex than those discussed in the preceding sections. The frame investigated, denoted as L30, is one of the frames subjected to shake-table excitation in a research programme concerned with the verification of the validity of the European code provisions for the design of earthquake-resistant structures (Carydis 1997). The numerical results obtained are then compared with those resulting from the analysis of the same frame, but without allowing cracking within the beam–column joints of the frame, in an attempt to establish the effect of cracking of the joint on the overall behaviour of the frame, thus complementing previous work on the subject by Cotsovos and Kotsovos (2008) described in Section 5.3.

7.6.1 Design details

The frame was designed in accordance with EC2/EC8 for low ductility and its design (both geometric and reinforcement) details are shown in Figure 7.44. The uniaxial cylinder compressive strength of the concrete used was $f_c = 50$ MPa, whereas the yield stress of the

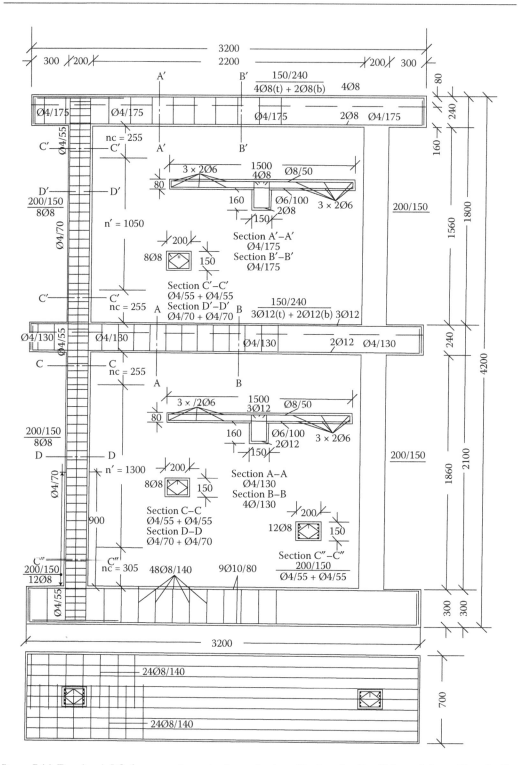

Figure 7.44 Two-level RC frame under seismic excitation. Design details. (Adapted from Carydis P., 1997, ECOEST PPREC8, Report 8, 182; Comite Europeen de Normalisation, ENV–1992–1, Eurocode No. 2 (EC2), 2004, *Design of Concrete Structures. Part I: General Rules and Rules of Building.* CEN, Brussels.)

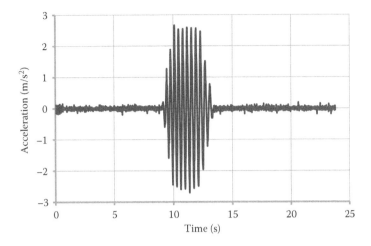

Figure 7.45 Two-level RC frame under seismic excitation. Acceleration record used in the numerical and experimental investigation.

reinforcement was $f_y = 500$ MPa. The frame was loaded with masses of 2.87 and 2.62 tons at the lower and upper girders, respectively, and subjected to the horizontal motion described by the acceleration record shown in Figure 7.45. This was one of the loading regimes adopted in the present work. The behaviour of the frame was also investigated under the action of two equal horizontal point loads exerted at the levels of the lower (level 1) and upper (level 2) girders. These loads either increased monotonically to failure or were applied in two cycles with maximum values approximately equal to ±40% and ±70% of the maximum sustained load under monotonic loading P_u, before increasing monotonically to failure.

7.6.2 FE modelling

The FE model adopted for the frames is shown in Figure 7.46, together with the location of the additional masses imposed. It should be noted that in order to simplify the analysis and reduce its cost, the flanges of the girders (see Figure 7.44) were not included in the FE model

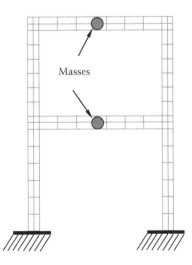

Masses

Figure 7.46 Two-level RC frame under seismic excitation. FE model.

representing the frame. Although this simplification may affect the predicted deformational response of the frame, the results of the comparative study (as will be seen later) remain essentially unaffected.

The frames were subdivided into 92 27-node brick elements representing concrete, whereas the 3-node truss elements representing the steel reinforcement bars were placed along successive series of nodal points in both vertical and horizontal directions. Since the spacing of these truss elements was predefined by the location of the brick elements' nodes, their cross-sectional area was adjusted so that the total amount of both longitudinal and transverse reinforcement to be equal to the design values.

The size of the 27-node Lagrangian brick finite elements (FEs) used is dictated by the philosophy upon which the FE model adopted in the present work is based, which does not employ small FEs (Kotsovos and Pavlovic 1995). It is reminded that the material model adopted is based on data obtained from experiments in which concrete cylinders were subjected to various triaxial loading conditions. Consequently, these cylinders are considered to constitute a 'material unit' for which average material properties are obtained and hence the volume of these specimens provides a guideline to the order-of-magnitude of the size of the FE used for the modelling of concrete structures.

7.6.3 Results

The results of the analyses are shown in Figures 7.47 through 7.56 and Tables 7.1 and 7.2. Figures 7.47 through 7.49 provide a comparative study of the dynamic response of frame L30 established experimentally and its counterpart predicted by means of non-linear FE analysis for the case of the seismic excitation depicted in Figure 7.45. The response is expressed in the form of curves describing the time history of the numerically predicted values of displacement and acceleration of the masses of levels 1 and 2 and of the base shear.

In Figure 7.50, the behaviour of frame L30 predicted numerically as described in Figures 7.47 through 7.49 is compared with that predicted for the same frame, but, without allowing cracking to occur within its beam–column joint regions. Thereafter, in the former case the frame is designated as 'F-C' and in the latter as 'F-E'; suffix 'C' denotes 'concrete' and this is intended to indicate that the behaviour of concrete throughout the frame, joints included, is modelled as described in the preceding section, whereas suffix 'E' denotes 'elastic' in order to indicate that not allowing cracking to occur within the joint is equivalent to adopting elastic properties for concrete in this region. It should also be noted that the letters 'C' and 'E' are also used as subscripts in Figures 7.51, 7.53, 7.54, 7.57 and 7.58 with the same meaning.

More specifically, Figure 7.50a shows the load–drift curves of frames F-C and F-E under monotonic loading, together with the load–drift curve obtained by assuming rigid joint behaviour; in the latter case the frame is designated as 'F-R' with 'R' denoting 'rigid joint'. Moreover, Figure 7.50b provides an indication of the frame crack patterns at various stages of the monotonically applied load. Figure 7.51 shows the variation with load of the differences in the drift values calculated for the first and second levels of frames F-C and F-E under monotonically increasing load; the values are expressed in a normalised form by dividing them with their counterparts for frame F-E. The load–drift curves of frames F-C and F-E under cyclic loading are shown in Figure 7.52, the normalised differences in the drift values with load being shown in Figures 7.53 and 7.54.

The response of the frames under the base acceleration record shown in Figure 7.45 is described in Figures 7.55 and 7.56 in the form of drift versus time and base shear vs. time diagrams, whereas a comparison of the numerically predicted dynamic responses of frames F-C and F-E are shown in Figures 7.57 and 7.58. In the latter figures, the frame response is

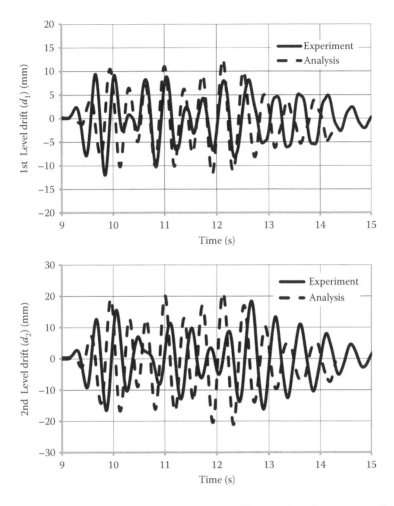

Figure 7.47 Two-level RC frame under seismic excitation. Numerical and experimental displacement response of levels 1 and 2.

expressed in a normalised form by dividing the differences in response with the corresponding maximum values for frame F-E.

Finally, Table 7.1 shows the maximum values of displacement and acceleration at levels 1 and 2 and the maximum value of base shear predicted by analysis together with their counterparts established experimentally. The predicted values shown in Table 7.1 are also included in Table 7.2 together with their counterparts predicted for frame F-E.

7.6.4 Discussion of the results

An indication of the model's validity may be obtained by reference to Table 7.1 and Figures 7.47 through 7.49, the latter describing the response of frame L30 in the form of displacement, acceleration and base shear versus time graphs under the seismic excitation shown in Figure 7.45. From the figures, it appears that the correlation between predicted and experimental values is significantly closer in the cases of the base shear (see Figure 7.49) and the acceleration of the masses at levels 1 and 2 (see Figure 7.48) than it is for the case of the

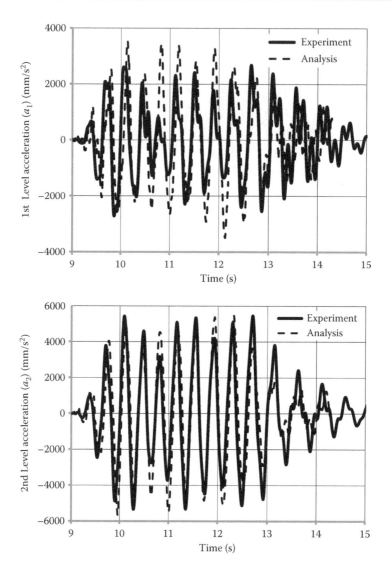

Figure 7.48 Two-level RC frame under seismic excitation. Numerical and experimental acceleration response of levels 1 and 2.

drift values (see Figure 7.47). Moreover, Table 7.1 shows that the maximum values obtained numerically for the drift, acceleration and base shear differ from their experimentally established counterparts by amounts ranging between approximately 3% and 30%.

Figure 7.50 shows the load–drift curves predicted for levels 1 and 2 of frames F-C and F-E under monotonic loading together with their counterparts for the case of frame F-R characterised by rigid joint behaviour. From the figure, it can be seen that the curves associated with elastic and rigid joint behaviour essentially coincide and this confirms the view that it is cracking that causes the deviation of the true joint response (response of frame F-C) from that predicted by assuming rigid joint behaviour (response of frame F-R). The differences in drift between frames F-C and F-E are shown in Figure 7.51 from which it appears that cracking of the joints causes the drift to increase significantly with the applied load up

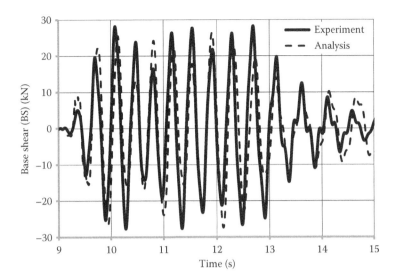

Figure 7.49 Two-level RC frame under seismic excitation. Comparison of the experimental and numerically predicted values of base shear.

to nearly 50% at peak load level (P_u). It is interesting to note in Figure 7.50b that cracking within the joint regions (indicated by the dashes within the joint elements) initiates at a value of the applied load of the order of 25 kN, which is less than $0.3P_u$.

The load–drift curves predicted for the frames under cyclic loading are shown in Figure 7.52 with the differences in drift between the two frames being shown in Figures 7.53 and 7.54 for levels 1 and 2, respectively. The latter figures show that the effect of cracking is more pronounced in the first load cycle rather than the second one. However, during the third stage of the loading process, when the imposed load increases monotonically to failure, the difference in drift increases to a value of over 50% that of frame F-E, for which cracking was not allowed to form within the joint region.

It is interesting to note in Figures 7.55 and 7.56 that crack formation within the joint regions affects mainly the amplitude of the drift and base shear oscillations of frame F-C under seismic excitation, as frames F-C and F-E appear to oscillate in phase with essentially the same period. Moreover, from Figures 7.57 and 7.58, it can be seen that the differences in the values of both drift and base shear between the frames can be larger than about 30% the maximum value of their counterparts for frame F-E.

As the maximum values of drift and base shear predicted by the analysis are important for design purposes, these are included in Table 7.2. From this table, it can be seen that the maximum values of drift predicted by assuming rigid joint behaviour differ from their counterparts predicted by analysis when allowing for the effect of joint cracking by amounts of the order of 19% and 13% for the first and second levels, respectively. From Table 7.2, it can also be seen that the maximum values of base shear predicted when assuming rigid joint behaviour differs from its counterpart predicted when allowing for the effect of joint cracking by an amount of the order of 26%. Since the assumption of rigid joint underlies practical structural analysis, it appears that such analyses may underestimate the values of both design loads and displacements. As a result of this, current design practice appears to be unable to safeguard the margins of safety and the structural performance requirements specified by the codes.

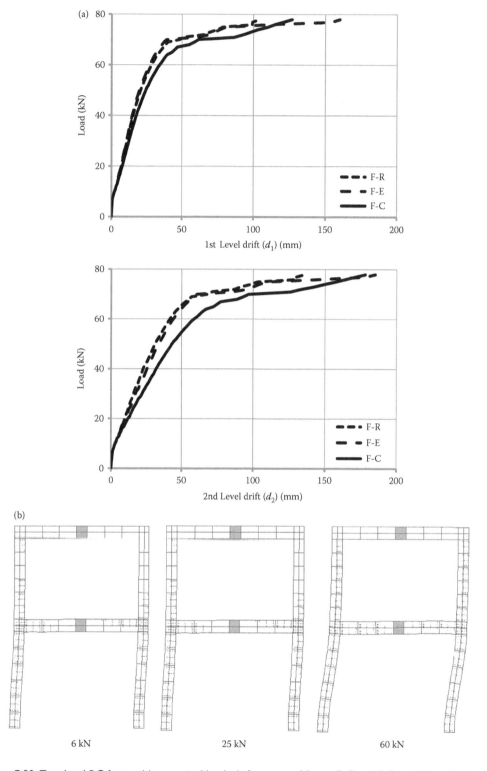

Figure 7.50 Two-level RC frame: (a) numerical load–drift curves of frame F-C and F-E and (b) typical crack patterns under monotonic loading.

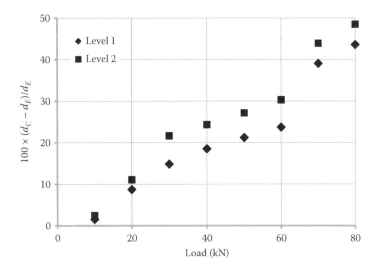

Figure 7.51 Two-level RC frame under monotonic loading. Difference in drift between frames F-C and F-E normalised with respect to the drift of frame F-E.

7.6.5 Concluding remarks

The NLFEA model, already found to yield close predictions of the behaviour of a wide range of RC structural forms under arbitrary static (monotonic and cyclic) loading conditions, is also shown to yield realistic predictions of the behaviour under seismic excitation of the two-level frame investigated in the present work.

   Using this model for investigating the effect of cracking within the joint regions forming at the beam–column intersections on structural behaviour, the effect of such cracking is found to be significant. As a result, practical structural analysis based on the assumption of rigid joints yields results that cannot always safeguard the code-specified margins of safety and structural performance requirements. It appears, therefore, that there is an urgent need for improving current design methods so as to allow for the inconsistencies resulting from the condition for rigid joint behaviour that underlies current methods for frame analysis.

7.7 EFFECT OF THE CONFINEMENT OF REINFORCEMENT IN BOUNDARY-COLUMN ELEMENTS ON THE BEHAVIOUR OF STRUCTURAL-CONCRETE WALLS UNDER SEISMIC EXCITATION

7.7.1 Background

Current code [American Concrete Institute (ACI) 2006, EC8 2004, Earthquake Planning and Protection Organization (EKOS) 2001] provisions for the design of earthquake-resistant RC structural walls (SW) specify reinforcement arrangements comprising two parts: one part forming 'boundary column (BC)' elements (usually extending between the ground and first-floor levels of buildings) along the two vertical edges of the walls; the other consisting of a set of grids of uniformly distributed vertical and horizontal bars, within the wall web, arranged parallel to the wall side faces. The BC elements are intended to impart the walls the code-specified ductility, whereas the wall web is designed against the occurrence of 'shear' failure, before the wall flexural capacity is exhausted. The specified ductility is considered to

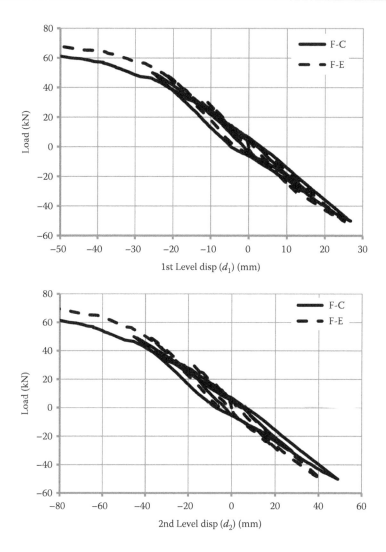

Figure 7.52 Two-level RC frame under cyclic loading. Numerically established load–drift curves of frames F-C and F-E.

be achieved by confining concrete within the BC elements through the use of a dense stirrup arrangement, thus increasing both the strength and the strain capacity of the material; on the other hand, shear failure is mainly prevented by providing horizontal web reinforcement capable of sustaining the portion of the shear force in excess of that which can be sustained by concrete. It is also important to add that the calculation of the wall flexural capacity allows for the contribution of all vertical reinforcement, within both the BC elements and the web.

The above design procedure, however, has a significant drawback: the dense spacing of the stirrups often results in reinforcement congestion within the BC elements and this may cause difficulties in concreting and, possibly, incomplete compaction of the concrete. Although it is widely recognised that there is a need for confining reinforcement within the BC elements in order to increase the ductility of SW, it has been argued that such reinforcement, but in a significantly smaller amount, is only required for the case of SW with a shear span-to-depth ratio larger than 2.5 (Kotsovos and Pavlovic 1999).

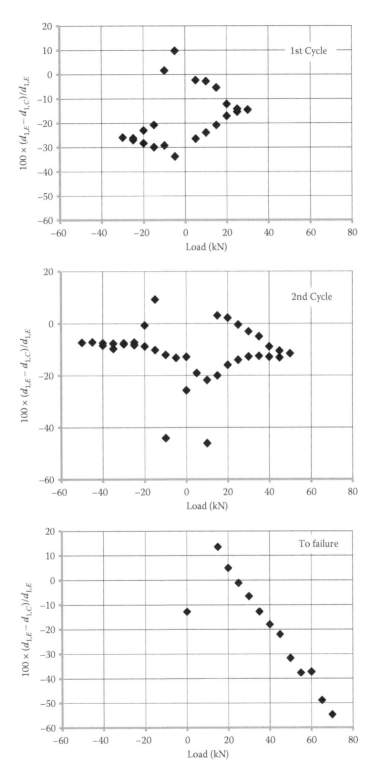

Figure 7.53 Two-level RC frame under cyclic loading. Normalised differences in drift of the 1st level between frames F-C and F-E.

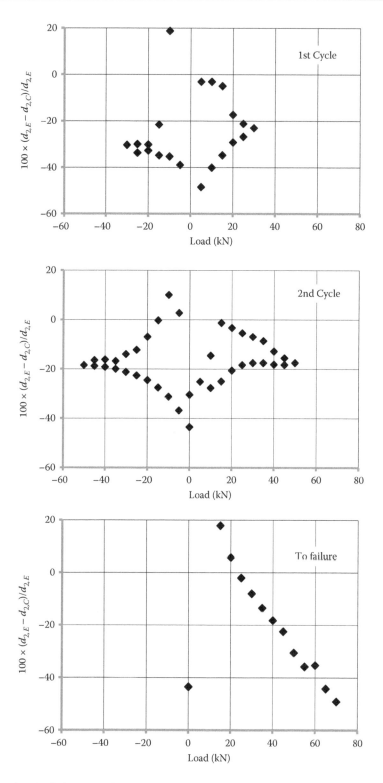

Figure 7.54 Two-level RC frame under cyclic loading. Normalised differences in drift of the 2nd level between frames F-C and F-E.

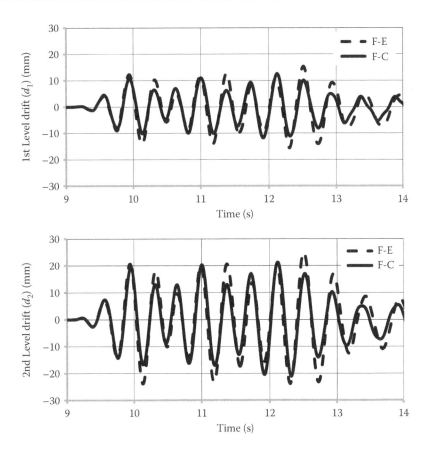

Figure 7.55 Two-level RC frame under seismic excitation. Time history of first and second level drift of frames F-C and F-E.

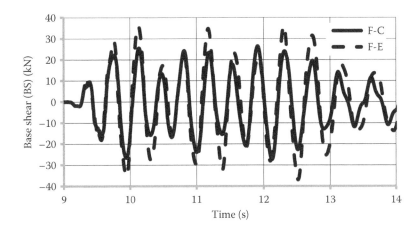

Figure 7.56 Two-level RC frame under seismic excitation. Time history of the base shear of frames F-C and F-E.

Table 7.1 Two-level RC frame under seismic excitation

Response	Experiment (1)	Analysis (2)	$100 \times [(2)-(1)]/(1)$
maxD$_1$ (mm)	12	12.6	5
maxD$_2$ (mm)	18.4	21.2	6.3
maxA$_1$ (mm/s²)	2713	3540	30.5
maxA$_2$ (mm/s²)	5455	5660	3.8
maxBS (kN)	28.3	27.4	−3.2

Experimentally established and numerically predicted maximum (numerical) values of the drift (maxD$_1$ and maxD$_2$) and the acceleration (maxA$_1$ and maxA$_2$) of the masses at levels 1 and 2 and base shear (maxBS)

Table 7.2 Two-level RC frame under seismic excitation

Response	F-C (1)	F-E (2)	$100 \times [(1)-(2)]/(2)$
maxD$_1$ (mm)	12.6	15.6	−19.2
maxD$_2$ (mm)	21.2	25	−12.9
maxBS (kN)	27.4	37	26

Numerically predicted maximum (numerical) values of the drift (maxD$_1$ and maxD$_2$) of the masses at levels 1 and 2 and the base shear (maxBS) of frames F-C and F-E.

Evidence in support of the above argument has recently been obtained from numerical experiments carried out, as for the case of the flat slabs discussed in Section 5.5, through the use of the proposed FE model (Cotsovos and Kotsovos 2007). The work involved a comparative study of the behaviour of SW designed by using two distinctly different approaches: that of the compressive force path (CFP) method (Kotsovos and Pavlovic 1999) and that of the methods adopted by codes (ACI 318 2006, EC2 2004, EC8 2004, EKOS 2001). Two types of SW were investigated under static (both monotonic and cyclic) and dynamic (seismic) loading conditions: Walls W1 with a height-to-length ratio of 1 and walls W2 with a height-to-length ratio of 2. A concise description of the above numerical-test programme is presented in what follows, with full details being provided elsewhere (Cotsovos and Kotsovos 2007). It should be noted that the conclusions drawn from the comparative study reported herein have been subsequently verified with results recently obtained by experiment and reported elsewhere (Kotsovos ct al. 2011, Zygouris et al. 2013).

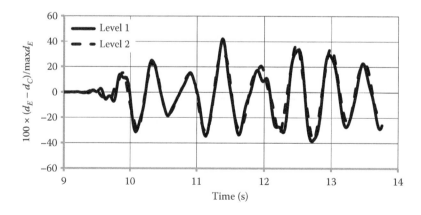

Figure 7.57 Two-level RC frame under seismic excitation. Time history of the normalised differences in drift between specimens F-C and F-E.

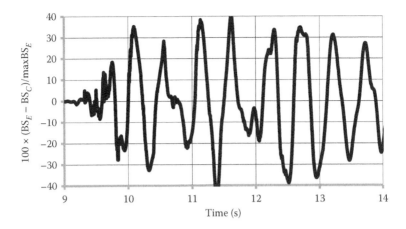

Figure 7.58 Two-level RC frame under seismic excitation. Time history of normalised differences in base shear between specimens F-C and F-E.

7.7.2 Design details

The design details of the SW investigated are shown in Figure 7.59. The figure indicates that walls W1 have a length $l = 3000$ mm (see Figure 7.59a), whereas for walls W2 $l = 1500$ mm (see Figure 7.59b); the height and width are equal to $h = 3000$ mm and $b = 250$ mm, respectively, for all walls. The longitudinal reinforcement of both types of wall comprises six 20 mm diameter bars arranged in pairs at 150 mm centre-to-centre spacing within the BC elements, and 12 mm diameter bars within the web, also arranged in pairs, with 11 pairs being placed in walls W1 at 200 mm spacing and three in walls W2 at 250 mm spacing. In all cases, it was assumed that the steel bars had a yield stress $f_y = 500$ MPa and strength $f_u = 600$ MPa, whereas the concrete (cylinder) compressive strength was $f_c = 30$ MPa.

In contrast with the longitudinal reinforcement, the horizontal reinforcement placed in the walls depends on the method of design employed. Within the BC elements of the walls, the horizontal reinforcement comprises 8 mm diameter stirrups at centre-to-centre spacing of 62 mm (ACI 318), 82 mm (EC2) and 26 mm (EKOS) for walls W1 (the CFP method does not specify any stirrups for this type of walls), whereas the stirrup spacing for walls W2 was 62 mm (ACI 318), 58 mm (EC2), 44 mm (EKOS) and 263 mm (CFP). The horizontal reinforcement of the web comprises 8 mm diameter straight bars that form a grid with the longitudinal bars at each side of the wall. The horizontal bars are placed at a centre-to-centre distance of 160 mm (ACI and EC2) and 250 mm (EKOS and CFP) for walls W1, and 160 mm (ACI), 135 mm (EC2), 185 mm (EKOS) and 343 mm (CFP) for walls W2. (It should be noted that no attempt was made to choose practical values for the diameter and spacing of the transverse reinforcement, as the choice of particular values does affect the analysis results; moreover, the calculated values show very clearly the differences resulting from the application of the different codes.) As for the case of vertical reinforcement, the values of yield stress and strength of the transverse reinforcement were assumed to be equal to 500 and 600 MPa, respectively. Depending on the method employed to design the transverse reinforcement the walls are classified as W1–ACI, W1–CFP, W1–EC, W1–EKOS, W2–ACI, W2–CFP, W2–EC and W2–EKOS.

It is interesting to note in Figure 7.59 that for walls W1 the CFP method does not specify any stirrups, while the number of horizontal straight bars is equal to that specified by EKOS. On the other hand, for the walls W2, the CFP method specifies nearly five times

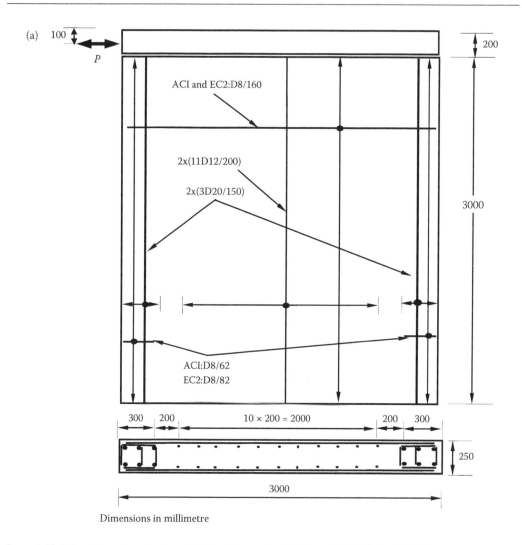

Figure 7.59 RC walls under seismic excitation. Design details: (a) walls W1; (b) walls W2.

(Continued)

fewer stirrups and approximately half the number of horizontal straight bars specified by the codes.

7.7.3 Loading regimes

The walls were subjected to three types of horizontal loading applied along the horizontal axis of symmetry of a rigid prismatic element monolithically connected to the walls at their top face (see Figure 7.59):

a. Static load monotonically increasing to failure
b. Static cyclic load also increasing to failure from a value of approximately 50% of the wall load-carrying capacity, in increments of 10% with each load reversal, as indicated in Figure 7.60

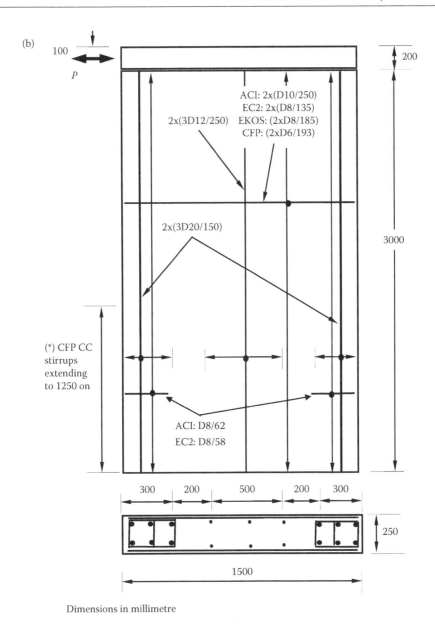

Dimensions in millimetre

Figure 7.59 (Continued) RC walls under seismic excitation. Design details: (a) walls W1; (b) walls W2.

    c. Seismic excitation as indicated in Figure 7.61 assuming a 'weightless' mass of 100 tonnes concentrated at the geometric centre of the prismatic element.

7.7.4 Mesh discretisation adopted

The walls were subdivided into $6 \times 1 \times 6 = 36$ (walls W1) and $4 \times 1 \times 6 = 24$ (walls W2) brick elements as shown in Figure 7.62. As discussed in Section 7.7.3, the load was applied through a prismatic member monolithically connected to the walls at the upper face; this member was subdivided into $6 \times 1 \times 1 = 6$ brick elements, as also indicated in Figure 7.62. The line elements representing the steel reinforcement were placed along successive series of

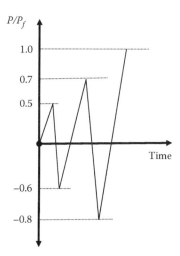

Figure 7.60 RC walls. Loading history adopted for the cyclic tests.

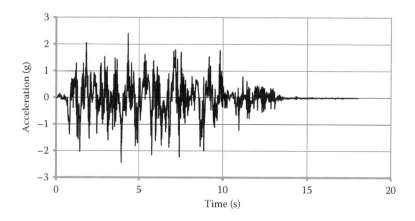

Figure 7.61 RC walls under seismic excitation. Acceleration record.

nodal points in both vertical and horizontal directions. Since the spacing of these line elements was predefined by the location of the brick element nodes, their cross-sectional area was adjusted so that the total amount of reinforcement within both the BC elements and the wall web was equal to the design values shown in Figure 7.59.

7.7.5 Results

The main results of the work are given in Figures 7.63 through 7.68. Figures 7.63 and 7.66 show the load–deflection curves obtained under statically applied monotonic loading, whereas the load–deflection curves obtained under statically applied cyclic loading are shown in Figures 7.64 and 7.67. Finally, Figures 7.65 and 7.68 show the response of the walls to seismic excitation, the wall response being expressed in the form of variations of displacement, velocity, acceleration and base shear with time.

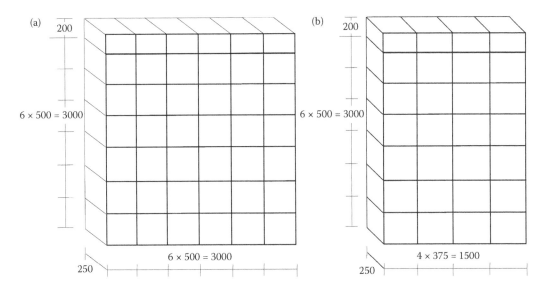

Figure 7.62 RC walls under seismic excitation. FE mesh: (a) walls W1; (b) walls W2.

7.7.6 Discussion of results

7.7.6.1 Walls W1

From Figure 7.59a, it can be seen that, of the walls W1, those designed in accordance with (a) EKOS (W1–EKOS) and (b) the CFP method (W1–CFP) differ in that the reinforcement of W1–EKOS includes a dense stirrup arrangement within its BC elements, whereas the reinforcement of W1–CFP does not. And yet, as indicated in Figures 7.63 through 7.65, these two walls exhibited small differences in behaviour under the loading regimes investigated. Under static loading (both monotonic and cyclic) wall W1–EKOS yielded at a slightly higher load, whereas wall W1–CFP exhibited a slightly higher load-carrying capacity and larger maximum horizontal displacements. Under dynamic loading simulating seismic

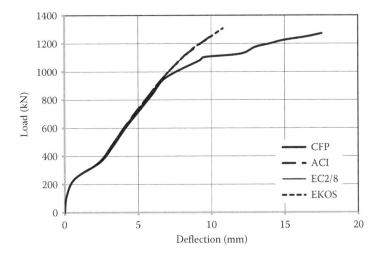

Figure 7.63 RC walls under monotonic loading. Load–deflection curves for walls W1.

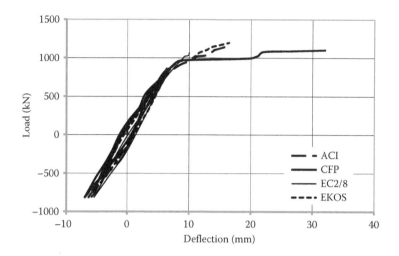

Figure 7.64 RC walls under cyclic loading. Load–deflection curves for walls W1.

excitation both walls exhibited similar behaviour. Such behaviour clearly demonstrates that, for these types of wall, the presence of stirrups within the BC elements does not affect the structural behaviour.

A comparison of the response of walls W1–ACI, W1–EC and W1–EKOS – that is, the walls designed in accordance with current codes – indicates that all walls exhibited similar behaviour. All these walls have similar reinforcement arrangements characterised by the presence of stirrups within their BC elements, with the spacing of the stirrups depending on the code adopted for their design. It is interesting to note that EKOS specifies spacing of only 26 mm, as compared with the 62 mm spacing specified by ACI 318 and the 82 mm specified by EC2. Moreover, ACI 318 and EC2 specify web horizontal reinforcement with 160 mm spacing, as compared with the 250 mm spacing specified by EKOS.

It is noted that for all types of SW the codes considered specify (a) stirrups in order to provide confinement to concrete within the BC elements and (b) horizontal reinforcement that would increase the wall shear capacity beyond the value of the shear force corresponding to flexural capacity. On the other hand, for RC structural members with a shear span-to-depth ratio between 1 and 2.5 (with 'depth' being the distance from the extreme compressive fibre of the resultant of the forces sustained by the flexural reinforcement in tension when flexural capacity is attained), such as walls W1, the CFP method specifies only horizontal reinforcement, uniformly distributed along the wall height, in an amount sufficient to enhance the bending moment corresponding to a non-flexural mode of failure to the value of flexural capacity (Kotsovos 2014). It appears from Figures 7.63 through 7.65 that the CFP method satisfies the code requirements for load-carrying capacity and ductility in a more efficient manner than the methods incorporated in the codes (see Figure 7.59a).

It may also be interesting to note in Figures 7.63 through 7.65 that the values of maximum load and horizontal displacement exhibited by all walls under both cyclic and dynamic loading were similar to those attained under monotonic loading.

7.7.6.2 Walls W2

In contrast with walls W1, all methods of design investigated specify stirrups within the BC elements of walls W2. However, unlike the code methods which, as discussed in the preceding section, specify stirrups for confining concrete within the BC elements, the CFP method

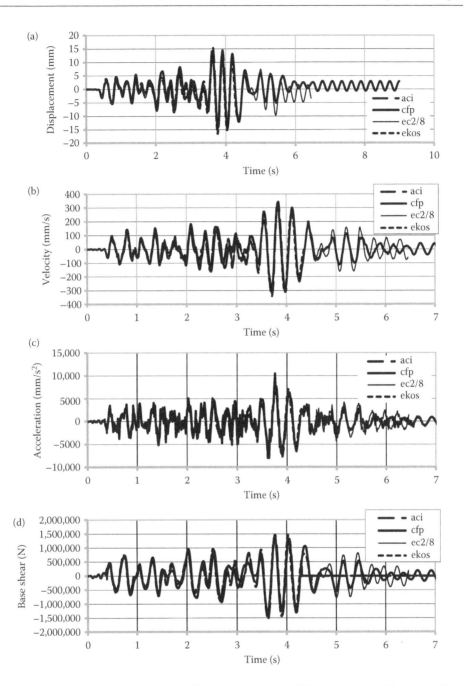

Figure 7.65 RC walls under seismic excitation. Variation with time of (a) displacement, (b) velocity, (c) acceleration and (d) base shear for walls WI.

specifies such reinforcement in order to sustain tensile stresses developing within the compressive zone at right angles to the vertical direction as a result of the loss of bond between concrete and the flexural reinforcement in regions, such as the base of the walls, under the action of a large bending moment combined with a large shear force (Kotsovos 2014).

Figures 7.66 through 7.68 show that variations of the order of 40% in either the stirrups of the BC elements or the horizontal bars of the web have no apparent effect on the

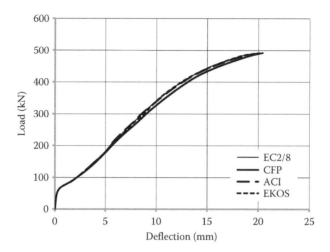

Figure 7.66 RC walls under monotonic loading. Load–deflection curves for walls W2.

behaviour of the walls designed in compliance with the code specifications for the loading regimes considered. More importantly, the wall behaviour appears to be unaffected by the large reductions of the order of 80% for the stirrups and 60% for the horizontal bars resulting from the application of the CFP method for the design of the walls. In fact, the reduction in stirrups is even larger when considering that such reinforcement is placed over the lower portion of the BC elements extending to approximately one-third of the wall height, as opposed to the total wall height specified by the codes (see Figure 7.59b).

It appears from the above, therefore, that, as for the case of walls W1, walls W2 exhibit similar behaviour under the loading regimes investigated. Such similarity in behaviour indicates that designing the walls in accordance with the CFP method yields more efficient

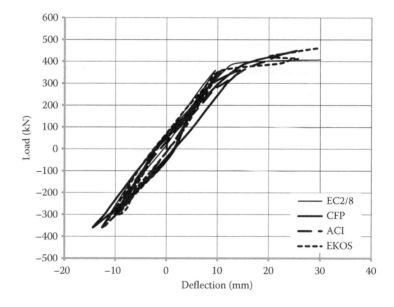

Figure 7.67 RC walls under cyclic loading. Load–deflection curves for walls W2.

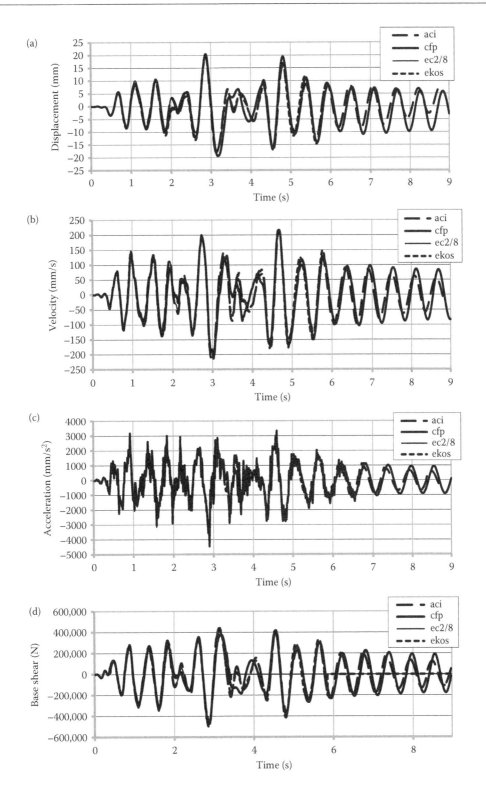

Figure 7.68 RC walls under seismic excitation. Variation with time of (a) displacement, (b) velocity, (c) acceleration and (d) base shear for walls W1.

design solutions (see Figure 7.59b) without compromising the code performance require-ments for strength and ductility (see Figures 7.66 through 7.68).

7.7.7 Conclusions

Within the scope of this study, designing in accordance with the CFP method leads to sig-nificant savings in horizontal reinforcement for both types of SW investigated, without com-promising the code performance requirements.

More specifically, in contrast with the methods incorporated in codes, the CFP method does not specify any stirrups within the boundary-column elements of walls with shear span-to-depth ratios between 1 and 2.5, whereas the amount of horizontal web reinforce-ment is similar to that specified by the codes.

On the other hand, for walls with a shear span-to-depth ratio larger than 2.5, the amount of stirrup reinforcement specified by the CFP method is significantly lower than that speci-fied by current codes. Moreover, such reinforcement is placed within a portion of the boundary-column elements extending to just over one-third of the wall height, as compared with the full element height recommended by the codes. As for the case of the stirrups, the amount of horizontal web reinforcement specified by the CFP method is also significantly smaller (nearly half) than that specified by the codes.

7.8 CONCLUDING REMARKS

From the case studies presented in the present chapter, it can be concluded that the proposed non-linear FE solution procedure is numerically stable and capable of not only producing realistic predictions of structural behaviour under seismic excitation, but, also, identifying the causes and mechanisms of failure. Moreover, the close correlation between predicted and experimental results indicates that the effect of damping is effectively accounted for by the adopted constitutive models of concrete and steel and the numerical description of the crack-ing processes.

It is important to note, however, that the predicted loss of load-carrying capacity occurs earlier than what the experimental results indicate. This difference in behaviour is consid-ered to reflect indefinable effects, stemming from the interaction between specimen and test-ing set-up. Such effects become dominant when disruption becomes excessive and enable the specimen to resort to alternative, unstable in nature, mechanisms of load transfer. It is con-sidered, therefore, that, during the last stage of the testing procedure, the results obtained describe the post-ultimate interaction between the specimen and the testing set-up and, hence, are of no practical significance.

In all but one of the case studies presented in the present chapter, the interaction between the specimen and the testing set-up was ignored, since its effect on the structural behav-iour became pronounced after the occurrence of considerable disruption of the specimen. However, for the case of the three-level wall discussed in Section 7.5, the interaction between specimen and testing set-up affected structural behaviour throughout the test duration, since the masses interacting with the wall were supported by an auxiliary frame structure. In this case, the energy dissipated during the seismic excitation could not be entirely attrib-uted to the damage reflected in the non-linear structural response; a part of it, therefore, was allowed for through the use of a dumping matrix expressing the interaction between the specimen and the experimental set-up.

REFERENCES

American Concrete Institute (ACI), 2006, *Building Code Requirements for Reinforced Concrete (ACI 318R-06) and Commentary*. ACI, Detroit.

Carydis P., 1997, Shaking table tests of R.C. frames, *ECOEST PPREC8*, Report 8, 182.

Comite Europeen de Normalisation, ENV–1992–1, Eurocode No. 2 (EC2), 2004, *Design of Concrete Structures. Part 1: General Rules and Rules of Building*. CEN, Brussels.

Comite Europeen de Normalisation, ENV–1998–1, Eurocode No. 8 (EC8), 2004, *Design of Structures for Earthquake Resistance. Part 1: General Rules, Seismic Actions and Rules for Buildings*, CEN, Brussels.

Cotsovos D. M., 2004, Numerical investigation of structural concrete under dynamic (earthquake and impact) loading, PhD thesis, University of London, UK

Cotsovos D. M., 2013, Cracking of RC beam/column joints: Implications for the analysis of frame-type structures, *Engineering Structures*, 52, 131–139.

Cotsovos D. M. and Kotsovos M. D., 2007, Seismic design of structural concrete walls: An attempt to reduce reinforcement congestion, *Magazine of Concrete*, 59, 9, 627–637.

Cotsovos D. M. and Kotsovos M. D., 2008, Cracking of RC beam/column joints: Implications for practical structural analysis and design, *The Structural Engineer*, 86(12), 33–39.

Earthquake Planning and Protection Organization (EKOS), 2001, *Greek Code for Reinforced Concrete*, Ministry of Environment, Planning and Public Works, Athens, Greece, April 2001 (in Greek).

Kotsovos M. D., 2014, *Compressive Force-Path Method: Unified Ultimate Limit-State Design of Concrete Structures*, Engineering Materials Series, Springer, London, 221pp.

Kotsovos G. M., Cotsovos D. M., Kotsovos M. D. and Kounadis A. N., 2011, Seismic behaviour of RC walls: An attempt to reduce reinforcement congestion, *Magazine of Concrete Research*, 63(4), 235–246.

Kotsovos, M. D. and Pavlovic, M. N., 1995, *Structural Concrete: Finite-Element Analysis and Design*, London, Thomas Telford, 550pp.

Kotsovos M. D. and Pavlovic M. N., 1999, *Ultimate Limit-State Design of Concrete Structures: A New Approach*, Thomas Telford (London), 164pp.

Lestuzzi P., Wenk T. and Bachmann H., 1999, Dynamic tests of RC structural walls on the ETH earthquake simulator, *IBK Report No. 240*, Institüt für Baustatik und Konstruktion, ETH, Zurich.

Minowa C., Ogawa N., Hayashida T., Kogoma I. and Okada T., 1995, Dynamic and static collapse tests of reinforce-concrete columns, *Nuclear Engineering and Design*, 156, 269–276.

Takeda T., Sozen M. A. and Nielsen N. N., 1970, Reinforced concrete response to simulated earthquakes, *Journal of Structural Engineering*, ASCE, 96, 2557–2573.

Zygouris N. St., Kotsovos G. M. and Kotsovos M. D., 2013, Effect of transverse reinforcement on short structural wall behavior, *Magazine of Concrete Research*, 67(17), 1034–1043.

Chapter 8

Structural concrete under impact loading

8.1 INTRODUCTION

The work presented in this chapter is intended to demonstrate that the use of the FE model presented in Chapter 4, as complemented in Chapter 6 so as to cater for dynamic problems, is also capable, through the use of the constitutive model presented in Chapter 3, of predicting the behaviour of structural concrete under high loading rates. As the use of this model implies that the mechanical properties of concrete are independent of the rate of loading, the numerical investigation presented is also aimed at investigating the effects of the rate of loading on structural behaviour and, in doing so, to explain the causes of these effects.

8.2 STRUCTURAL CONCRETE UNDER COMPRESSIVE IMPACT LOADING

8.2.1 Background

Over the past century, a large number of experiments have been carried out on the behaviour of prismatic or cylindrical concrete specimens under high rates of uniaxial compressive loading (for a thorough bibliography of the copious laboratory data encompassing hundreds of tests see e.g., Bischoff and Perry 1991, 1995, Ross et al. 1995, 1996, Gary and Bailly 1998, Grote et al. 2001). The primary objective of such experiments is to investigate the behaviour of concrete under such extreme loading conditions at a material level, since the response exhibited by these specimens during dynamic tests has been shown to differ from that of their counterparts tested under static conditions. This difference primarily takes the form of an increase in the specimens' load-carrying capacity and maximum sustained axial strain, a difference which becomes more apparent as the loading rate increases.

The experimental data obtained from these tests are usually processed in order to derive laws which are incorporated into already existing material models aiming to enhance them by making them sensitive to the rate of loading to enable them to describe the behaviour of concrete under high loading rates. The majority of these models are then incorporated into various finite-element (FE) packages (e.g., LS-DYNA, ABAQUS, ADINA, etc.) aiming at predicting accurately the behaviour of RC structures under extreme loading conditions such as those encountered in impact and explosion situations.

The formulations of such material models have been based on a variety of theories, including plasticity (Malvar et al. 1997, Thabet and Haldane 2001), viscoplasticity (Cela 1998, Winnicki et al. 2001, Gomes and Awruch 2001, Georgin and Reynouard 2003, Barpi 2004), continuum damage mechanics (Cervera et al. 1996, Hatzigeorgiou et al. 2001, Koh et al. 2001, Lu and Xu 2004) or a combination of these theories (Dube et al. 1996, Faria et al. 1998). Models such as these are referred to as phenomenological since they are based on theories capable of providing a close fit to experimental information without taking into consideration the causes of the observed material behaviour. Regardless of the theory upon which their formulation is based, these models share a number of fundamental assumptions, the validity of which is inherently questioned herein: strain softening, stress-path dependency and loading-rate sensitivity are the most common among such assumptions.

In order for such material models to be used in problems involving the investigation of structural concrete under static or dynamic loading, one must initially calibrate them carefully (based on the available experimental data) by assigning certain values to a number of parameters which are essential to fully define the material model. However, the use of such parameters usually makes the FE packages which incorporate them case-sensitive owing to the fact that their ability to produce accurate predictions is limited only to certain problem types. Applying them to different problem types often means that they have to be recalibrated. Therefore, it is not surprising that these existing FE models for structural concrete cannot provide a phenomenological explanation for the behaviour of concrete (at a material level) under high rates of impact: instead, when faced with this type of problem, all the existing FE packages for structural concrete merely adapt their material properties on the basis of the experimental high-rate loading data mentioned above.

The work presented herein, extracted from Cotsovos and Pavlovic (2008a), follows the opposite approach: it employs the proposed FE model to reproduce the experimental data and, in so doing, also aims to provide a fundamental explanation for the sudden increase in concrete strength and the overall change in the specimen behaviour as a certain value of the loading rate is exceeded. As already shown in Chapters 5 and 7, the FE model is capable of yielding realistic predictions of the response of a wide range of RC structures under static (monotonic and cyclic) and dynamic (earthquake) loading, respectively. It should also be reminded that the material model presently adopted for describing the behaviour of concrete is fully defined only by one parameter: the uniaxial compressive strength f_c; moreover, in compliance with the experimental information presented in Chapters 2 and 3, it does not account for strain softening and stress-path dependency, but places special emphasis on the response of concrete to multiaxial (i.e., triaxial) stress conditions ignoring loading-rate sensitivity.

The aim of the numerical investigation is to demonstrate the reliability of the proposed FE model in realistically predicting the behaviour of structural concrete under extreme loading conditions, such as those encountered during impact and explosion situations, despite its reliance on purely static material properties. This is achieved by comparing the results obtained from the numerical investigation with available experimental data. Such a comparative study allows for the validation of the assumptions, based upon the FE model's formulation, and provides a simple explanation as to the causes of loading-rate effects in impact situations.

8.2.2 Experimental information

The results obtained from experiments investigating the behaviour of prismatic and cylindrical concrete specimens under high rates of uniaxial compressive loading presented graphically in Figure 8.1 clearly indicated that the increase of the loading rate beyond certain levels leads to a substantial increase in the specimen strength. Furthermore, the majority of experimental data depicted in Figures 8.2 and 8.3 (based on the data presented by Bischoff and Perry 1995) suggest that the maximum axial strain exhibited prior to failure and the axial stiffness of the specimen are also characterised by a significant increase with respect to their counterparts under static loading. However, while it is evident that the concrete

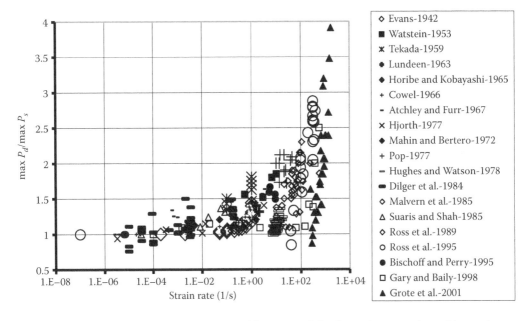

Figure 8.1 Concrete in uniaxial compression. Variation of load-carrying capacity with strain rate (max P_d = load-carrying capacity, max P_s = load-carrying capacity under static loading). (Data from Cotsovos D. M. and Pavlovic M. N., 2008a. *Computers and Structures*, 86 (1–2), 145–163.)

characteristics under high-rate loading differ significantly from those under static loading, what the experimental data cannot provide are the reasons for these differences.

By inspecting the experimental data it is clear that they are characterised by considerable scatter and, therefore, it is extremely difficult to derive a law able to describe realistically the change in the specimen strength and its overall behaviour under high rates of loading. Furthermore, the existence of a number of parameters (such as the experimental technique adopted for the tests, the shape, size and moisture content of the specimens and the different

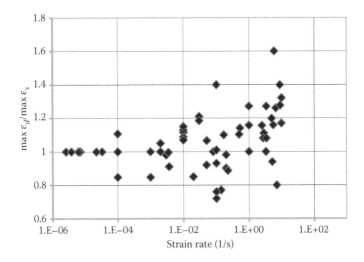

Figure 8.2 Concrete in uniaxial compression. Variation of maximum exhibited strain with strain rate (max ε_d = maximum strain, max ε_s = maximum strain under static loading).

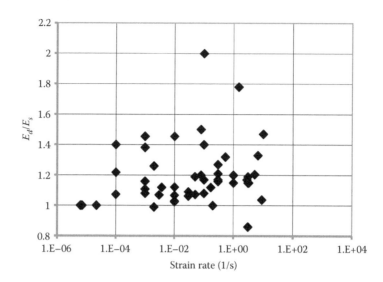

Figure 8.3 Concrete in uniaxial compression. Variation of specimen axial stiffness with strain rate (E_d = specimen axial stiffness, E_s = specimen axial stiffness under static loading).

types of concrete used), which vary from experiment to experiment, and the absence of laws able to quantify accurately their individual effect on the specimen behaviour add to the uncertainty and difficulty in interpreting the experimental data. Additionally, the difficulty inherent in dynamic tests (in the sense of being able to obtain accurate and meaningful measurements) due to their extremely short duration, combined with the fact that many of these tests were performed before the 1980s when the equipment used was not as advanced as it is today, are also significant factors which undermine the validity of the data obtained from these tests. Therefore, it is clear that the experimental data can only qualitatively describe the effect of the loading rate on the specimens' behaviour.

8.2.3 Structural form investigated

The structural form which provides the basis of this investigation is a concrete prism, similar to the concrete specimens used in various experimental investigations carried out to date on this subject (e.g., Bischoff and Perry 1995). The prism is assumed to be fixed at its bottom face, and to be subjected to an axial load applied at its upper face through a rigid element with the same cross section (see Figure 8.4) in order for the external load to be distributed uniformly on the upper face of the concrete prism. It is assumed that concrete and the rigid element on the top are fully bonded at their interface. The prism height is 253 mm and its cross section forms a square with a side of 100 mm, whereas the rigid element has a height of 200 mm. The uniaxial compressive strength of concrete f_c is assumed to be 30 MPa, a fairly typical value in practice.

In order to explain the behaviour exhibited by the concrete specimen, the dynamic problem must be viewed as a wave-propagation problem. By applying the external load on an upper area of the specimen, the stress wave created is transferred initially from the area where the external load is imposed towards the lower part of the specimen. Because the latter is fixed at the bottom, the stress wave bounces off its bottom surface and moves back upwards. Once the stress wave reaches the top of the specimen, it bounces back down again. The wave is, therefore, trapped by the boundary conditions imposed on the specimen and

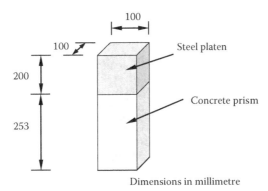

Dimensions in millimetre

Figure 8.4 Concrete in uniaxial compression. Details of specimens (dimensions in millimetre).

moves from top to bottom and vice versa. The velocity of the stress wave is given by equation $\dot{u}_w = \sqrt{G/\rho}$ (Hinton and Owen 1980) where $G = E/(2 + 2\nu)$ is the shear modulus of concrete, and ρ is the density of the concrete specimen. By assuming that the Poisson ratio ν, the shear modulus G and the density ρ of concrete used are 0.2, 13.3 GPa and 2,400 kg/m³, respectively (which are the average values for concrete with an f_c of 30 MPa), the velocity of the stress wave is found to be $\dot{u}_w = 2354$ m/s. Based on the fact that the height of the concrete specimen used in the present numerical investigation is 253 mm, the time needed for the stress wave to move from top to bottom is approximately 0.0001 s.

8.2.4 Numerical modelling of the dynamic problem

Both the concrete prism and the rigid element (Figure 8.5) are modelled by using the 27-node Lagrangian brick element. Meshes consisting of $3 \times 1 \times 1$ and $1 \times 1 \times 1$ elements are adopted in order to model the concrete prism and the rigid element, respectively (see Figure 8.5a and b depicting possible models described below). The use of a sparse FE mesh contrasts with what other investigators have used previously. Usually, a dense FE mesh is preferred in order to model the concrete specimen with several investigations adopting FEs as small as 2–3 mm (Tedesco et al. 1989, 1991, 1997, Thabet and Haldane 2001, Koh et al. 2001, Li and Meng 2003). However, the philosophy upon which the FE model

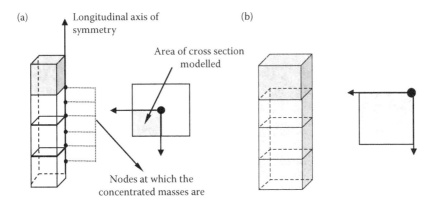

Figure 8.5 Concrete in uniaxial compression. FE models: (a) model A; (b) model B.

adopted in the present work is based is different and does not employ small FEs (Kotsovos and Pavlovic 1995). This is because, as discussed in Section 4.4.2, the material model now adopted is based on data obtained from experiments in which cylindrical concrete specimens (subjected to various triaxial loading conditions) constituted a 'material unit' for which average material properties were obtained: thus, the volume of these specimens provides a guideline to the order-of-magnitude of the size of the FE which should be used for the modelling of the concrete structures (in the present instance the structure is a prism specimen). Furthermore, each Gauss point within an FE should correspond to a volume having a size which must be at least three times the size of the largest aggregate used in the concrete mix in order to provide a realistic representation of concrete rather than a description of its constituent materials.

The mass of the specimen is modelled as concentrated masses either located on the FE nodes situated along the longitudinal axis of symmetry of the specimen (model A) or distributed to all FE nodes (model B). In the case of model A, only mass displacement in the direction of the applied load (i.e., along the axis of symmetry) is allowed and hence it is sufficient to analyse only one quarter of the specimen (see Figure 8.5a). In the case of model B, the mass is allowed to have all the three degrees of freedom of the nodes and hence, in order to avoid wave deflection problems on the boundaries of the prism, the whole structure is analysed (see Figure 8.5b). On the other hand, the external load is imposed as a force incrementally at the beginning of each time step. In order to vary the rate of loading, the load increments are kept constant and the time step is varied. For the case of model A, because of the fourfold symmetry, only one quarter of the specimen is modelled and a load increment equal to 0.4 kN is used for each time step. For the case of model B, however, the fourfold symmetry does not hold as the masses have three degrees of freedom, and hence the whole specimen must be modelled, thus making it necessary to impose four times the load used in the previous case (i.e., 1.6 kN). The numerical investigation consists of two case studies. Case study 1 adopts model A whereas case study 2 relies on model B.

At this point, it is important to stress that there is a considerable confusion in the literature regarding the way the loading is applied and described, and also how it is measured. For this reason, several possibilities are investigated. To remove any uncertainty, their definitions are as follows:

1. *Average strain rate*: calculated as the average rate of displacement exhibited by the very top of the specimen divided by the length of the whole specimen.
2. *Maximum value of average strain rate*: calculated by dividing the specimen into zones by using the nodes along the axis of symmetry and by evaluating which one of these zones exhibits the largest average strain rate.
3. *Mid-height strain rate*: evaluated at the mid-height region of the specimen.
4. *Applied stress rate*: defined as the load increment applied in each time step divided by the cross-sectional area of the specimen and the length of the time step used.

8.2.5 Numerical predictions

The results obtained from case studies 1 and 2 show that the behaviour of the concrete specimen under high rates of uniaxial compressive loading differs considerably compared to that exhibited under static loading. This change in the specimen behaviour takes the form of an increase in load-carrying capacity and sustained maximum axial strain exhibited by the specimen as the rate of loading becomes higher. This can be seen by observing the numerical results of Figures 8.6 and 8.7. Figure 8.6 shows the variation of the load-carrying capacity max P_d (normalised with respect to its value under static loading max P_s) with

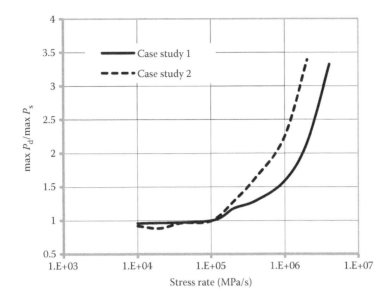

Figure 8.6 Concrete in uniaxial compression. Variation of load-carrying capacity (max P_d), normalised with respect to its value under static loading (max P_s), with applied stress rate for case studies 1 and 2.

the applied stress rate. Figure 8.7 depicts the maximum strain max ε_d (normalised with respect to its values under static loading max ε_s) at the maximum sustained load (which, in these case studies, also implies maximum strain) with the applied stress rate. Such gradual changes in the specimen's behaviour with the rate of loading can also be observed by inspecting the imposed stress–displacement curves presented in Figure 8.9a–f obtained from case studies 1 and 2 for various rates of loading: it can be seen that, for low rates of loading (lower than 100,000 MPa/s), the behaviour predicted by the two case studies is similar, but, as the applied loading rate increases above a value of about 100,000 MPa/s, the

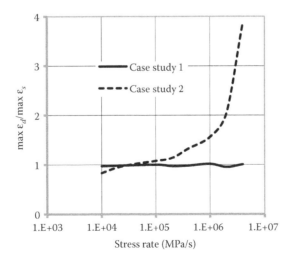

Figure 8.7 Concrete in uniaxial compression. Variation of maximum axial strain (max ε_d), normalised with respect to its value under static loading (max ε_s), with applied stress rate for case studies 1 and 2.

stress–displacement curves obtained from case studies 1 and 2 begin to differ from their static counterparts, as evidenced by an increase in axial stiffness and load-carrying capacity of the specimen and a decrease in its axial deformation.

A comparison of the numerical results of case studies 1 and 2 reveals significant differences between the two FE models for the concrete prism. The results in Figure 8.6 show that, for stress rates up to about 100,000 MPa/s, the predictions obtained from the two case studies are similar; however, the differences in these predictions begin to occur for stress rates over around 200,000 MPa/s. In particular, the results obtained from case study 2, which adopts model B for the modelling of the specimen, predict a higher increase in the specimens' load-carrying capacity compared to that predicted by case study 1 (which adopts model A). Moreover, the data in Figure 8.7 reveal that case study 2 predicts an increase of the maximum value of axial strain (max ε_d) in the specimen prior to failure for stress rates above approximately 200,000 MPa/s, whereas case study 1 predicts that this value remains constant, being practically unaffected by the rate of loading. Such differences in the predictions of case studies 1 and 2 increase as the rate of loading becomes higher. A further comparison of the predicted behaviours obtained from case studies 1 and 2 is available in the plots of Figure 8.8. Up to about 100,000 MPa/s, both models seem to predict approximately the same behaviour. However, as the applied stress rate increases above 200,000 MPa/s, the response predicted by the two case studies begins to differ, with the specimen in case study 2 exhibiting larger load-carrying capacities and larger axial deformations prior to failure than its counterpart in case study 1 for the same rate of loading.

Based on the predictions obtained from the numerical investigation, it can be concluded that inertia has a significant effect on the specimen behaviour under high rates of loading. However, it is interesting to note the very considerable effect of the modelling of the specimen's mass, as evidenced by the different results stemming from the two case studies (1 and 2). It will be recalled that, in case study 1, the specimen mass was lumped at the mesh nodes situated on the longitudinal axis of symmetry and was allowed to have only one degree of freedom in this longitudinal direction (model A). On the other hand, in case study 2, the mass was equally distributed at all the mesh nodes and was allowed to have three degrees of freedom (model B). Clearly, the effect of inertia in the lateral direction has a significant effect on the predicted behaviour of the prisms, especially during the final stages of the loading procedure (i.e., prior to failure). Evidently, case study 2 allows for a more precise modelling of the problem since it is able to account for the effect of inertia in both axial and lateral directions, as well as providing a more refined discretised version of the actual (i.e., continuously distributed) mass of the specimen.

An interesting observation emerges from the numerical results by reference to the values of the reactions at the bottom of the prism. Thus, for high rates of loading, where the duration of the loading procedure is less than 0.0001 s (which corresponds approximately to the earlier estimate for the time needed for the stress wave to travel from the top to the bottom of the specimen), the numerical results obtained from case study 2 (presented in Figure 8.9) reveal that the failure of the concrete prism precedes the development of substantial reactions since the stress wave is unable to reach the bottom of the specimen within the time over which the loading procedure lasts. Therefore, it is realistic to assume that, in such cases, the external load does not affect the whole specimen but only a part of it extending to a level which the stress wave is able to reach in the time that the loading procedure lasts. Based on this, it is possible to conclude that higher concentration of stresses (and strains) develop in the upper part of the specimen whereas, in the bottom part, these concentrations of stresses and strains are much lower. On the other hand, for low rates of loading, where the duration of the loading procedure exceeds 0.0001 s, the stress wave reaches the bottom of the specimen, bounces off it and starts to travel backwards and forwards along the length of the specimen, trapped by the imposed boundary conditions. Because of this, the stress

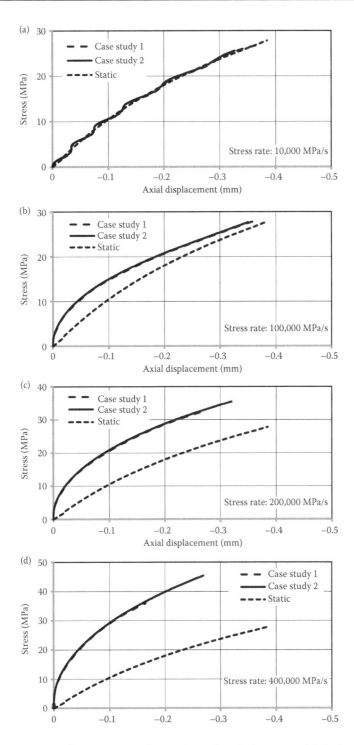

Figure 8.8 Concrete in uniaxial compression. Comparison of applied stress–axial displacement curves for case studies 1 and 2 under different stress rates with their static counterparts. Values of stress rate (MPa/s) equal to: (a) 10,000; (b) 100,000; (c) 200,000; (d) 400,000; (e) 2,000,000 and (f) 4,000,000.

(Continued)

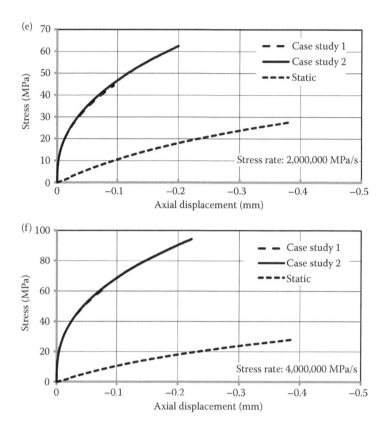

Figure 8.8 (Continued) Concrete in uniaxial compression. Comparison of applied stress–axial displacement curves for case studies I and 2 under different stress rates with their static counterparts. Values of stress rate (MPa/s) equal to: (a) 10,000; (b) 100,000; (c) 200,000; (d) 400,000; (e) 2,000,000 and (f) 4,000,000.

wave affects the whole of the concrete specimen, adding to the complexity of the stress field which develops within it (except, of course, for static or quasi-static loadings). In such cases, it is difficult to predict where the highest concentrations of stresses (and strains) will develop because of the continuous travelling of the stress wave which causes the internal stress field within the concrete prism to constantly change.

By further analysing the numerical data obtained from case study 2 and by examining the displacements exhibited by the specimen during the duration of the loading procedure at certain nodes (shown in Figure 8.10) it is possible to make an assessment of the part of the specimen affected by the external loading. The ensuing numerical results presented in Figure 8.11 show the axial displacements exhibited by those nodes located on the longitudinal axis of symmetry of the specimen (nodes 30–35) and the lateral displacements exhibited by the nodes situated at the middle of one of the four side faces of the specimen (nodes 9–14) with respect to the level of stress applied on the upper face of the concrete prism for different rates of loading. These figures indicate that, as the rate of loading increases (over the value of 200,000 MPa/s), the displacements (both axial and lateral) of the nodes close to the bottom of the specimen gradually decrease. In the case of very high rates of loading, these displacements become very small (even negligible) compared with the displacements exhibited within the upper part of the specimen. It appears, therefore, that the numerical results presented in Figure 8.11 indicate that the upper part of the specimen deforms far more than its lower part when subjected to high rates of compressive loading, as shown qualitatively in Figure 8.12.

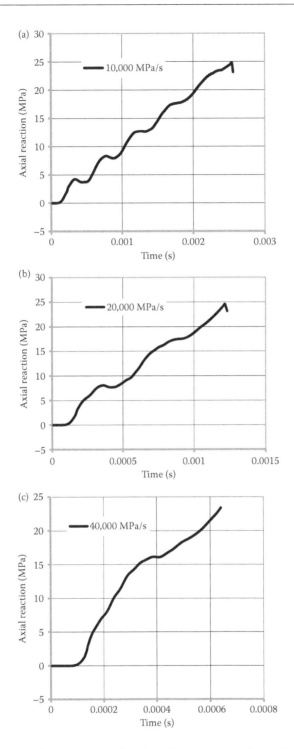

Figure 8.9 Concrete in uniaxial compression. Variation of the axial reaction with time for various stress rates. Values of stress rate (MPa/s) equal to: (a) 10,000; (b) 20,000; (c) 40,000; (d) 100,000; (e) 200,000; (f) 400,000 and (g) 1,000,000.   (*Continued*)

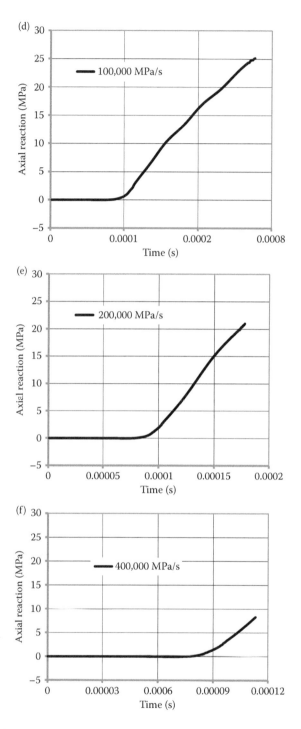

Figure 8.9 (Continued) Concrete in uniaxial compression. Variation of the axial reaction with time for various stress rates. Values of stress rate (MPa/s) equal to: (a) 10,000; (b) 20,000; (c) 40,000; (d) 100,000; (e) 200,000; (f) 400,000 and (g) 1,000,000. *(Continued)*

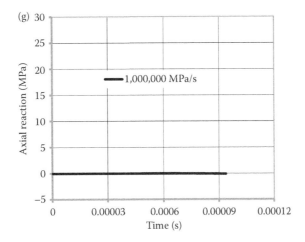

Figure 8.9 (Continued) Concrete in uniaxial compression. Variation of the axial reaction with time for various stress rates. Values of stress rate (MPa/s) equal to: (a) 10,000; (b) 20,000; (c) 40,000; (d) 100,000; (e) 200,000; (f) 400,000 and (g) 1,000,000.

Finally, Figure 8.13 shows the cracking patterns for the above specimen prior to failure for different rates of loading. During the final stages of the loading procedure and at high rates of loading, cracking occurred within the upper part of the specimen. This agrees with previous conclusions since it implies the development of high concentrations of stresses (and, therefore, strains) in this upper region of the specimen. However, in the case of low loading rates, cracking may form in any area of the specimen owing to the fact that the stress wave trapped within the specimen causes continuous changes in the internal stress field within the prism, so that high concentrations of stress leading to macro-cracking may form in any area of the specimen.

8.2.6 Validation of numerical predictions

In order to establish the validity and accuracy of the predictions obtained from the proposed FE model, the results from case studies 1 and 2 are now compared with published

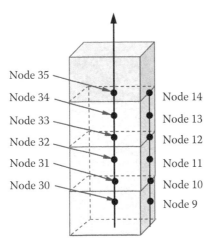

Figure 8.10 Concrete in uniaxial compression. Nodes for which the values of axial and lateral displacements are calculated.

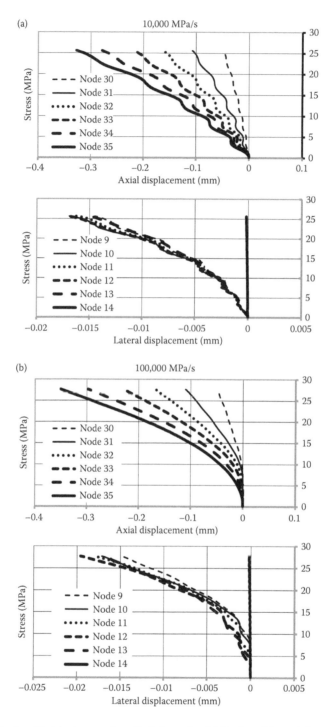

Figure 8.11 Concrete in uniaxial compression. Variation of axial (top) and lateral (bottom) displacements with applied stress at various distances from the top face of the specimen. Values of stress rate (MPa/s) equal to: (a) 10,000; (b) 100,000; (c) 200,000; (d) 400,000; (e) 1,000,000; (f) 2,000,000 and (g) 4,000,000. *(Continued)*

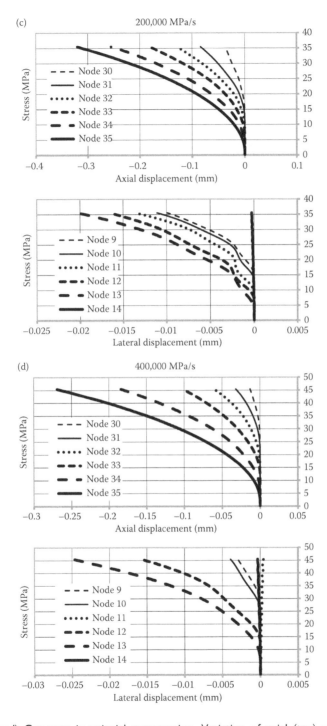

Figure 8.11 (*Continued*) Concrete in uniaxial compression. Variation of axial (top) and lateral (bottom) displacements with applied stress at various distances from the top face of the specimen. Values of stress rate (MPa/s) equal to: (a) 10,000; (b) 100,000; (c) 200,000; (d) 400,000; (e) 1,000,000; (f) 2,000,000 and (g) 4,000,000.   (*Continued*)

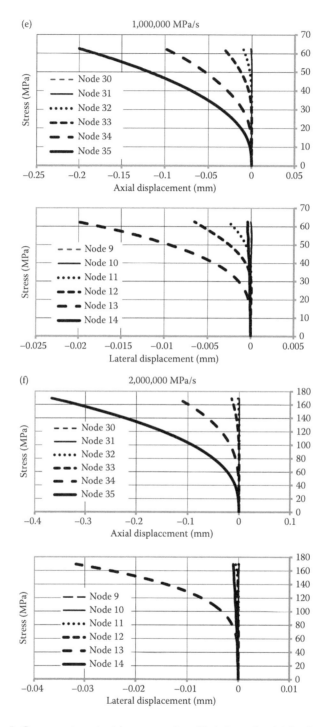

Figure 8.11 (Continued) Concrete in uniaxial compression. Variation of axial (top) and lateral (bottom) displacements with applied stress at various distances from the top face of the specimen. Values of stress rate (MPa/s) equal to: (a) 10,000; (b) 100,000; (c) 200,000; (d) 400,000; (e) 1,000,000; (f) 2,000,000 and (g) 4,000,000.   *(Continued)*

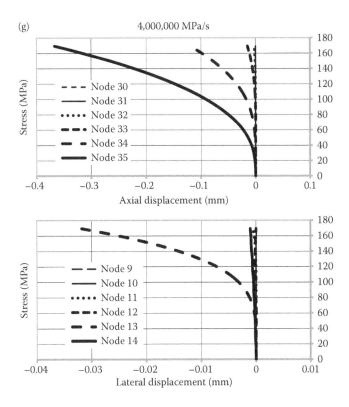

Figure 8.11 (*Continued*) Concrete in uniaxial compression. Variation of axial (top) and lateral (bottom) displacements with applied stress at various distances from the top face of the specimen. Values of stress rate (MPa/s) equal to: (a) 10,000; (b) 100,000; (c) 200,000; (d) 400,000; (e) 1,000,000; (f) 2,000,000 and (g) 4,000,000.

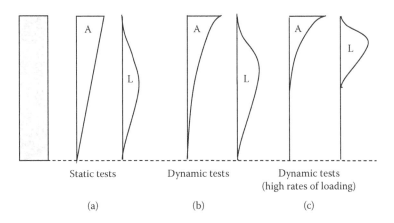

Figure 8.12 Concrete in uniaxial compression. Qualitative distribution of axial displacement (A) and lateral displacement (L) exhibited prior to failure along the longitudinal axis for (a) static tests, (b) dynamic tests and (c) dynamic tests with high rates of loading.

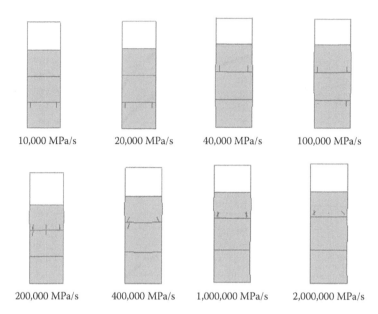

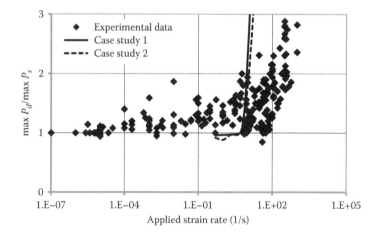

Figure 8.13 Concrete in uniaxial compression. Predicted crack patterns prior to failure for different rates of loading.

experimental data. Such a comparison reveals good agreement between the two (see Figures 8.14 through 8.17). However, the scatter that characterises the experimental data does not allow the immediate identification of which of the two case studies provides the closer fit to the experimental data. Both case studies yield results well within the scatter. However, if one accepts that, at high loading rates, the apparent strength of the specimen increases sharply, then it is quite evident from Figures 8.16 and 8.17 that case study 1 yields a poorer model (as expected) since it does not mimic the sharp strain increase of most experiments (while agreeing with some of the data, likely to be less accurate), which records either no strain increase or even a strain decrease in this range).

Figure 8.14 Concrete in uniaxial compression. Comparison of numerical (assuming average strain rate) and experimental results describing the variation of load-carrying capacity with the applied strain rate.

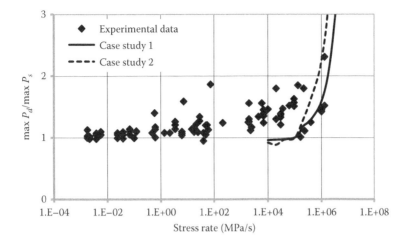

Figure 8.15 Concrete in uniaxial compression. Comparison of numerical and experimental results describing the variation of load-carrying capacity with the applied stress rate.

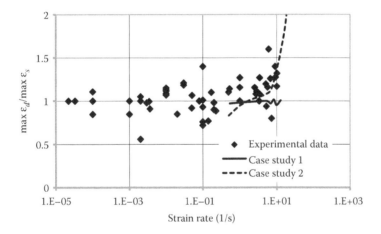

Figure 8.16 Concrete in uniaxial compression. Comparison of numerical (assuming average strain rate) and experimental results describing the variation of maximum exhibited strain with applied strain rate.

For extremely high rates of loading, there appears to be some divergence between numerical and experimental data. This divergence may be partly due to the fact that the numerical investigation is carried out by using a specimen with specific dimensions, f_c, and method of loading characteristics which vary in different experimental investigations. However, this deviation is only clearly observed in Figure 8.14 which shows the variation of max P_d/max P_s with the applied strain rate, whereas in Figures 8.15 through 8.17 this deviation is not apparent.

When comparing numerical and experimental results it is important to keep in mind that a high-rate loading test has an extremely short duration, so that it is difficult to obtain accurate measurements. This is particularly relevant for the case of the impact loading tests that were carried out from the early 1940s to the 1980s, which may not have been as accurate as more recent ones benefiting from modern equipment. Furthermore, measuring strains in the mid-height region of the specimen is based on the assumption that the specimen behaves as in the case of static loading. This may explain why many experimental investigations carried

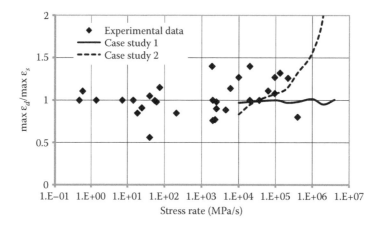

Figure 8.17 Concrete in uniaxial compression. Comparison of numerical and experimental results describing the variation of maximum exhibited strain with applied stress rate.

out before the 1980s report a significant increase in their specimen strength at low strain rates. Whereas in static tests the specimen has a more uniform distribution of strain along its height, for the case of high loading rate the lower part of the specimen deforms significantly less than the upper part. Therefore, the strain measured in the mid-height region of the specimen may be considerably less than that which is actually exhibited in the upper part.

In Figure 8.18 a relationship is presented between max P_d/max P_s and strain rates evaluated by using different methods or, rather, criteria as defined at the end of Section 8.2.4 (namely average strain rate, maximum value of average strain rate and mid-height strain rate). By comparing the three resulting relationships (Figure 8.18) it is obvious that they are significantly different. The use of mid-height strain rate calculated in the mid-height region of the specimen may lead to misleading conclusions due to the fact that, for high rates of loading, only the upper part of the prism is affected and, therefore, the mid-height region exhibits less deformation, so that the strain rate calculated in this region is much less than the strain rate exhibited in the upper area of the specimen. Comparing these relationships with the experimental data

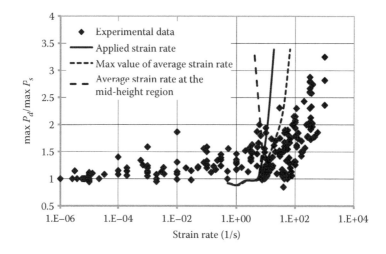

Figure 8.18 Concrete in uniaxial compression. Relationship between max P_d/max P_s and differently evaluated strain rates and their comparison with experimental data.

in Figure 8.18, it is obvious that the relationship between the max P_d/max P_s and the maximum value of average strain rate is the relationship which is closest to the experimental data.

8.2.7 Causes of the loading-rate effect on the behaviour of the specimen

The numerical investigation carried out proves that the change in the behaviour of concrete prismatic elements when subjected to high rates of compressive loading can be attributed primarily to the effect of inertia. Furthermore, it has been proved that the inertia effect in both axial and lateral directions is significant for an accurate description of specimen response. However, in order to understand more comprehensively how inertia affects the behaviour of concrete under high rates of loading, its behaviour must be investigated at a material level so as to fully comprehend the cracking procedure that concrete undergoes and how this is affected by the loading rate.

In a static test, the application of a uniaxial compressive load onto a concrete cubical or cylindrical specimen results in the development of an internal complex stress field. As discussed in Section 2.3, the complexity of the stress field is due to the non-homogeneous nature of the material, characterised by the existence of micro-cracks (Kotsovos 1979, Kotsovos and Newman 1981). At the tips of the micro-cracks, high concentrations of tensile stresses form and, once the tensile ultimate strength of concrete is overcome, the cracks extend in the direction of the maximum principal compressive stress (Griffith 1921, Kotsovos 1979, Kotsovos and Newman 1981). This extension offers relief to concrete, as it is followed by a decrease of the value of the tensile stresses acting at the crack tips (see Figure 8.19). The extension of the cracks continues as the applied load increases until, at some stage, the edges of the micro-cracks meet and larger cracks (macro-cracks) begin to form. Under a high rate of loading the procedure becomes even more complicated since the effect of the inertia forces must be also taken into consideration. The effect of these forces is dual since, on the one hand, they affect the overall response of the concrete specimen and, on the other hand, they have a local effect on the area of the concrete specimen where cracks form.

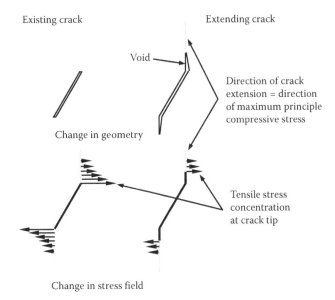

Figure 8.19 Micro-cracking sequence.

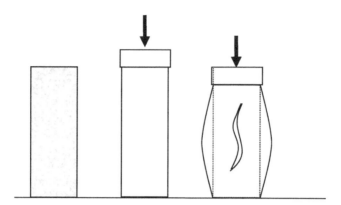

Figure 8.20 Concrete in uniaxial compression. Lateral deformation just prior to failure.

At the final stages of macro-cracking the specimen exhibits an increase in lateral strain. In the static tests this lateral deformation is exhibited in the mid-height region of the specimen (see Figure 8.20); however, as already discussed, this is not the case in the dynamic tests, especially when high rates of loading are involved. Because of the high rate of compressive loading imposed on the specimen, the lateral strain rate is also high. In case study 2, because the mass is distributed throughout the FE mesh and is active in all three directions, it reacts to the lateral deformation trying to restrict it. Hence, the reaction of the mass slows down the cracking process at these final stages and, in doing so, allows the specimen to increase its strength. In case study 1, on the other hand, the mass is not distributed throughout the FE mesh and it is active only in the direction of the external load. Because of this, the mass is unable to respond to the lateral deformation of the specimen and, therefore, failure occurs earlier, the deformation being smaller than that exhibited in case study 2.

As the rate of applied loading becomes higher and the duration of the loading procedure becomes less than 0.001 s, the height of the specimen affected (effective height h_{eff}) decreases. This decrease of the effective height coincides with the increase of the specimen strength. Here, it is interesting to note that, from experimental investigations of the behaviour of cylindrical concrete specimens under uniaxial-compressive loading, it is known that the maximum sustained load depends on the height-to-diameter (h/d) ratio of the specimen. In fact, it has been found that the specimens are characterised by a substantial increase in strength when the h/d ratio decreases below 1.0 (see Figure 8.21) (Neville 1973). The cause of this increase is the interaction which occurs at the interface between the specimen and the steel platens (Kotsovos 1983, Van Mier 1984, Van Mier et al. 1997, Zissopoulos et al. 2000) which are usually employed to apply the load. This interaction is due to the different material properties of concrete and steel, and is caused by the development of frictional forces.

The above static problem is now also investigated numerically. Concrete prisms, with a square cross section of 100×100 mm, are subjected to uniaxial compressive loading. The FE modelling of the specimen is the same as that used in case study 1 (since, under static conditions, the results of case studies 1 and 2 coincide). The compressive load is applied monotonically through a rigid steel platen situated on the top of the concrete specimen (a similar confinement is provided at the bottom, fixed face). The numerical results obtained are presented in Figure 8.22 in the form of a relationship between strength and h/d ratio. The increase in strength is found to become substantial once h/d becomes less than 1. This increase is caused by the interaction between the rigid-steel platens (or similar boundary conditions) used to apply the external load and the concrete specimen, which results in the development of a triaxial compressive state of stress which leads to an increase in concrete strength.

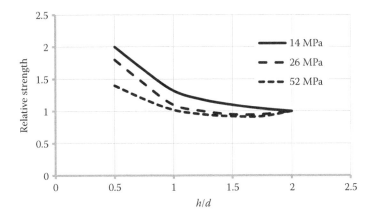

Figure 8.21 Concrete *n* uniaxial compression. Experimental relationship between the strength increase and the *h/d* ratio (*note*: relative strength = actual strength/strength of 'long' specimens).

A similar phenomenon occurs when subjecting concrete prism specimens to uniaxial compression at high-loading rates. The analysis of the numerical predictions revealed that, at high rates of loading, only the upper portion of the specimen is affected by the external load, whereas the rest of the specimen remains practically unaffected. Based on the numerical predictions, the deformation exhibited by the upper part of the specimen (where the external loading is imposed), under high loading rates, is significantly greater than that exhibited by the lower part. As the rate of loading increases, the height of the portion of the specimen (effective height) h_{eff} affected by the external load decreases. However, when comparing the relationship between the increase of strength with the *h/d* ratio for the static loading case (Figures 8.21 and 8.22) with the relationship between the increase in strength obtained from case study 2 (Figure 8.23), it is obvious that the increase in strength under dynamic loading is much larger. The height h_{eff} is assumed to be the distance that the stress wave travels inside the specimen during the loading procedure and is calculated by multiplying the length of time that the loading procedure lasts with the velocity of the stress wave. Based on these results, it can be concluded that the inertia of mass (both axial and lateral) under dynamic loading has a dominant effect on the specimen strength.

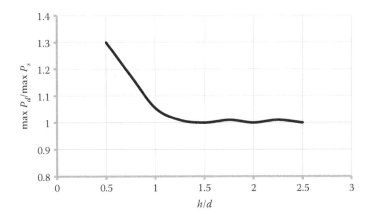

Figure 8.22 Concrete in uniaxial compression. Relationship between the static strength increase and the *h/d* ratio obtained from the numerical investigation.

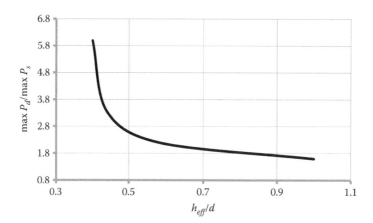

Figure 8.23 Concrete in uniaxial compression. Relationship between the strength increase and the h_{eff}/d ratio obtained from case study 2.

8.2.8 Conclusions

The constitutive model used by the FE program in this study to describe the behaviour of concrete is based on the static properties of the material, which remain constant and independent of the strain rate. The comparative study between numerical and experimental data revealed that the static brittle material model used is capable of providing realistic predictions of the behaviour of concrete at high rates of loading. In view of this, it can be concluded that the effect of loading rate on the specimen's behaviour reflects the effect of the inertia loads that reduce both the rate of cracking of the specimen and its effective height, factors which, in turn, lead to an increase in its load-carrying capacity.

At high rates of loading the numerical results reveal that only the upper region of the concrete specimen deforms whereas the rest remains practically unaffected by the application of the external load. This region is situated under the rigid element used to apply the external load and its height becomes smaller as the rate of loading increases. As a result, its behaviour is also affected by the interaction between the rigid element and the specimen which leads to the conclusion that the experimental and numerical data describe specimen (structural) response, rather than concrete (material) behaviour. On the other hand, for low loading rates, the deflection of the stress waves on the boundary surfaces of the specimen causes a different type of interaction between loading mechanism and concrete specimen. The fact that the stress waves, which arise from the continuous application of the external load during the loading procedure, are trapped by the boundary conditions imposed on the specimen, leads to a non-homogeneous distribution of stress inside the specimen in which it is extremely difficult (if not impossible) to predict where high concentrations of stress and strain will develop. It can, therefore, be concluded that the specimen behaviour cannot be considered to represent the behaviour of concrete as a material but represents its response as a structure.

To this end, concrete specimens under dynamic loading cannot be used to describe concrete behaviour (as usually assumed) since, in contrast with static loading, they do not constitute a material unit from which average material properties may be obtained. Under dynamic tests, concrete specimens must be viewed as structures since their response is directly linked to the inertia effect of their mass (and, of course, boundary conditions). Therefore, the use of experimental data from dynamic tests in order to develop constitutive models of concrete behaviour under dynamic loading is questionable. Moreover, the experimental data suggest that the behaviour of a concrete specimen under impact loading depends on a number of parameters, the effect of which has not been quantified so far. The success of the study reported so far suggests that the model used can serve as a reliable basis for elucidating the effect of these various parameters, and this forms the basis of ongoing research work, the first results of which have already been reported (Cotsovos and Pavlovic 2008b). An indication of the effect of these parameters on specimen behaviour is provided in the following section concerned with the behaviour of structural concrete under tensile impact loading.

8.3 STRUCTURAL CONCRETE UNDER TENSILE IMPACT LOADING

8.3.1 Background

The work discussed in what follows has been extracted from Cotsovos and Pavlovic (2008c); it is concerned with the numerical investigation of the effect of the rate of loading on the behaviour of concrete specimens subjected to uniaxial tension, complementing the work described in the preceding section. It is primarily intended to establish whether the proposed FE package is able to realistically model the material behaviour of concrete and the structural response of prismatic elements under high rates of uniaxial tension. The present investigation provides additional evidence of the generality of the proposed FE model in realistically predicting the behaviour of structural concrete under such extreme loading conditions (i.e., encountered during impact and explosion situations) despite its reliance on purely static material properties. This is achieved through a comparative study between the numerical predictions and available experimental data. Finally, the effect of various parameters on the structural response of the concrete prisms under high-rate uniaxial tensile loading has also been investigated in order to explain the causes of the observed scatter that characterises the available experimental data and to also show that experimental and numerical data describe structural, rather than material, behaviour.

8.3.2 Review of experimental data

Over the past few decades, a large number of experiments have been carried out on the behaviour of concrete (prismatic or cylindrical) specimens, not only under high rates of uniaxial compressive loading (see Section 8.2.2), but also under tensile loading (Malvar and Crawford 1998, ACI Committee 446 (ACI 446.4R-04) 2004, Cotsovos 2004, Wu et al. 2005, Brara and Klepaczko 2006, Schuler et al. 2006). A summary of the available experimental data for the case of tensile loading is presented in graphical form in Figure 8.24 expressing the relationship between the load-carrying capacities of the specimen (with respect to its counterpart under static loading) and the exhibited axial strain rate. The primary objective of such experiments is to investigate the behaviour of concrete at a material level under high-rate loading, since, as already pointed out in Section 8.2.2, the response exhibited by these specimens during the dynamic tests has been shown to differ considerably from that of their counterparts tested under static conditions. This difference primarily takes the form of an increase in the specimens' load-carrying capacity in both compression and tension, a difference which becomes more apparent as the rate of loading increases. Although all the experimental investigations arrive at the conclusion that there is a definite link between the loading rate and the exhibited response of the specimens, the considerable scatter that characterises the experimental data – which is evident from the data depicted in Figure 8.24 – makes it extremely difficult to derive a law that is able to accurately quantify this change of behaviour.

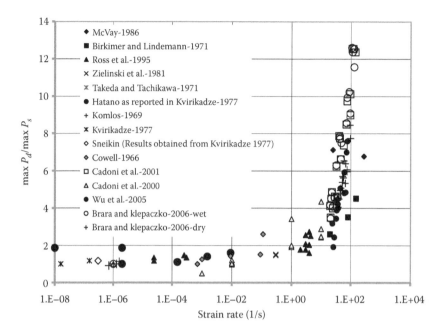

Figure 8.24 Concrete in uniaxial tension. Variation of load-carrying capacity with strain rate (max P_d = load-carrying capacity, max P_s = load-carrying capacity under static loading). (Data from Cotsovos D. M. and Pavlovic M. N., 2008c, *International Journal of Impact Engineering*, 35(5), 319–335.)

When taking a closer look at the experimental investigations carried out over the years, many parameters (such as the experimental techniques used for the tests, the shape, size and moisture content of the specimens, the different types of concrete used) vary from experiment to experiment. This variety of parameters contributes to the scatter which characterises the experimental data, and adds to the difficulty in interpreting the latter in a manner that leads to a clear understanding of the behaviour of concrete at high rates of loading.

It appears from the above, therefore, that, although the experimental data are able to qualitatively describe the effect that the rate of the applied loading has on the prism behaviour/response, they are unable to identify the causes of the gradual change exhibited in its structural response as described above. For this reason, FE analysis has been resorted to in order to investigate in more detail the response of concrete under high rates of uniaxial tensile loading.

8.3.3 Structural form investigated and FE modelling

The concrete prism shown in Figure 8.4 has also been used for investigating the effect of the loading rate of uniaxial tension on structural concrete behaviour. As for the case of the compressive load, the uniaxial compressive strength of concrete f_c is assumed to be 30 MPa. Through the reasoning followed in Section 8.2.3, the time needed for the stress wave, created when the external load on the upper area of the specimen, to move from top to bottom is approximately 0.0001 s. This value of 0.0001 s was used as reference in determining small time steps in the non-linear incremental analysis described in Chapter 6, so as to ensure sufficient accuracy for the numerical results. The time step adopted for these results varied between 0.000016 and 0.00000004 s depending on the stress rate under investigation and, it was clearly always a very small fraction of the time needed for the stress wave to travel along the specimen. For each stress rate different values for the time step were tried in order to ensure that predictions obtained were reliable.

The FE model of the concrete prism is that described in Section 8.2.4 (Figure 8.5), with the mass of the specimen being modelled as concentrated masses either located on the FE nodes situated along the longitudinal axis of symmetry of the specimen (model A) or distributed to all FE nodes (model B). As discussed in Section 8.2.4, in the case of model A, only mass displacement in the direction of the applied load (i.e., along the axis of symmetry) is allowed and hence it is sufficient to analyse only one quarter of the specimen (see Figure 8.5a). In the case of model B, the mass is allowed to have all the three degrees of freedom of the nodes and, hence, in order to avoid lateral wave deflection problems on the central planes of symmetry imposed in model A, the whole structure is analysed (see Figure 8.5b). The external load is imposed as a force incrementally at the beginning of each time step. In order to vary the rate of loading, the load increments are kept constant and the time step is varied. For the case of model A, because of the fourfold symmetry, only one quarter of the specimen is modelled and a load increment equal to 50 N is used for each time step. For the case of model B, however, the fourfold symmetry does not hold as the masses have three degrees of freedom, and hence the whole specimen must be modelled, thus making it necessary to impose four times the load used in the previous case (i.e., the load increment now becomes 200 N). The numerical investigation consists of two case studies. Case study 1 adopts model A whereas case study 2 relies on model B.

In order to remove any uncertainty regarding the terms 'average strain rate', 'maximum value of average strain rate', 'mid-height strain rate' and 'applied stress rate' the definitions provided in Section 8.2.4 have been adopted.

8.3.4 Presentation and discussion of the numerical predictions

The results obtained from case studies 1 and 2 show that in the case of the compressive loading (see Section 8.2.5), the behaviour of the concrete structural forms presently investigated under high rates of uniaxial tensile loading differs considerably from that exhibited under static loading. This change in the specimen's behaviour can be observed by inspecting the results presented in Figure 8.25 in the form of imposed stress–displacement curves, each one of which corresponds to different loading rates. By investigating the data presented in Figure 8.25 it

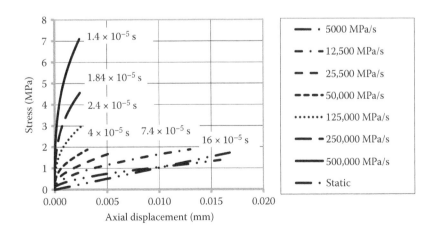

Figure 8.25 Concrete in uniaxial tension. Applied stress–axial displacement curves for case studies 1 and 2; the duration of each loading is also indicated for each of the curves.

becomes clear that the change in the specimen behaviour is gradual and primarily takes the form of an increase in load-carrying capacity exhibited by the specimen as the applied loading rate becomes higher, as was observed in the case of uniaxial compressive loading (see Section 8.2.5). The same conclusion can also be drawn from the inspection of the numerical results of Figure 8.26 which shows the variation of the load-carrying capacity max P_d (normalised with respect to its value under static loading max P_s) with the applied stress rate which is similar to that predicted for the case of compressive loading shown in Figure 8.6. However, in contrast with the case of compressive loading (see Section 8.2.5), where an increase in the axial maximum strain (max ε_d) exhibited by the specimen under high-rate loading was predicted when employing the more realistic model B (see Figure 8.7), in the present case of tensile loading the numerical results predict that the value of max ε_d with the applied stress rate slightly decreases (see Figure 8.27), although the trend is slight and hence inconclusive (in the sense that one could say that the value max ε_d/max ε_s is insensitive to the rate of loading).

The main reason for this difference is the fact that concrete has a small tensile strength (f_t), compared to its compressive strength (f_c) – i.e., $f_t = 0.1f_c$ – and brittle failure occurs prior to the specimen exhibiting practically any lateral deformation in the form of necking – as is the

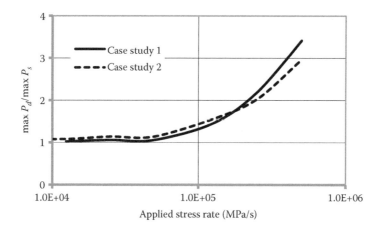

Figure 8.26 Concrete in uniaxial tension. Variation of max P_d/max P_s with the applied stress rate.

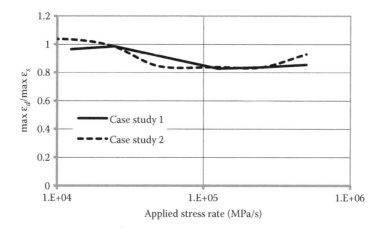

Figure 8.27 Concrete in uniaxial tension. Variation of the max ε_d/max ε_s with the applied stress rate.

case for other more ductile materials such as steel – in contrast with the case of compressive loading where lateral deformation is exhibited by the concrete prisms in the form of bulking.

Unlike the case of compressive loading for which the effect of inertia in the lateral direction has a significant effect on the predicted behaviour of the prisms, especially during the final stages of the loading procedure (i.e., prior to failure) since it was found to have a confining effect on the specimen, thus slowing down the cracking procedure, in the case of tensile loading, case studies 1 and 2 yield approximately the same results (see Figures 8.26 and 8.27). This leads to the conclusion that, in direct tension tests, the different distribution of the mass used in models A and B does not have a significant effect on the numerical predictions obtained for the specimen's behaviour. Therefore, an important observation, based on the present numerical investigation, is that in tension, unlike compression, the concrete specimen does not exhibit significant lateral deformation in the final stages of the test. This is prevented by the brittle nature of concrete and, also, by the fact that the tensile strength of concrete is only a small fraction of its uniaxial compressive strength (approximately 10%). Failure, therefore, occurs before any significant lateral deformation can be exhibited: thus, the effect of the inertia of the mass in this (lateral) direction is insignificant. It can therefore be concluded that only the inertia effect along the height of the specimen affects its behaviour under high rates of tensile loading.

Another interesting observation emerges from the numerical results by reference to the values of the reactions at the bottom of the prism. The results in Figure 8.28, presented in the form of curves describing the variation of the values of the applied stress and the axial reaction at the bottom of the specimen with time, show that for high rates

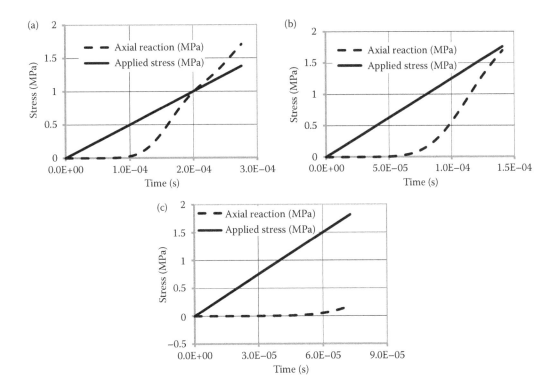

Figure 8.28 Concrete in uniaxial tension. Variation of the values of the applied stress and the axial reaction exhibited at the bottom of the concrete prism with time values of the applied loading rate (MPa/s) equal to: (a) 5000; (b) 12,500 and (c) 25,000.

of loading (i.e., for values over 12,500 MPa/s) the reaction which develops prior to the failure of the specimen (or the end of the loading procedure) is small compared to the stress applied on the top of the prism. In particular, the results show that, for high rates of loading (for values over 12,500 MPa/s), where the duration of the loading procedure is less than 0.0001 s (which corresponds approximately to the earlier estimate for the time needed for the stress wave to travel from the top to the bottom of the specimen), the numerical results reveal that the failure of the concrete prism precedes the development of substantial reactions (see Figure 8.28, and also Figure 8.29 which summarises the results of Figure 8.28 through a maximum axial reaction versus applied stress rate plot). This is due to the fact that the stress wave – which is generated on the top of the concrete prism by the applied load – is unable to reach the bottom of the concrete prism within the time over which the loading procedure lasts. Therefore, it is realistic to assume that, in such cases, the external load does not affect the whole specimen but only a part of it extending to a level which the stress wave is able to reach in the time that the loading procedure lasts. Based on this, it is possible to conclude that, beyond a threshold value of applied loading rate, higher concentrations of stresses (and strains) develop in the upper part of the specimen whereas, in the bottom part, the values of stresses and strains are much lower, and often negligible.

For low rates of loading, where the duration of the loading procedure exceeds 0.0001 s, the stress wave reaches the bottom of the specimen, bounces off it and starts to travel backwards and forwards along its height, trapped by the imposed boundary conditions. Because of this, the stress wave affects the whole of the concrete specimen, adding to the complexity of the stress field which develops within it (except, of course, for static or quasi-static loadings). In such cases, it is difficult to predict where the highest concentrations of stresses (and strains) will develop because of the continuous travelling of the stress wave which causes the internal stress field within the concrete prism to constantly change. It is important to note that for such cases the concrete prisms exhibit load-carrying capacities and maximum values of strain close to their counterparts obtained for the case of static loading (see Figures 8.26 and 8.27). These values begin to be affected by the applied stress rate as the duration of the loading procedure becomes less than 0.0001 s, resulting in the stress wave being unable to reach the bottom of the concrete prism.

In Figure 8.30 a relationship is presented between max P_d/max P_s and strain rates evaluated by using different methods for calculating the exhibited strain rate (namely average strain rate, maximum value of average strain rate, and mid-height strain rate). By comparing the three resulting relationships, it becomes clear that they are significantly different and that the use

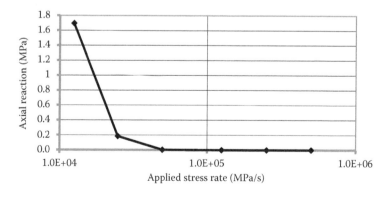

Figure 8.29 Concrete in uniaxial tension. Variation of the maximum axial reaction exhibited at the bottom of the concrete prism with the applied loading rate.

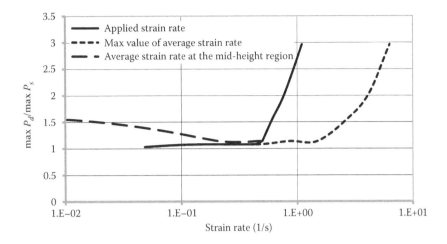

Figure 8.30 Concrete in uniaxial tension. Variation of max P_d/max P_s with the applied strain rate (variously defined).

of the average strain rate at the mid-height region leads to unrealistic results; hence the latter definition of strain rate will no longer be considered. When the two remaining curves are compared with the available experimental data (see Figure 8.31 which refer to experimental data), the one that describes the variation of max P_d/max P_s with the maximum value of average strain rate exhibited by the specimen appears to provide a closer fit to the experimental data. The use of the average strain rate and the mid-height strain rate calculated in the mid-height region of the specimen seems to lead to misleading conclusions due to the fact that, for high rates of loading, only the upper part of the prism is affected and, therefore, the mid-height region exhibits less deformation and strain, so that the strain rate calculated in this region is much less than the strain rate exhibited in the upper area of the specimen. Based on these

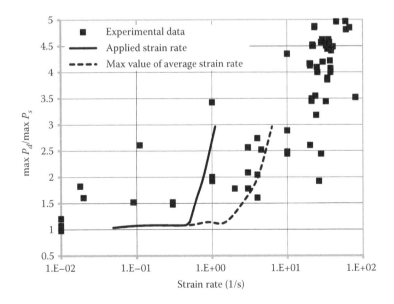

Figure 8.31 Concrete in uniaxial tension. Comparison of numerical and experimental data between max P_d/max P_s and differently evaluated strain rates.

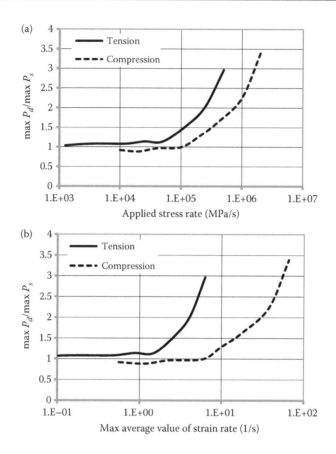

Figure 8.32 Comparison of numerical and experimental results established from tensile and compressive tests with the (a) applied stress rate and (b) maximum average value of strain rate.

results, it becomes apparent that, in order to calculate the maximum values of strain exhibited by the concrete prism, one has to identify with accuracy the exact location where such strains and their corresponding stresses will occur. This task is difficult to carry out in an experimental investigation, from which only estimate values of data can be hoped for.

Figure 8.32a and b show the curves obtained from the numerical investigation which describe the variation of the load-carrying capacity $\max P_d$ (normalised with respect to its value under static loading $\max P_s$) with (i) the applied stress rate (see Figure 8.32a) and (ii) the maximum value of average strain rate (see Figure 8.32b). By comparing both the cases of tensile and compressive loading it becomes apparent that the change in the specimen behaviour under tensile loading – which is primarily expressed by an increase in its load-carrying capacity – begins to occur at loading rates much lower than in the case of compression. The same conclusion is also drawn from the analysis of the available experimental data (Bischoff and Perry 1991, ACI 446 2004).

8.3.5 Validation of the numerical predictions

In Figure 8.31, the results obtained from the numerical investigation are compared with available experimental data which describe the increase of the specimen load-carrying capacity with increasing rates of tensile loading (more comprehensive experimental data are contained in Figure 8.24). The experimental data are obtained from a number of direct tensile and splitting tests. Such experimental data reveal that, for strain rates up to 0.1–1 s⁻¹, the specimen load-carrying capacity is similar to its counterpart obtained under static loading. However, when strain rates become larger than approximately 1 s⁻¹, a sudden increase in the specimen strength occurs. This observation agrees with the numerical predictions obtained from the numerical investigation presently reported.

As for the case of compressive loading, the experimental data presented in Figure 8.31 are characterised by considerable scatter. This scatter does not allow for very accurate predictions of the behaviour of concrete under high rates of tensile loading. As already mentioned, the large number of factors vary from experiment to experiment such as, for example, the size and shape of the specimens tested, the loading techniques used, the moisture content of the specimens, the different mixes of concrete used, the different types of experiments carried out for investigating the behaviour of concrete in tension (direct tension tests, splitting tests, etc.). Nevertheless, based on the comparison of the predictions obtained from the numerical investigation with the relevant experimental data, presented in Figure 8.31, one can conclude that, as in the case of compressive loading (see Section 8.2.5), the FE model's predictions are reasonable and provide a realistic phenomenological description of the behaviour of concrete in tension.

8.3.6 Parametric investigation

The available test data do not allow accurate conclusions to be drawn regarding the individual effect that the variation of each of the parameters referred to in the preceding section has on the specimen response. As for the case of compressive loading, this is due to both the lack of experimental data investigating the individual effect of each parameter and to the scatter that characterises all existing data often resulting in conflicting conclusions from investigation to investigation. Clearly, however, the variation of these parameters must have an effect on specimen behaviour. The majority of these parameters are linked to the dynamic (structural) characteristics of the specimen structure, such as the stiffness, the mass and the boundary conditions. Keeping in mind that the structural forms (i.e., concrete cylinders or prisms) used in tests must be viewed as structures, changes – even small ones – in the values of these various characteristics may (potentially) result in significant changes in the specimen's response under dynamic loading. Such changes clearly show that the data obtained from such experiments describes structural response rather than material behaviour.

The parametric study presented in what follows focuses on an investigation, through numerical analysis (adopting the more formal and accurate whole-specimen modelling with three degrees of freedom for the mass at each node), of the effects that the variation of some of these parameters have on specimen response and their contributions to the scatter in the experimental data. At this point, it is important to mention that one of the parameters that has been shown to affect the behaviour of concrete under high loading rates is the moisture content of concrete. However, it is impossible to include this parameter in the present investigation as the material model adopted for concrete is limited to the type of concrete used in RC structures, being usually air cured and, therefore, possessing low moisture content. The parameters that are presently investigated are the uniaxial compressive strength and density of the concrete used as well the height and cross-sectional area of the specimen (the effect of the measuring technique has already been considered – see e.g., Figure 8.30). In each of these cases only one of the parameters of the concrete prism investigated initially is changed at a time in order to investigate its individual effect. The results obtained are summarised in Figure 8.33a and b which show the variation of the load-carrying capacity

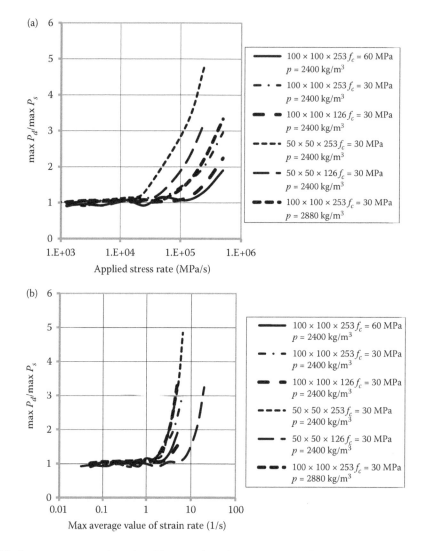

Figure 8.33 Concrete in uniaxial tension. Numerical predictions obtained from the parametric investigation describing the variation of max P_d/max P_s with (a) the applied stress rate and (b) the maximum value of exhibited average strain rate.

max P_d (normalised with respect to its value under static loading max P_s) with the applied stress rate as well as the maximum average value of strain rate.

Initially, the value of the uniaxial compressive strength of concrete was increased from 30 through 60 MPa. The predictions obtained are consistent with the CEB (1988) recommendations which suggest that the increase in ultimate strength of concrete specimens with loading rate depends on the value of f_c of the concrete used; in particular, the higher the value of f_c, the lower the increase of the specimen's strength when subjected to high rates of tensile loading.

In order to investigate numerically the effect of specimen cross-sectional area and length on specimen behaviour under high rates of tensile loading, three additional case studies are carried out. The specimens are similar to that shown in Figure 8.4 except for the changes described below. In the first case, the side of the cross section of the specimen is set to

50 mm (instead of 100 mm) ($50 \times 50 \times 253$) in order to reduce the cross-sectional area to a quarter of that used originally. In the second case, the length of the specimen is half that of the original (126 mm instead of 253 mm) ($100 \times 100 \times 126$). Finally, in the third case, both the side of the cross section and the length of the specimen are half of those in the original specimen ($50 \times 50 \times 126$). By examining the results in Figure 8.33a and b, it is apparent that the variation in cross-sectional area and height has some effect on specimen response with increasing stress rate. In particular, it seems that, the smaller the cross-sectional area and the longer the specimen, the larger the increase in strength.

Finally, the use of different mixes as well as the use of different types of components (aggregate, cement paste, etc.) may result in a variation of the density of the concrete specimen. Though these differences are expected to be rather small, the application of high rates of uniaxial loading onto a concrete specimen becomes a dynamic problem and, therefore, the inertia of its mass plays a significant role in its response under such loading conditions. In order to investigate the effect of the variation of density, the density of concrete (ρ) was increased by 20% of that originally used (2400 kg/m³). The numerical data obtained revealed that an increase of the density resulted in the specimen exhibiting a higher increase of load-carrying capacity under tensile loading with increasing loading rates. This can be explained by the fact that an increase in density results in an increase of the inertia forces that develop during the application of the external load, thus resulting in a further increase of the load-carrying capacity of the concrete prisms.

8.3.7 Conclusions

The comparative study between numerical and experimental data revealed that the brittle material model used and non-linear FE strategy adopted are capable of providing realistic predictions of the behaviour of concrete at high rates of uniaxial tensile loading. In view of this (static) material model's ability to describe the behaviour at high rates of loading, it can be concluded that the effect of the rate of loading on the specimen behaviour reflects the effect of the inertia loads that reduce the rate of cracking of the specimen, a fact which, in turn, leads to an increase of its load-carrying capacity. Furthermore, the numerical results obtained reveal that the scatter caused by varying individual parameters (such as the f_c and density of the concrete, the cross-sectional area and length of the specimen, and even the testing/measuring technique adopted) can account for the magnitude of the scatter that characterises the available experimental data. This conclusion becomes more apparent when considering the fact that the combined effect of a number of such parameters is certain to have an even more substantial effect on the overall prism response.

As for the case of compressive loading, at high rates of tensile loading the numerical results reveal that only the upper region of the concrete specimen deforms whereas the rest remains practically unaffected by the application of the external load. For the case of low rates of loading the deflection of the stress waves on the boundary surfaces of the specimen caused from the continuous application of the external load is trapped by the boundary conditions leading to a non-homogeneous distribution of the stress throughout the specimen. In neither of the above cases, the specimen behaviour can be considered to represent the behaviour of concrete as a material. This is further supported by results obtained from the parametric investigation which clearly suggests that the behaviour of the concrete prisms under high rates of uniaxial tensile loading are affected by a number parameters related to the structural characteristics of the prisms (i.e., mass, strength, stiffness).

Therefore, it is confirmed that a concrete specimen under dynamic loading cannot be used to describe concrete-material behaviour since, in contrast with static loading, it cannot constitute a material unit from which average material properties may be obtained. Under dynamic tests the concrete specimen must be viewed as a structure since its behaviour is directly linked to the inertia effect of its mass and the boundary conditions. Therefore, the use of experimental data from dynamic tests in order to develop constitutive models for concrete behaviour under dynamic loading is questionable.

Moreover, the experimental data suggest that the behaviour of a concrete specimen under impact loading depends on a number of parameters, the effects of which cannot be quantified due to the significant scatter that characterises the experimental data.

Finally, there are differences in the responses exhibited by the specimen under tensile and compressive impact. First, tensile specimens show rate sensitivity at lower applied stress rates. Secondly, tensile specimens appear to exhibit little, if any, changes in the maximum strain (compared with its static counterpart), which contrasts with the findings for compressive impact where a significant increase in maximum strain is recorded beyond a certain loading rate. Thirdly, there is a negligible confinement in the direction perpendicular to the loading (again, in contrast to compressive tests).

8.4 RC BEAMS UNDER IMPACT LOADING

8.4.1 Background

In what follows, it is shown that that the proposed FE model is also capable of providing a realistic description of the behaviour of RC beams subjected to concentrated load at mid-span applied at high rates. The results obtained have been extracted from Cotsovos and Pavlovic (2012) where full details of the work can be found.

8.4.2 Review of experimental data

A significant number of experiments have been carried out to date in order to investigate the behaviour of RC beams under high rates of concentrated loading (Hughes and Spiers 1982, Miyamoto et al. 1989, Kishi et al. 2001, 2002, 2006, May et al. 2006). In the majority of these cases, the load was applied by means of a steel mass (impactor) that was allowed to fall onto the specimens' mid-span from a certain height depending on the desired rate of loading. A simplified version of the experimental set-up used is presented in Figure 8.34. Special provisions were usually taken in order to avoid uplift at the supports and to moderate damage in the area of contact between the steel impactor and the concrete medium (usually in the form of pads made from various materials such as steel, ply and rubber) – see also Figure 8.34. The duration of loading in these tests was extremely short (of the order of some milliseconds [ms]) and the intensity of the applied load increased rapidly from zero to a maximum value. RC beams similar to those tested under high rates of loading were usually also tested under static loading, for purposes of comparison.

Typical results obtained from the tests are presented in Figures 8.35 through 8.39. Load–deflection curves describing the response of RC beams tested by Miyamoto et al. (1989) under static and impact loading at mid-span are shown in Figure 8.35, with the corresponding deflected shapes at the ultimate limit state of the beams being depicted in Figure 8.36. Typical crack patterns at the failure of RC beams tested by Hughes and Spiers (1982) are presented in Figure 8.37, whereas Figure 8.38 shows the measured values of the beam load-carrying capacity for various rates of loading and velocities of the steel impactor at the moment of impact. Finally, the measured values of the mid-span deflection at the failure of RC beams tested by Hughes and Spiers (1982) under various rates of loading are shown in Figure 8.39.

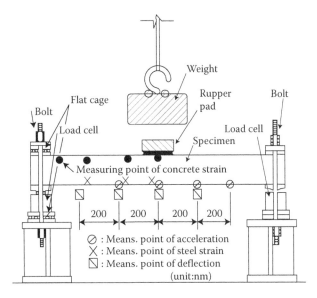

Figure 8.34 RC beams in impact. (Simplified representation of experimental set-up used by Miyamoto A., King M. W. and Fuji M., 1989, *Bulletin of the New Zealand National Society for Earthquake Engineering*, 22, 98–111.)

From Figures 8.35 and 8.36, the loading rate appears to affect both the load-carrying capacity and the deflected shape of the beams tested: Load-carrying capacity increases with the loading rate (see Figure 8.35), the latter also causing the formation of a 'discontinuity' point which marks the start of an abrupt increase of the slope of the deflected shape (see Figure 8.36). The formation of this 'discontinuity' point is compatible with the formation of near-vertical cracking which, as indicated at the bottom photograph of Figure 8.37, initiates

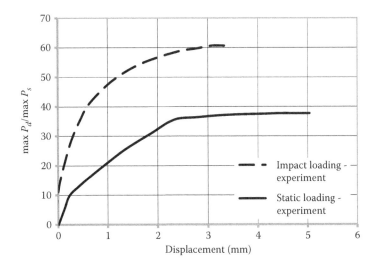

Figure 8.35 RC beams in impact. Typical curves expressing the relationship between applied load and deflection under static and impact loading.

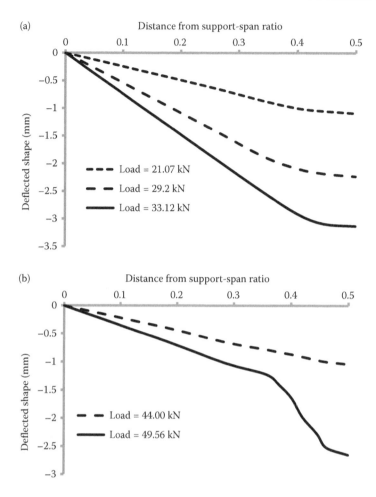

Figure 8.36 RC beams in impact. Typical deflected shapes under increasing load of beams tested by Miyamoto et al. (1989) under (a) static loading and (b) impact loading.

in the upper face of the beam and moves downwards indicating the development of negative bending moments under impact loading. Such cracking is preceded by the development of flexural cracking initiating at the bottom face of the beam and clustering within the mid-span region rather than spreading throughout the beam span as in the case of static loading indicated at the top photograph of Figure 8.37.

An important feature of the experimental results obtained from testing RC beams under load exerted at a high rate is the large scatter exhibited by the measured values of both load-carrying capacity and corresponding mid-span deflection shown in Figures 8.38 and 8.39, respectively. Although this may be partly due to the wide range of parameters linked with the experimental set-up adopted for testing (i.e., size and mass of the striker in relation to the beam length, the type of pad used, the boundary conditions, the beam dimensions, the grade of concrete, the amount, the spacing and strength properties of the longitudinal and transverse reinforcement), it is considered that the scatter predominantly reflects the difficulty of the experimental techniques adopted to accurately correlate the measured value of load-carrying capacity to the physical state of the specimens (Figure 8.40).

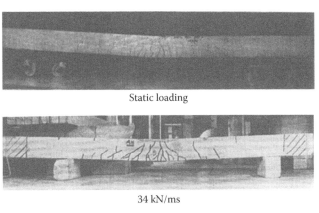

Static loading

34 kN/ms

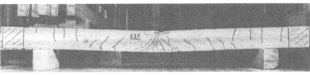

36 kN/ms

Figure 8.37 RC beams in impact. Typical crack patterns at failure for various rates of loading. (From Hughes G. and Spiers D. M., 1982, *An investigation on the beam impact problem, Cement and Concrete Association*, Technical Report 546.)

Loss of load-carrying capacity of an RC beam under static loading occurs when the compressive zone suffers longitudinal cracking close to the extreme compressive fibre after the occurrence of flexural or inclined cracking; this is consistent with both analytical and experimental findings which show that the occurrence of such cracking marks the start of the loss of load-carrying capacity either in flexure or shear (Kotsovos and Pavlovic 1995). In fact, Figure 8.40 shows that horizontal splitting of the compressive zone combined with flexural and inclined cracking is one of crack patterns characterising the state of an RC beam under impact at various time instances after contact with the impactor. From the figure it can be seen that the crack pattern of the beam 12 ms after contact with the impactor, apart from the spalling of the concrete cover, is characterised by flexural and inclined cracking that penetrates deeply into the compressive zone, with the latter suffering horizontal splitting. This is the typical crack pattern corresponding to a loss of load-carrying capacity which appears to occur within a time less than half the test duration (the latter being more than 28 ms); beyond this time of 12 ms, the experimentally established behaviour describes post-failure phenomena of little, if any, practical significance.

In the proposed FE package, loss of load-carrying capacity occurs when the stiffness matrix becomes non-positive definite (this essentially occurring when concrete disintegrates). For a statically determinate structure, the load step preceding the loss of load-carrying capacity is essentially that at which the load-carrying capacity is attained. In what follows, it is shown that the proposed FE model is capable of predicting trends of behaviour similar to those experimentally established, as well as providing a lower-bound fit to test strength data.

8.4.3 Structural form investigated

From the test data on beams under impact loading available in the literature, attention was focused on the behaviour of simply supported RC beams (beams C2) tested by Hughes and

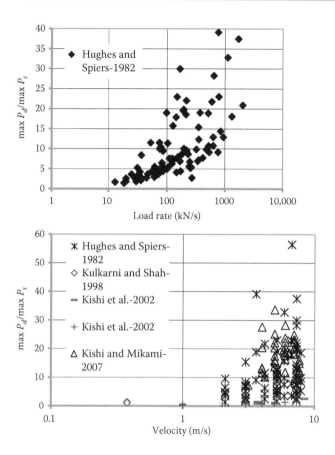

Figure 8.38 RC beams in impact. Measured values of load-carrying capacity obtained experimentally for various rates of loading and velocities of the impactor at the moment of impact (max P_d = load-carrying capacity under dynamic loading, max P_s = load-carrying capacity under static loading).

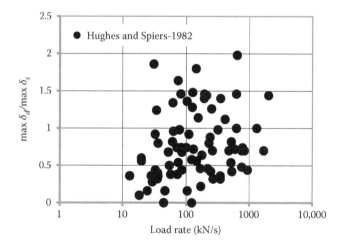

Figure 8.39 RC beams in impact. Measured values of the maximum deflection for various rates of loading (max δ_d = maximum deflection under dynamic loading, max δ_s = maximum deflection under static loading).

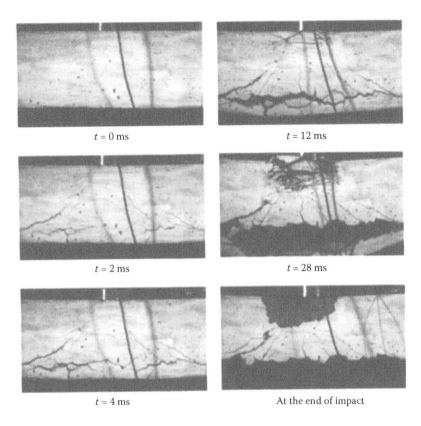

$t = 0$ ms $t = 12$ ms

$t = 2$ ms $t = 28$ ms

$t = 4$ ms   At the end of impact

Figure 8.40 RC beams in impact. Crack patterns characterising the response of an RC beam's under impact loading at various time instances after contact with the impactor. (From May I. M. et al., 2006, *Computers and Concrete* 3 (2/3), 79–90.)

Spiers (1982). The beams had a rectangular cross section with a height of 200 mm, a width of 100 mm and a clear span equal to 2,700 mm. The longitudinal reinforcement consisted of four bars: two 12 mm diameter bars placed at the two bottom corners of the beams cross section and two 6 mm diameter bars placed at the two top corners (see Figure 8.41). The transverse reinforcement comprised 6 mm diameter stirrups with an approximately 180 mm centre-to-centre spacing. The modulus of elasticity (E_S), the yield stress (f_y) and the ultimate strength (f_u) of both the longitudinal and transverse reinforcement bars used were equal to

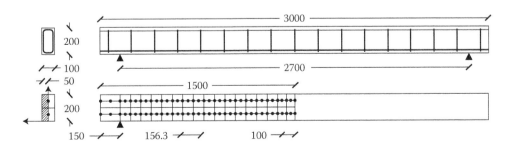

Figure 8.41 Beams in impact. Design details and FE model adopted for the analysis (dimensions in millimetre).

approximately 200 GPa, 460 and 560 MPa, respectively. The uniaxial cylinder compressive strength of the concrete used was $f_c = 45$ MPa. The RC beams were subjected to point load at mid-span by means of a steel mass allowed to fall onto the specimen from a certain height, depending on the desired rate of loading; several loading rates were considered in the test, as well as a quasi-static load application. Mild steel, rubber or ply pads are placed on the top face of the specimen in order to prevent or moderate local damage by the impact of the falling mass.

8.4.4 FE modelling of the problem

Concrete is modelled by using 27-node brick Lagrangian elements, whereas three-node isoparametric truss elements are used to model the steel reinforcement. As for all structural forms discussed herein, the size of the 27-node Lagrangian brick FEs used is dictated by the philosophy upon which the FE package adopted in the present work is based. It is reminded that the material model implemented in the package is based on data obtained from experiments on cylindrical specimens under triaxial loading conditions, and thus these cylinders may be assumed to constitute a 'material unit' for which average material properties are obtained and hence the volume of these specimens provides a guideline to the order-of-magnitude of the size of the FE that should be used for the modelling of concrete structures. Making use of the double symmetry that characterises the problem considered it is sufficient to model only a quarter of the RC beam shown in Figure 8.41. The truss elements representing the steel reinforcement were placed along successive series of nodal points in both vertical and horizontal directions. Since the spacing of these truss elements is predefined by the location of the brick elements' nodes, their cross-sectional area was adjusted so that the total amount of both longitudinal and transverse reinforcement is equal to the design values.

The beam mass was assumed to be lumped at the nodes along the longitudinal axes of symmetry of the two rows of brick elements comprising the beam FE mesh, each mass being allowed to move horizontally and vertically. The applied load increased linearly to failure at a constant rate which was varied between 10^3 and 10^6 kN/s.

8.4.5 Static loading

The main results obtained from the analysis of the beam under static loading are presented in Figure 8.42 in the form of a load–deflection curve and crack patterns at various load levels. The load–deflection curve obtained experimentally is also included in the figure which shows a close correlation between predicted and experimentally established behaviour. The beam exhibited ductile behaviour due to yielding of the longitudinal tension bars in the mid-span region which suffered extensive cracking eventually leading to loss of load-carrying capacity due to horizontal splitting of the compressive zone. Flexural cracks first appeared in the mid-span region and, with increasing load, gradually spread towards the supports. The predicted crack pattern just before failure is found to correlate closely with its experimental counterpart in Figure 8.42.

8.4.6 Impact loading

In what follows, attention is focused, on the one hand, on the effect of the loading rate on beam behaviour and, in particular, on load-carrying capacity, load–deflection curves, deformation profile and cracking process, and, on the other hand, on the causes of the predicted behaviour.

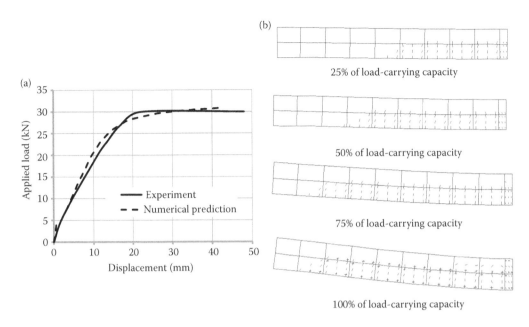

Figure 8.42 RC beam under static loading: (a) Comparison of load–deflection curves established by experiment and predicted by analysis; (b) deformed shape and crack patterns predicted by analysis for various stages of the applied load.

8.4.6.1 *Predicted behaviour*

The effect of the loading rate on the load-carrying capacity, max P_d, normalised with respect to its counterpart under static loading, max P_s, is described in Figure 8.43 which shows a close correlation between the numerically predicted values and their experimentally established counterparts. From the figure, it can also be seen that, after a small gradual increase, the load-carrying capacity increases sharply once a certain threshold of the loading rate is surpassed.

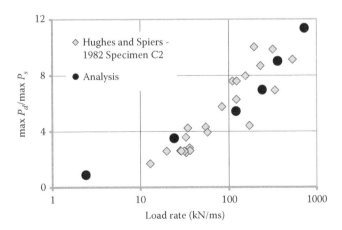

Figure 8.43 RC beams in impact. Experimental and numerical results expressing the variation of the load-carrying capacity with the loading rate (max P_d = load-carrying capacity under dynamic loading; max P_s = load-carrying capacity under static loading).

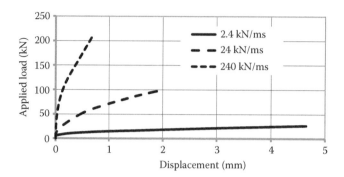

Figure 8.44 RC beams in impact. Predicted load–deflection curves for various rates of loading.

Figure 8.44 shows the numerically predicted load–mid-span deflection curves of the beams under load applied at various rates. From the figure, it can be seen that, under dynamic loading, there is a significant change in the RC beam's response when compared to its counterpart under static loading: as the rate of loading increases the beam response is characterised by a considerable increase in stiffness and load-carrying capacity and, at the same time, a decrease of the deflection at mid-span.

The numerically predicted and experimentally established values of the maximum mid-span deflection, max δ_d, normalised with respect to its counterpart under static loading, max δ_s, for various rates of loading are shown in Figure 8.45. The figure shows that, unlike the numerical values which show that the mid-span deflection decreases with increasing loading rate, the large scatter of the experimental values does not allow any conclusion to be drawn regarding the effect of the loading rate on deflection. The main cause of the scatter characterising the test values has already been discussed in Section 8.4.2: it reflects the inability of the measuring techniques to correlate with adequate accuracy the recorded values of deflection with the crack pattern corresponding to the loss of load-carrying capacity.

The numerically predicted deformed shapes of the RC beam analysed, shown in Figure 8.46, suggest that, for low rates of loading, the RC beam deforms in a manner similar to that of the static case (see Figure 8.42). More specifically, for rates of loading less than 24 kN/ms, the deflected shape of the beam has a near-parabolic form. For rates of loading beyond the above threshold, the numerical predictions indicate that the deflected shape progressively attains an inverted bell-like form which becomes more pronounced as its concave mid-span portion decreases in length with the loading rate increasing beyond the value of 120 kN/ms.

The numerically predicted crack formation and extension are also presented in Figure 8.46. The figure shows that, for loading rates less than 24 kN/ms, the crack pattern is similar to that occurring under static loading and gradually spreads throughout the beam span as the load increases. For higher rates of loading, cracking also forms in the upper part of the specimen, gradually extending downwards and this is indicative of the development of negative moments. The region of the deepest downward extending crack marks the transition (referred to as 'discontinuity' point in Figure 8.36) between the 'concave' and 'convex' portions of the inverted bell-like deflected shape of the beam. For loading rates higher than 120 kN/ms, the crack patterns exhibited by the beam indicate that the portion of the beam between the deepest downward extending crack and the mid-span may be viewed to behave as a fixed-end beam.

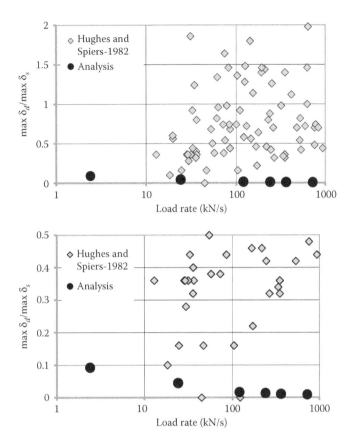

Figure 8.45 RC beams in impact. Variation of the numerical and experimental values of the maximum midspan deflection with the loading rate (max δ_d = maximum deflection under dynamic loading, max δ_d = maximum deflection under static loading).

8.4.6.2 Causes of beam behaviour

Figure 8.47 shows the variation with time of the numerically predicted values of the axial strain rate in the region of the compressive zone where the load is applied at a rate of 200 kN/ms. From the figure, it is seen that these values are in the range of ±1 s⁻¹, for which the variation of concrete strength with strain rate presented in Figure 8.48 shows that the increase in concrete strength is insignificant. Therefore, the increase in the beam's load-carrying capacity under impact loading (see Figure 8.43) cannot be attributed to an increase of the strength of concrete.

Figure 8.49 presents the time histories of (a) the applied load and (b) the corresponding support reactions predicted by analysis. From the figure, it can be seen that for rates of loading over 24 kN/ms, the support reaction is considerably less than the applied load; in fact, the higher the loading rate the larger the difference between the two values. Such behaviour indicates that the stress waves generated by the applied load (at the mid-span cross section) are practically unable to reach the supports either due to the short duration of the loading procedure, when this is smaller than the time required for the stress wave to travel from the mid-span to the beam ends, or due to the vertical cracking which initiates at the upper face of the beam due to the bell-shaped deflection of the beam indicated in Figure 8.46. Such behaviour indicates that, as the loading rate exceeds a certain threshold, the end portions of the beam, that is, the portions extending from the support to the location of the deepest

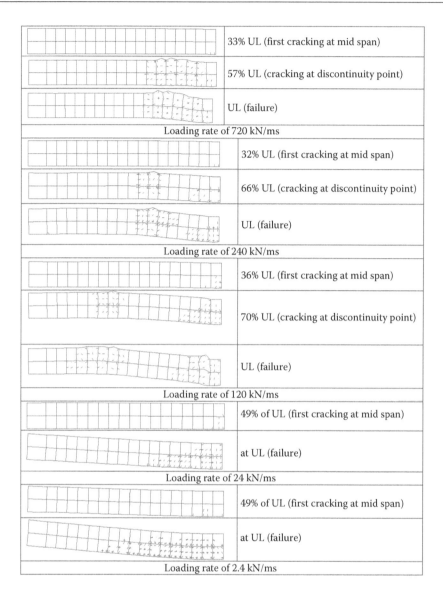

Figure 8.46 RC beams in impact. Numerically predicted deformation profiles and crack patterns under different rates of loading.

crack initiating at the upper face of the beam, remain essentially inert under impact loading. In contrast with the end portions, the middle portion of the beam which, as discussed earlier, may be viewed as a 'fixed-end beam', is the portion responding to this action. It appears, therefore, that the increase of the load-carrying capacity of the beam under impact loading reflects the increase of the load-carrying capacity of the 'fixed-end beam', as it reduces with increasing loading rate.

8.4.7 Conclusions

The constitutive models (for both steel and concrete) used for the numerical investigation of the behaviour of RC beams under impact loading are strain-rate independent. For the case

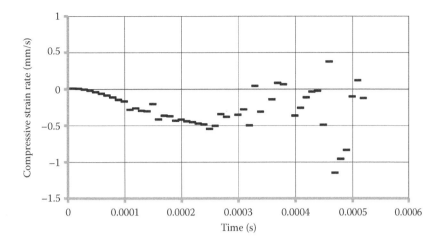

Figure 8.47 RC beams in impact. Variation with time of the numerically predicted values of the axial strain rates developing in the region of the compressive zone where the load is applied at a rate of 200 kN/ms.

of the beams considered, the comparative study between numerical predictions and experimentally obtained data revealed that through the use of these constitutive models, the proposed FE package is capable of providing realistic predictions of structural behaviour under increasing rates of loading. Such predictions encompass not only the beams' load-carrying capacity, but also the deformed shape and crack patterns at failure under impact. It was, in fact, found that the load-carrying capacity of the beam depends on the portion of its span

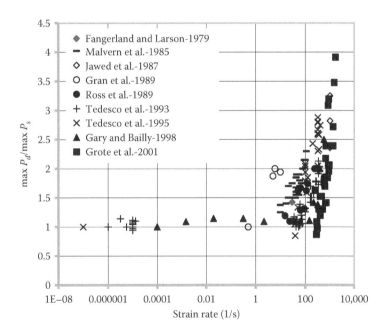

Figure 8.48 RC beams in impact. Variation of the load-carrying capacity of concrete specimens in compression with strain rate obtained from SHPB tests (max P_d = load-carrying capacity under dynamic loading; max P_s = load-carrying capacity under static loading). (Data from Cotsovos D. M. and Pavlovic M. N., 2008b, *Computers & Structures*, 86 (1–2), 164–180.)

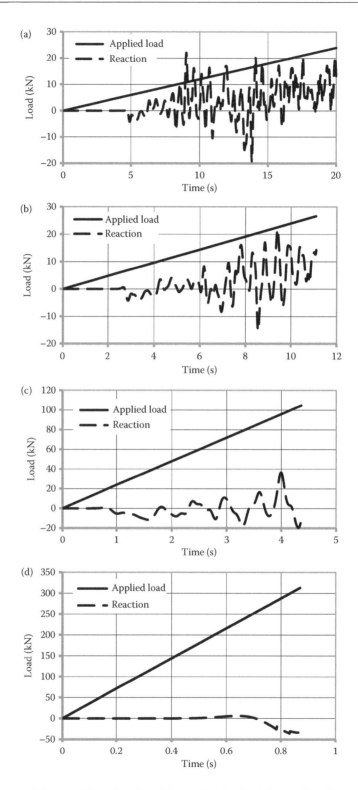

Figure 8.49 RC beams in impact. Time histories of the applied load and the predicted reaction forces under loading rates: (a) 1.2 kN/ms; (b) 2.4 kN/ms; (c) 24.0 kN/ms and (d) 240.0 kN/ms.

extending on either side of the load point at a distance equal to that travelled by the stress waves during impact: the smaller this portion the larger the beam load-carrying capacity. Moreover, the rate of the strain developing within the compressive zone was found to be insufficient to cause an increase in the strength of concrete.

8.5 CONCLUDING REMARKS

As for the case of structural behaviour under seismic excitation, the proposed FE model is found to produce, not only realistic predictions of structural behaviour under impact loading, but, also, an insight into the causes underlying such behaviour.

The close correlation between predicted and measured structural response shows that, in contrast with current code tenets, concrete behaviour is independent of strain rate effects and, thus, the increase in concrete strength with increasing loading rate reflects inertia effects. Moreover, as for the case of the flat slabs and structural walls discussed in Chapters 5 and 7, respectively, the FE program was found capable of identifying the causes of the observed RC beam behaviour, thus proving to be a valuable research tool.

REFERENCES

ABAQUS, Finite Element Software, Dassault Systems, www.3ds.com/products-services/simulia/portofolio/latest-release.

ACI Committee 446 (ACI 446.4R-04), 2004, *Report on Dynamic Fracture of Concrete.*

ADINA, FEA Software, ADINA, www.adina.com/index.shtml.

Atchley B. L. and Furr H. L., 1967, Strength and energy absorption capabilities of plain concrete under dynamic and static loadings, *ACI Journal*, 64, 745–756.

Barpi F., 2004, Impact behaviour of concrete: A computational approach, *Engineering Fracture Mechanics*, 71, 2197–2213.

Birkimer, D. L. and Lindemann, R., 1971, Dynamic tensile strength of concrete materials, *ACI Journal, Proceedings*, 47–49.

Bischoff P. H. and Perry S. H., 1991, Compressive behaviour of concrete at high strain rates, *Materials and Structures RILEM*, 24, 425–450.

Bischoff P. H. and Perry S. H., 1995, Impact behaviour of plain concrete loaded in uniaxial compression, *Journal of Engineering Mechanics*, 24, 425–450.

Brara A. and Klepaczko J. R., 2006, Experimental characterization of concrete in dynamic tension, *Mechanics of Materials*, 38, 253–267.

Cadoni E., Labibes K., Berra M., Giangrassos M. and Albertini C. 2000, High-strain-rate tensile behaviour of concrete, *Magazine of Concrete Research*, 52, 365–370.

Cadoni E., Labibes K., Berra M., Giangrassos M. and Albertini C. 2001, Strain-rate effect on the tensile behaviour of concrete at different relative humidity levels, *Materials and Structures*, 34 (235), 21–26.

CEB, 1988, *Concrete structures under impact and impulsive loading, Synthesis Report, Bulletin d' Information No. 187*, Comité Euro-International du Beton, Lausanne.

Cela J. J. L., 1998, Analysis of reinforced concrete structures subjected to dynamic loads with a viscoplastic Drucker-Prager model, *Applied Mathematical Modelling*, 22, 495–515.

Cervera, M., Oliver, J. and Manzoli, O., 1996, A rate-dependent isotropic damage model for the seismic analysis of concrete dams, *Earthquake Engineering & Structural Dynamics*, 25, 987–1010.

Cotsovos D. M., 2004, *Numerical Investigation of Structural Concrete under Dynamic (Earthquake and Impact) Loading, PhD thesis*, University of London, UK

Cotsovos D. M. and Pavlovic M. N., 2008a, Numerical investigation of concrete subjected to compressive impact loading, Part 1: A fundamental explanation for the apparent strength gain at high loading rates, *Computers and Structures*, 86 (1–2), 145–163.

Cotsovos D. M. and Pavlovic M. N., 2008b, Numerical investigation of concrete subjected to compressive impact loading, Part 2: Parametric investigation of factors affecting behaviour at high loading rates, *Computers & Structures*, 86 (1–2), 164–180.

Cotsovos D. M. and Pavlovic M. N., 2008c, Numerical investigation of concrete subjected to high rates of uniaxial tensile loading, *International Journal of Impact Engineering*, 35 (5), 319–335.

Cotsovos D. M. and Pavlovic M. N., 2012, Modelling of RC beams under impact loading, *Proceedings of the ICE-Structures and Buildings*, 165 (2), 77–94.

Cowell W. L., 1966, *Dynamic properties of plain concrete*, Technical Report No R447, DASA-13,0181, U.S. Naval Civil Engineering Laboratories, Port Hueneme, California.

Dilger W. H., Koch R. and Kowalczyk R., 1984, Ductility of plain and confined concrete under different strain rates, *ACI Journal*, 81, 73–81.

Dube J. -F., Pijaudier-Cabot G. and La Borderie C., 1996, Rate dependent damage model for concrete in dynamics, *Journal of Engineering. Mechanics Division ASCE*, 122, 359–380.

Evans R. H., 1942, Effect of rate of loading on the mechanical material of some materials, *ICE*, 18, 296–306.

Fangerlund G. and Larsson B., 1979, 'Betongs slaghallfasthet', (Impact strength of concrete), *Cement-och Betong Institutet (CBI), foskning FO4:79*, Stockholm.

Faria R., Olivera, J. and Cevera M., 1998, A strain-based plastic viscous-damage model for massive concrete structures, *International Journal of Solids and Structures*, 35, 1533–1558.

Gary G. and Bailly P., 1998, Behaviour of quasi-brittle material at high strain rate, *European Journal of Mechanics*, 17, 403–420.

Georgin J. F. and Reynouard J. M., 2003, Modeling of structures subjected to impact: Concrete behaviour under high strain rate, *Cement and Concrete Composites*, 217, 131–143.

Gomes H. M. and Awruch A. M., 2001, Some aspects on three-dimensional numerical modelling of reinforced concrete structures using the finite element method, *Advances in Engineering Software*, 32, 257–277.

Gran J. K., Florence A. L. and Colton J. D., 1989, Dynamic triaxial test of high strength concrete, *Journal of Engineering. Mechanics Division ASCE*, 115, 891–904.

Griffith A. A., 1921, The phenomenon of rupture and flow in solids, *Philosophical Transactions of the Royal Society*, London, Series A, 221, 163–198.

Grote D. L., Park S. W. and Zhou M., 2001, Dynamic behavior of concrete at high strain rates and pressures: I. Experimental characterization, *International Journal of Impact Engineering*, 25, 869–886.

Hatzigeorgiou G., Beskos D., Theodorakopoulos D. and Sfakianakis M., 2001, A simple concrete damage model for dynamic FEM applications, *International Journal of Computational Engineering Science*, 2, 267–286.

Hinton E. and Owen D. R. J., 1980, *Finite Elements in Plasticity: Theory and Application*, Pineridge Press, Swansea.

Hjorth O., 1977, *Ein beitrag zur frage der festigkeiten und des verbundverhaltens von stahl und betonbei hohen beanspruchungsgesehwindigkeiten (A contribution to the problem of strength and bond relationships for steel and concrete at high rates of loading)*, Dissertation, Universitat Carolo-Wilhelmina, Braunschweig, Heft 32.

Horibe T. and Kobayashi R., 1965, On mechanical behavior of rock under various loading rates, *Journal of the Society of Material Science (Japan)*, 14, 498–506.

Hughes B. P. and Watson A. J., 1978, Compressive strength and ultimate strain of concrete under impact loading, *Magazine of Concrete Research* 30, 189–199.

Hughes G. and Spiers D. M., 1982, *An investigation on the beam impact problem*, Cement and Concrete Association, Technical Report 546.

Jawed I., Childs G., Ritter A., Winzer S., Johnson T. and Barker D., 1987, High-strain rate behaviour of hydrad cement pastes, *Cement and Concrete Research*, 17, 433–440.

Kishi N. and Mikami H., 2007, A proposal for verification method of flexural-failure type RC beams under impact loading following performance-based design concept, *Structural Engineering Proceedings, Japan Society of Civil Engineers*, 53A, 1251–1260.

Kishi N., Mikami H. and Ando T., 2001, *An applicability of the FE impact analysis on shear-failure-type RC beams with shear rebars*, 4th Asia-Pacific Conference on Shock and Impact Loads on Structures, pp. 309–315.

Kishi N., Mikami H., Matsuoka K. G. and Ando T., 2002, Impact behaviour of shear-failure-type RC beams without shear rebar, *International Journal of Impact Engineering*, 27, 955–968.

Kishi N., Ohno T., Konno H. and Bhatti A. Q., 2006, Dynamic response analysis for a large-scale RC girder under a falling weight impact load, *Advances in Engineering Structures, Mechanics and Construction*, 140 (part 2), 99–109.

Koh C. G., Liu Z. J. and Quek S. T., 2001, Numerical and experimental studies of concrete under impact, *Magazine of Concrete Research*, 53, 417–427.

Komlos K., 1969, Some factors influencing the value of relative deformation of concrete axial tension, *Inzenyrske Stavby*, 17 (4), 177–182.

Kotsovos M. D., 1979, Fracture processes of concrete under generalized stress state, *Materials & Structures RILEM*, 12, 431–437.

Kotsovos M. D., 1983, Effect of testing techniques on the post-ultimate behaviour of concrete in compression, *Materials & Structures RILEM*, 16, 3–12.

Kotsovos M. D. and Newman J. B., 1981, Fracture mechanics and concrete behaviour, *Magazine of Concrete Research*, 31, 103–112.

Kotsovos M. D. and Pavlovic M. N., 1995, *Structural Concrete: Finite-element analysis and design*, Thomas Telford, London.

Kulkarni, S. M. and Shah, S. P., 1998, Response of reinforced concrete beams at high strain rates, *ACI Structural Journal*, 95 (6), 705–715.

Kvirikadze, O. P., 1977, Determination of the Ultimate Strength and Modulus of Deformation of Concrete at Different Rates of Loading, International Symposium, Testing In-Situ of Concrete Structures, Budapest, pp. 109–117.

Li Q. M. and Meng H., 2003, About the dynamic strength enhancement of concrete-like materials in a split Hopkinson pressure bar test, *International Journal of Solids and Structures*, 40, 343–360.

LS-DYNA, FEA Software, Livemore Software technology Corp (LSTC), http://www.lstc.com/products/ls-dyna

Lu Y. and Xu K., 2004, Modelling of dynamic behaviour of concrete materials under blast loading, *International Journal of Solids and Structures*, 41, 131–143.

Lundeen R. H., 1963, *Dynamic and Static tests of plain concrete specimens, Report I, Miscellaneous Paper No. 6-609*, U.S. Army Engineer Waterways Experiment Station, Corps of Engineers, Vicksburg, Mississippi.

Mahin S. A. and Bertero V. V., 1972, *Rate of loading effects on uncracked and repaired reinforced concrete members, Report No. OCB EERC 72-9*, Earthquake Research Center, University of California, Berkeley.

Malvar L. J. and Crawford J. E., 1998, *Dynamic increase factors for concrete, Twenty-Eighth DDESB Seminar*. Orlando, FL.

Malvar L. J., Crawford J. E., Wesevich J. W. and Simons D., 1997, A plasticity concrete material model for DYNA3D, *International Journal of Impact Engineering*, 19, 847–873.

Malvern L. E., Tang T., Jekins D. A. and Gong J. C. 1985, Dynamic compressive testing of concrete, *Proceedings of the 2nd Symposium on the Interaction of Non-Nuclear Munitions with Structures*, U.S. Department of Defence, Florida, pp. 119–138.

May I. M., Chen Y., Roger D., Owen J., Feng Y. T. and Thiele P. J., 2006, Reinforced concrete beams under drop-weight impact loads, *Computers and Concrete*, 3 (2/3), 79–90.

McVay, M. K., 1988, *Spall Damage of Concrete Structures*, Technical Report SL-88-22, U.S. Army Corps of Engineers, Vicksburg, Mississippi.

Miyamoto A., King M. W. and Fuji M., 1989, Non-linear dynamic analysis and design concepts for RC beams under impulsive loads, *Bulletin of the New Zealand National Society for Earthquake Engineering*, 22, 98–111.

Neville A. M., 1973, *Properties of Concrete*, Pitman, London, 2nd edition.

Popp C., 1977, *Unterschungen uber das vehalten von beton bei sclagartiger beansprufung (A study of the behaviour of concrete under impact loading)*, Deutscher Ausschuss fur Stahlbeton, No. 281.

Ross C. A., Jerome D. M., Tedesco J. W. and Hughes L. M., 1996, Moisture and strain effects on concrete strength, *ACI Materials Journal*, 93, 293–300.

Ross C. A., Tedesco J. W. and Kuennen S. T., 1995, Effects of strain rate on concrete strength, *ACI Materials Journal*, 92 (1), 37–47.

Ross C. A., Thompson P. Y. and Tedesco J. W., 1989, Split Hopkinson pressure bar tests on concrete and mortar in tension and compression, *ACI Materials Journal*, 86, 475–481.

Schuler H., Mayrhofer C. and Thoma K., 2006, Spall experiments for the measurement of the tensile strength and fracture energy of concrete at high strain rates, *International Journal of Impact Engineering*, 32, 1635–1650.

Suaris W. and Shah S. P., 1985, Constitutive model for dynamic loading of concrete, *Journal of Structural Engineering, ASCE*, 111, 563–576.

Takeda J., 1959, A loading apparatus for high speed testing of building materials and structures, *Proceedings of the Second Japan Congress on Testing Materials*, Japan Society for Testing Materials, Kyoto, pp. 236–238.

Takeda, J. and Tachikawa, H., 1971, Deformation and fracture of concrete subjected to dynamic load, *Mechanical Behavior of Materials, Proceedings of the International Conference*, Kyoto, Vol. IV.

Tedesco J. W., Powell J. C., Ross A. C. and Hughes M. L., 1997, A strain-rate-dependent concrete material model for ADINA, *Computers & Structures*, 64, 1053–1067.

Tedesco J. W., Ross A. C. and Brunair R. M., 1989, Numerical analysis of dynamic split cylinder tests, *Computers & Structures*, 32, 609–624.

Tedesco, J. W., Ross, C. A. and Kuennen, S. T., 1993, Experimental and numerical analysis of high strain rate splitting tensile tests, *ACI Materials Journal*, 90 (2), 162–169.

Tedesco J. W., Ross A. C., McGill P. B. and O'Neil B. P., 1991, Numerical analysis of high strain rate concrete tension tests, *Computers & Structures*, 40, 313–327.

Thabet A. and Haldane D., 2001, Three-dimensional numerical simulation of the behaviour of standard concrete test specimens when subjected to impact loading, *Computers & Structures*, 79, 21–31.

Van Mier J. G. M., 1984, *Strain-softening of concrete under multiaxial loading conditions*, PhD thesis, Eindhoven University of Technology.

Van Mier J. G. M., Shah S. P., Armand M., Balayssac J. P., Bascoul A., Choi S., Dasenbrock D. et al., 1997, Strain softening of concrete in uniaxial compression, *Materials and Structures RILEM*, 30, 195–209.

Watstein D., 1953, Effect of straining rate on the compressive strength and elastic properties of concrete. *ACI Journal*, 24, 735–744.

Winnicki A., Pearce C. J. and Bicanic N., 2001, Viscoplasic Hoffman consistency model for concrete, *Computers & Structures*, 79, 7–19.

Wu H., Zhang Q., Huang F. and Jin Q., 2005, Experimental and numerical investigation on the dynamic tensile strength of concrete, *International Journal of Impact Engineering*, 32, 605–617.

Zielinski A. J., Reinhardt H. W. and Kormeling H. A., 1981, Experiments on concrete under uniaxial impact tensile loading, *Materials and Structures*, 14 (2), 103–112.

Zissopoulos P. M., Kotsovos M. D. and Pavlovic M. N., 2000, Deformational behaviour of concrete specimens in uniaxial compression under different boundary conditions, *Cement and Concrete Research*, 30, 153–159.

Appendix A: Octahedral formulation of stresses and strains

A.1 OCTAHEDRAL COORDINATES

Consider the Cartesian stress space depicted in Figure A.1a, the coordinate axes of which are the principal stresses (σ_1, σ_2, σ_3), these being used to define any given state of stress as indicated generally by point P. The octahedral coordinates (z, r, Θ'), on the other hand, refer to a cylindrical coordinate system having its z axis coincide with the space diagonal ($\sigma_1 = \sigma_2 = \sigma_3$) while r and Θ' represent the radius and rotational variable, respectively: such a system is also shown in Figure A.1a. In addition, the cylindrical system may be viewed more clearly along the space diagonal or z axis, as in Figure A.1b, which thus lies in the plane normal to z (the deviatoric plane). Basic vector algebra enables the following relationships between octahedral and Cartesian coordinates to be established:

$$z = (1/\sqrt{3})(\sigma_1 + \sigma_2 + \sigma_3) \tag{A.1}$$

$$r = (1/\sqrt{3})[(\sigma_1 - \sigma_2)^2 + (\sigma_2 - \sigma_3)^2 + (\sigma_3 - \sigma_1)^2] \tag{A.2}$$

$$\cos\Theta' = (1/\sqrt{6}r)(\sigma_1 + \sigma_2 - 2\sigma_3) \tag{A.3}$$

A.2 OCTAHEDRAL STRESSES

The octahedral stress (σ_{oct}) acts on a plane orthogonal to the line that trisects equally the sets of axes defined by the principal stress directions. As stated above, such a plane is known as the deviatoric plane: in Figure A.2a, it is shown in the form of a triangle. With u_z (the unit vector along z) given by $(\sqrt{3})^{-1}$ (1, 1, 1), the octahedral stress is obtained through the following product:

$$\sigma_{oct} = \begin{bmatrix} \sigma_1 & 0 & 0 \\ 0 & \sigma_2 & 0 \\ 0 & 0 & \sigma_3 \end{bmatrix} \frac{1}{\sqrt{3}} \begin{bmatrix} 1 \\ 1 \\ 1 \end{bmatrix} = \frac{1}{\sqrt{3}} \begin{bmatrix} \sigma_1 \\ \sigma_2 \\ \sigma_3 \end{bmatrix} \tag{A.4}$$

This octahedral stress is defined fully by, and is usually more conveniently expressed in terms of, its direct and shear components, σ_o and τ_o, as well as the angle, Θ, that the shear octahedral stress vector forms with the projection of any given principal direction on the

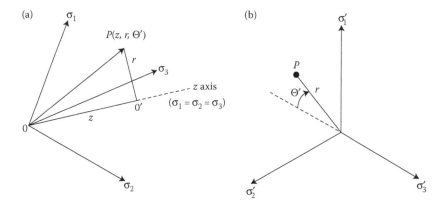

Figure A.1 Stress space: (a) Cartesian (σ_1, σ_2, σ_3) and cylindrical (z, r, Θ') coordinates; (b) view from the deviatoric plane.

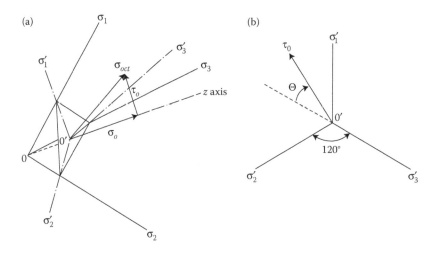

Figure A.2 Octahedral stress: (a) $\sigma_{oct} = \sigma_o + \tau_o$; (b) view along σ_o of τ_o and Θ.

deviatoric plane (e.g., σ_3' in Figure A.2b). The magnitudes of these parameters are denoted by σ_o, τ_o, Θ and are known as the hydrostatic stress, the deviatoric stress and the rotational angle, respectively. Simple vectorial operations lead to the following expressions which show how these octahedral quantities are related to the principal stresses and the octahedral coordinates:

$$\sigma_o = (1/3)(\sigma_1 + \sigma_2 + \sigma_3) = (1/\sqrt{3})z \tag{A.5}$$

$$\tau_o = (1/3)[(\sigma_1 - \sigma_2)^2 + (\sigma_2 - \sigma_3)^2 + (\sigma_3 - \sigma_1)^2]^{(1/2)} = (1/\sqrt{3})r \tag{A.6}$$

$$\cos\Theta = (1/\sqrt{2}\tau_o)(\sigma_o - \sigma_3) = \cos\Theta' \tag{A.7}$$

A.3 INVARIANTS OF THE STRESS TENSOR AND THEIR RELATION TO THE OCTAHEDRAL PARAMETERS

As is well known, the search for principal stresses $(\sigma_1, \sigma_2, \sigma_3)$ – and associated principal directions – corresponding to an arbitrary stress state $(\sigma_x, \sigma_y, \sigma_z, \tau_{xy}, \tau_{xz}, \tau_{yz})$ leads to an eigenvalue formulation that results in the setting of the determinant of the matrices of coefficients of the three homogeneous equations to zero. The resulting cubic equation may be written as

$$\sigma^3 - I_1\sigma^2 + I_2\sigma - I_3 = 0 \tag{A.8}$$

where all three roots $(\sigma_1, \sigma_2, \sigma_3)$ are always real. Since these principal stresses are unique physical quantities, they cannot depend on the (arbitrary) system of axes chosen originally (x, y, z): thus it follows that the coefficients I_1, I_2 and I_3 in Equation A.8 must be the same in all coordinate systems, and this is why they are known, respectively, as the first, second and third *invariants* of the stress tensor. The two instances of particular interest are the arbitrary coordinates (x, y, z) and the coordinate system defined by the principal directions; for these cases, the expressions for the three invariants become

$$I_1 = \sigma_x + \sigma_y + \sigma_z = \sigma_1 + \sigma_2 + \sigma_3 \tag{A.9}$$

$$I_2 = \sigma_x\sigma_y + \sigma_y\sigma_z + \sigma_x\sigma_z - \left(\tau_{xy}^2 + \tau_{yz}^2 + \tau_{zx}^2\right) = \sigma_1\sigma_2 + \sigma_2\sigma_3 + \sigma_3\sigma_1 \tag{A.10}$$

$$I_3 = \sigma_x\sigma_y\sigma_z + 2\tau_{xy}\tau_{yz}\tau_{zx} - \left(\sigma_x\tau_{yz}^2 + \sigma_y\tau_{zx}^2 + \sigma_z\tau_{xy}^2\right) = \sigma_1\sigma_2\sigma_3 \tag{A.11}$$

Of interest here are the relations between octahedral quantities $(\sigma_o, \tau_o, \Theta)$ and the stress invariants, so that the former may be computed directly from the given stress state σ_{ij} in arbitrary Cartesian coordinates x_i without the necessity to calculate the principal stresses as an intermediate step. It is easily verified that the relevant expressions are

$$\sigma_o = (1/3)I_1 \tag{A.12}$$

$$\tau_o = \left[2\sigma_o^2 - (2/3)I_2\right]^{(1/2)} \tag{A.13}$$

$$\cos 3\Theta = -\sqrt{2}(J_3/\tau_o^3) \tag{A.14}$$

where J_3 is the third invariant of the deviatoric stress tensor $s_{ij} = \sigma_{ij} - \sigma_o\delta_{ij}$ (and the use was made of the trigonometric relation $\cos 3\Theta = 4\cos^3\Theta - 3\cos\Theta$).

A.4 OCTAHEDRAL STRAINS

All the preceding definitions related to the octahedral stress and its component quantities $(\sigma_o, \tau_o, \Theta)$ apply equally to their strain counterparts, as would be expected on account of the essentially identical mathematical nature of stresses and strains. Therefore, denoting the three principal strains by $(\varepsilon_1, \varepsilon_2, \varepsilon_3)$, the following definitions for the hydrostatic and deviatoric strains (ε_o and γ_o, respectively), as well as the angle between the vector γ_o and the projection of the ε_3 axis (δ), hold

$$\varepsilon_o = (1/3)(\varepsilon_1 + \varepsilon_2 + \varepsilon_3) \tag{A.15}$$

$$\gamma_o = (1/3)[(\varepsilon_1 - \varepsilon_2)^2 + (\varepsilon_2 - \varepsilon_3)^2 + (\varepsilon_3 - \varepsilon_1)^2]^{(1/2)} \tag{A.16}$$

$$\cos\delta = (1/\sqrt{2}\gamma_o)(\varepsilon_o - \varepsilon_3) \tag{A.17}$$

A.5 ELASTIC CONSTITUTIVE RELATIONS IN TERMS OF OCTAHEDRAL STRESSES AND STRAINS

Since all the octahedral parameters may be expressed in terms of principal stresses/strains, only three of the six constitutive relations linking strains and stresses need to be invoked, namely those dealing with normal or direct quantities. These are

$$\varepsilon_1 = (1/E)[\sigma_1 - \nu(\sigma_2 + \sigma_3)] \tag{A.18}$$

$$\varepsilon_2 = (1/E)[\sigma_2 - \nu(\sigma_3 + \sigma_1)] \tag{A.19}$$

$$\varepsilon_3 = (1/E)[\sigma_3 - \nu(\sigma_1 + \sigma_2)] \tag{A.20}$$

Combining the above gives

$$\varepsilon_o = [(1 - 2\nu)/E]\sigma_o = \sigma_o/(3K) \tag{A.21}$$

$$\gamma_o = [(1 + \nu)/E]\tau_o = \tau_o/(2G) \tag{A.22}$$

where K and G are the bulk and shear moduli, respectively, defined in the usual way

$$K = E/[3(1 - 2\nu)] \tag{A.23}$$

$$G = E/[2(1 + \nu)] \tag{A.24}$$

Appendix B: Coordinate transformations

Transformation of stresses and strains from one coordinate system to another is a recurrent operation in assessing the critical stress/strain state in a structure, one of its familiar forms being the well-known Mohr's circle construction. In finite element analysis, transformations of constitutive ($[D]$) and/or stiffness ($[k]$) matrices occur repeatedly as, for example, in expressing stiffness properties from local to global coordinates, or the imposition of constraints which do not coincide with global axes. As indicated in Section 4.1.3, coordinate transformation is implicit in the isoparametric formulation of stiffness matrices; however, an explicit transformation is required for purposes of expressing cracked D matrices – set up initially in temporary local axes that follow the direction of cracking – in (overall) structure coordinates.

B.1 STRAIN TRANSFORMATIONS

Consider two sets of Cartesian axes $(x, y, z \equiv x_i)$ and $(x', y', z' \equiv x_i')$ sketched in Figure B.1 with the direction cosines that define their relative orientations given by (l_1, m_1, n_1) (for x' relative to x, y, z, respectively), (l_2, m_2, n_2) (for y') and (l_3, m_3, n_3) (for z'). Through the use of these direction cosines, the relations linking displacements in the two coordinate systems can readily be formulated, and then, in combination with the chain rule of differentiation and the strain–displacement expressions of three-dimensional elasticity (with the engineering definition of shear strains applicable), the transformation of strains ε (in the x_i system) into strains ε' (in the x_i' system) follows

$$
\begin{bmatrix} \varepsilon_x' \\ \varepsilon_y' \\ \varepsilon_z' \\ \gamma_{xy}' \\ \gamma_{xz}' \\ \gamma_{yz}' \end{bmatrix} = \begin{bmatrix} l_1^2 & m_1^2 & n_1^2 & l_1 m_1 & m_1 n_1 & n_1 l_1 \\ l_2^2 & m_2^2 & n_2^2 & l_2 m_2 & m_2 n_2 & n_2 l_2 \\ l_3^2 & m_3^2 & n_3^2 & l_3 m_3 & m_3 n_3 & n_3 l_3 \\ 2 l_1 l_2 & 2 m_1 m_2 & 2 n_1 n_2 & l_1 m_2 + l_2 m_1 & m_1 n_2 + m_2 n_1 & n_1 l_2 + n_2 l_1 \\ 2 l_3 l_1 & 2 m_3 m_1 & 2 n_3 n_1 & l_3 m_1 + l_1 m_3 & m_3 n_1 + m_1 n_3 & n_3 l_1 + n_1 l_2 \\ 2 l_2 l_3 & 2 m_2 m_3 & 2 n_2 n_3 & l_2 m_3 + l_3 m_2 & m_2 n_3 + m_3 n_2 & n_2 l_3 + n_3 l_3 \end{bmatrix} \begin{bmatrix} \varepsilon_x \\ \varepsilon_y \\ \varepsilon_z \\ \gamma_{xy} \\ \gamma_{xz} \\ \gamma_{yz} \end{bmatrix}
$$

$$(B.1)$$

that is,

$$\varepsilon' = [T_\varepsilon] \varepsilon \tag{B.2}$$

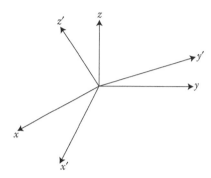

Figure B.1 Two sets of Cartesian coordinate systems arbitrarily oriented with respect to each other.

B.2 STRESS TRANSFORMATIONS

The relations governing the transformations of stresses σ (system x_i) into stresses σ' (system x_i') are derived by straightforward equilibrium considerations, and may be written as

$$\sigma' = [T_\sigma]\sigma \tag{B.3}$$

where the ordering of the elements of the stress vectors is compatible with that of their strain counterparts as listed in Equation B.1. It is interesting that $[T_\sigma]$ may be obtained from $[T_\varepsilon]$ by simply moving the factors 2 symmetrically with respect to the diagonal. It may also be shown that

$$[T_\varepsilon]^{-1} = [T_\sigma]^T \tag{B.4}$$

$$[T_\sigma]^{-1} = [T_\varepsilon]^T \tag{B.5}$$

B.3 TRANSFORMATIONS OF MATERIAL PROPERTIES

Consider the constitutive matrices $[D]$ and $[D']$ in the two coordinate systems x_i and x_i', respectively. Their relationship can be established by invoking the invariance of the strain energy stored (as a result of strains caused by the application of virtual displacements) in either system of axes. Here, however, a more direct derivation may be achieved through expressions such as Equations B.4 and B.5. Thus, in the system x_i'

$$\sigma' = [D]\varepsilon' \tag{B.6}$$

which, upon use of Equations B.2 and B.3, may be written as

$$[T_\sigma]\sigma = [D'][T_\varepsilon]\varepsilon$$

so that

$$\sigma = [T_\sigma]^{-1}[D'][T_\varepsilon]\varepsilon$$

On account of Equation B.5,

$$\sigma = [T_\varepsilon]^T [D'][T_\varepsilon]\varepsilon$$

and hence firmly

$$[D] = [T_\varepsilon]^T[D'][T_\varepsilon] \tag{B.7}$$

Index